HERANT KATCHADOURIAN

The Way It Turned Out

Published by

Pan Stanford Publishing Pte. Ltd.
Penthouse Level, Suntec Tower 3
8 Temasek Boulevard
Singapore 038988

Email: editorial@panstanford.com
Web: www.panstanford.com

British Library Cataloguing-in-Publication Data
A catalogue record for this book is available from the British Library.

The Way It Turned Out

ISBN 978-981-4364-75-1 (Hardcover)
ISBN 978-981-4364-76-8 (eBook)

Printed in the USA

For my parents,
my nanny, and
my first grandchild

Contents

Preface

I have been always fascinated by the lives of others. As a youngster, I read avidly about nineteenth-century Armenian revolutionaries. During adolescence, romantic lives in novels captured my interest. As an adult, the accounts of historical figures helped me understand the past. After I became a psychiatrist, my interests became more clinical. I wanted to know what was underneath the surface of people's lives. Understanding others helped me to understand myself. However, it never occurred to me that one day I would write my own memoir.

Nabokov called his memoir *Speak Memory*. Yet our memories are mute. It is we who speak through them by looking at the past through the lens of the present and the benefit of hindsight. And we do it in order to serve the purposes of the present. Autobiographies are suspect because they can be so easily falsified. Even when authors do not willfully distort the truth, to glorify themselves or vilify their enemies, they may inadvertently misrepresent it through their failure of memory in the reconstruction of events. However, accounts of one's life may also serve a nobler purpose. Near the end of the fourth century, St. Augustine wrote arguably the first autobiography, which actually was the account of his spiritual journey from paganism to Christianity and which has been a source of inspiration to so many others since then. However, since I am not a St. Augustine, why would I want to write an account of my life? Why would I want to revive the pain of the past and the regrets of missed opportunities? What would be the cost of revealing my inner thoughts and feelings? Why would I want to expose myself after living a guarded and private life? Why distress my family and friends by public disclosures and invite the *schadenfreude* of those who might think ill of me?

It was my mother who provided the antidote for these qualms. When she was in her eighties, she wrote the story of her extraordinary life. My mother's account of her early life contained revelations that I could not have imagined and that opened new vistas into her and my own life. Having learned so much from her, I deeply regretted that I knew so little about my father's past. He rarely talked about it and I never asked. My wife, Stina, persuaded me that my own children would also know very little about my earlier life unless I revealed myself to them. A good friend, Reva Tooley, further convinced me that the story of my unforeseen life would be of interest to others beyond my family and friends.

I was forewarned by a literary agent that the only autobiographies that would be widely read were those by celebrities and by individuals with pervasive trauma in their lives. I could understand how the accounts of eminent figures would provide material for historians and how damaged lives would be of interest to clinicians. But what could most people find in the lives of the famous and the dysfunctional to identify with? Is it voyeuristic gratification? Instead, as I have grown older, I have become increasingly interested in the lives of ordinary people with extraordinary lives—lives that are not the stuff of fame or abuse.

My life has been a variegated cultural tapestry woven from the diverse threads of my personal experiences. There are themes in my life that readers can readily identify with and that may help them to make better sense of their own lives. For instance, the role of chance in shaping one's life. Serendipity plays a part in every life but it may have played a greater role in my life than those of many others. Moreover, the way we interpret events at the time they occur does not always hold up over time. For each of us to look back upon our lives honestly is painful, but also gives us the strength to look forward with honesty.

Our sense of identity is at the core of who we are. The struggles with my sense of identity, the question of who am I, is a recurrent theme in my story. The circuitous course of my life shows the more unusual ways we may define ourselves. I am an immigrant in the quintessential nation of immigrants. Accounts of that experience vary greatly but they also point to important common elements at various times in the history of the nation. Many of us may still think of immigrants to America in the terms of Emma Lazarus's famous words ("Give me your tired, your poor/Your huddled masses yearning to breathe free") inscribed under the Statue of Liberty. However, these characterizations hardly apply to me, or to many other more recent immigrants to the United States. (The tired, the poor, and the huddled masses may have trouble getting even a visitor's visa these days.) Yet the question of identity remains a crucial issue for immigrants of every stripe.

No life exists in isolation and a personal account will by necessity involve the lives of others: my parents and my nanny, as well as my wife and our children, play key roles in the story of my life. However, this book is about me and not about them, and therefore my references to them will be discreet and, I hope, nonintrusive. A few key individuals have played a critical role in my professional development as mentors and colleagues and they too figure prominently in my account. Then there are failed relationships—lingering memories of unrequited love, disappointing friendships, and the scars of my first, failed marriage. These make the telling of the tale more difficult. I will try keep the focus on myself, but even then, not everyone may be pleased with what I have to say. It is not my intention to embarrass or upset anyone, but if telling the truth requires it, then so be it.

Writing the story of one's life is like undressing in public. How many layers of clothing should one take off? Self-exposure has now become a form of public entertainment, but some of us prefer to stay buttoned up. A memoir is even more self-referential than an autobiography because it reveals more of one's personal reflections and feelings, in addition to the story of one's life. How far should I go telling the world things that even my own family and friends might not know? Will I be able to walk the thin line between being truthful and embarrassing myself and others I care about? How can I give myself due credit without appearing smug and boastful?

The fact that I have been reticent all my life to reveal myself to others became, surprisingly, a spur for me to come out of my shell. Countless people have opened their hearts and minds to me—family, friends, lovers, colleagues, students, and patients. It was time for me to reciprocate. My life may not be exemplary enough to serve as a model for others but it might help them to learn from my strengths and weaknesses, my successes and failures.

None of the half-dozen other books that I have written provided a useful model for writing this book. Nor did the autobiographies of others that I read. Two of these were by well-known persons of shared background—one by Edward Said, the other by Vartan Gregorian. They were particularly interesting, but their focus was different from my own. Unlike the other books I have written, this one did not require extensive research. The material had to come not from the library, but from within my own mind. There was no crutch to lean on. Yet despite the passage of seven decades, I was surprised by how much I could remember. One recollection seamlessly led to another, although some had to be dug out laboriously. Old family photographs triggered many memories. An important source of information was my diaries. The first set of diaries was from my college years at the American University of Beirut between 1950 and 1953. The second set came twenty-seven years later, beginning in 1980, when I was a university administrator at Stanford, and ended in 2005 shortly after my retirement. (Its first entry is a quotation from John Berryman's poem, "Eleven Addresses to the Lord": "Surprise me on some ordinary day with a blessing gratuitous.")

My family and friends were another important source. My mother's memoir provided invaluable insights into her life and details about my childhood. Stina and our children, Nina and Kai, had their own recollections and observations to contribute. My cousin Nora, who is several years older, was an important link to our earlier shared lives. My good friend since college, John Racy, provided valuable information and observations for important periods when our lives overlapped during the past five decades.

The publication of this book is the result of a yet another serendipitous encounter. Two years ago, I was to board the plane at Paris Charles de Gaulle Airport and I was carrying my viola with me. A vivacious young lady approached me and asked if I was a famous violinist she had seen in a newspaper. It was tempting to say I was, but I did not. Instead, we introduced ourselves to each other. Her name was Jenny Rompas and we were soon joined by her husband, Stanford Chong. I had taught at Stanford for four decades. What was his connection with Stanford? There was no connection. It was coincidentally his name. They and a few other publishing professionals had founded a publishing house in Singapore called Pan Stanford Publishing. We kept in touch. I told them I was writing my memoir and looking for a publisher. Would they be interested? Yes, they would. So here we are.

To thank those who have helped me with this venture, I would like to acknowledge them in this preface rather then in a separate section since their participation has been an integral part of my writing this book. I am especially grateful to my wife, Stina, who was a source of unfailing support. She patiently read successive versions of the manuscript. As a writer and translator, she offered critical observations and constructive suggestions with respect to both content and language, as well as came up with the title of the book. My daughter Nina provided an insightful overview and as a conceptual artist helped with the design of the cover. My son Kai and his wife, Anni, read the manuscript together and made their own astute observations. A number of friends and colleagues read the entire text and made invaluable comments and suggestions. I extend my sincere appreciation to David and Susan Abernethy, Sanford Gifford, Richard Gunde, David Hamburg, Esther Hewlett, Sheila Melvin, John Racy, Reva Tooley, Nora Tour-Sarkissian, and Ernle Young. Select chapters were read by Robert Gregg, Walter Hewlett, Brent Sockness, Scotty McLennan, Paul Minus, John Reynolds, Steve Toben, Christine Tour-Sarkissian, and Abraham Verghese. The careful editing by Sarabjeet Garcha and Richard Gunde helped greatly to clarify and improve the text. The enthusiastic responses of these readers helped allay my concerns in exposing my life to public scrutiny and offered the prospect that this memoir may be of interest to others as well. I hope that is the way it will turn out.

1

The Legacy of Alexander

IN THE FALL OF 333 BCE, Alexander the Great inflicted a crushing defeat on Darius III, the King of Kings of Persia. The battle of Issus took place at the entrance of the Beylan Pass in the Amanus mountain range at the northeast corner of the Mediterranean. Subsequently, a settlement, called Alexandretta (now Iskenderun, Turkey), was established at the site. It was one of many such towns that sprang up in the wake of Alexander's conquests and a pale version of the great Hellenistic city of Alexandria in Egypt.

For the next almost two thousand years nothing noteworthy took place in the sleepy little port town—until my birth on January 23, 1933. That, at least, was my mother's version of history. Moreover, this geographical kinship made me in her eyes heir to the great Alexander and destined for greatness—if not on the battlefield, at least in the marketplace. I was well poised for such a prospect. My father was the richest Armenian merchant in town—a big fish in a small pond. I was his only child, destined to inherit his mini-empire. That, however, is not what happened. Why it did not happen, and what happened instead, is the story of this book.

~

My father and mother came from the provincial town of Aintab (now Gaziantep, Turkey) in south-central Anatolia.[1] My father and his two older brothers had established a successful business in Iskenderun before World War I. Beyond the fact that it involved importing construction materials, I never learned the details of my father's business.

I recall as a child the cavernous warehouse next to his office with huge piles of iron rods, barrels full of metal parts, and sacks of cement. I learned later from my mother that my father also acted as a customs broker and invested in real estate. Being on the fringes of Anatolia, rather than in the heartland of Turkey, saved him from the murderous deportations that virtually destroyed the Armenian population in Ottoman Turkey during World War I.

My parents were married in Aleppo, Syria, on November 7, 1919. It was a marriage of minds, not of hearts, but a good marriage nonetheless. My father, Aram, was a successful, highly respected, and a formidably intelligent thirty-five-year-old man from an old and prominent family. My mother, Efronia, at twenty-five, was a beautiful, thoughtful, and self-possessed young woman with a keen intellect who came from a respectable family of more modest means. Under normal circumstances, my father and mother could have been married earlier, not necessarily to each other, but the turmoil of World War I and the Armenian genocide in Ottoman Turkey made that difficult.

At the time of my parents' wedding, World War I had ended but the Armenian community of Aintab was again fighting for its life. My parents were so upset and concerned over the safety of their families that they didn't have the heart to have their wedding portrait taken. In other photographs taken at the time my father appears as an imposing, heavy-set man with a serious expression. My mother is a lovely woman whose warm brown eyes convey a wistful and subtle melancholy mood. (Photographs taken at the time are in the next chapter.)

My parents were eager to have children, but for the next fourteen years my mother could not conceive. They went on a pilgrimage to Jerusalem to pray for a child. My mother then went on another pilgrimage, walking barefoot to the shrine of a local saint near Iskenderun. They finally consulted the renowned obstetrician Dr. John Dorman, at the American University of Beirut. Dr. Dorman assured my parents that there was no medical reason why my mother should not be able to get pregnant, and he counseled patience. So, when she did get pregnant, at nearly the age of forty, it was hailed as a miracle. I was to be my parents' only child. My mother became pregnant once more, but it ended in a miscarriage. She told me years later in a voice heavy with regret that it was a boy. I have wondered how my life would have been different had I had a younger brother.

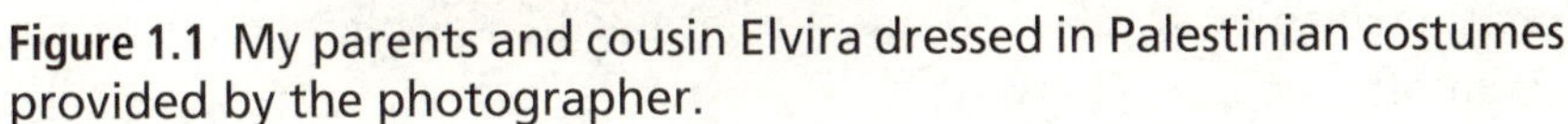

Figure 1.1 My parents and cousin Elvira dressed in Palestinian costumes provided by the photographer.

My birth was difficult. My mother was in labor for three days, and it was not certain whether we would both survive. The attending French physician, Dr. Bruce, had to resort to a forceps delivery. Given the risks, he wanted to get my father's consent, but my mother told him to say nothing to my father and to go ahead with it. When my father got to the hospital, the procedure was already under way. On hearing the baby's cry, he began to sob thinking that if the child survived, then his wife must have died.

My mother was too ill to see me for several days. She developed pneumonia that kept us in the hospital for forty days. The hospital bill came to forty Ottoman sovereigns. A nurse-midwife and family friend, Mademoiselle Marie, stayed with us in the hospital to provide special care. My mother could not nurse me, and I had to be bottle-fed. Cow's milk was considered unsafe, so they used condensed milk from Nestlé. With no refrigeration available, each can got used for a single feeding and then was discarded. It was a needless extravagance, since the sweetened condensed milk caused a constant diarrhea. To spare her waking up for nightly feedings, I slept in Marie's bed. This delightful arrangement nearly came to grief when one night she almost rolled over me and would have suffocated me had she not woken up in the nick of time.

On the night of my birth, relatives and friends gathered at our house to await the news. Finally my teenage cousin Puzant raced from the hospital on his bicycle, flung it aside, and ran into the house, yelling, "It's a boy." Lucine, who was to become my nanny, was standing at the top of the stairs holding a tray full of cups of Turkish coffee. She was so startled that she dropped the tray, sending the cups scattering down the steps. As she stood mortified, women ran up to her to hug and kiss her, since the breaking of china brought good luck.

Our homecoming caused great rejoicing. My mother's oldest brother, Yacoub, had a sheep slaughtered as a sacrificial offering on our doorstep. Holding me in her arms, my mother had to step over the bleeding animal to get into the house. I received many pieces of baby jewelry, including gold bracelets with tiny pearls and blue beads set in silver, to protect me against the evil eye. Such amulets continued to be stitched to my clothes until I was old enough to rip them off.[2]

The next big event would have been my baptism. Unfortunately, my father's older brother Sarkis was terminally ill, and my father had to go Beirut to be with him in his final days. With my uncle's death, the family went into mourning. My baptism had to be scaled down to a simpler affair that took place at home, instead of the splendid occasion it would have been in church. My baptismal dress, nonetheless, was spectacular with its intricate Aintab lacework. (Both of our children were to wear it at their own baptisms.)

I was named Hrant (from *hur*, Armenian for fire) after an ancient patriarch called Hrand. The spelling of my name was later changed to Herant.[3] Marie *Nanoug* ("Aunt") had seen the name in a newspaper while visiting my mother at the hospital, and suggested it to her.[4] Traditionally, first-born Armenian boys would be named after their paternal grandfather. That would have made me Adour, but the name

Figure 1.2 With my mother in my baptismal dress.

sounded old-fashioned to my mother's ears, and my father agreed to her choice.

~

We lived in Iskenderun until I was six years old. These uneventful and happy years came to an abrupt end when political circumstances made it necessary for us to leave for Beirut, with far-reaching consequences for our family. It was an event that would leave an indelible impression on me for the rest of my life.

Four persons occupied a central place in my young life. In their order of importance, they were my mother, my nanny Lucine, my father, and my cousin Elvira. At the age of four, Elvira had come to live with my parents, although she was never formally adopted by them. She was thirteen when I was born. Lucine was brought to us from the orphanage when she was about twelve (years later, she told me with much embarrassment that she was not yet menstruating at the time); she was eighteen when I was born. (Lucine became "Lucy" five decades later, when she came to America with my mother.)

I was the only child in our household and have no recollection of there being other children around. I remember being alone a good deal of the time, but I did not feel lonely; nor have I ever felt lonely since or bored by myself. I do not recall ever acting silly or being rambunctious. This may have been part of a precocious and misplaced sense of maturity that turned me to an adult in the form of a little boy.

We lived in one of largest houses in Iskenderun—a two-story mansion built to my father's exacting specifications. My mother would relate, with a mixture of admiration and exasperation, how during the construction of the house my father had the stone balustrade torn down and rebuilt because of a slight irregularity. Of the many rooms in our house, I remember well the salon that was used only for formal guests. Its curtains were usually drawn, which made the room dark and forbidding. When I occasionally peeked in from the doorway, Lucine would lead me away even though there was no likelihood of my touching the fine furniture or the delicate curtains.

Figure 1.3 Our house in Iskenderun.

The room I was most fond of was the far more intimate bedroom of my parents where I also slept. I remember vividly their massive gleaming twin brass beds, but I barely remember my own wooden bed, custom-made in Aleppo. I often started sleeping in my mother's bed or would go to her if I woke up at night. I was not afraid sleeping alone. I simply felt happier snuggling against my mother's bosom, and I was sure she too liked having me close to her. I was intrigued by the two cabinets next to my parents' beds, but I never touched them until early one morning, when I picked up a strange odor coming from one of them. I opened its door gingerly and recoiled at the pungent smell of dark urine in a large potty.

My earliest memory also goes back to my mother's bed. I must have been about four. My mother had put me down in her bed for my afternoon nap. It was a stiflingly hot summer day with harsh sunlight pouring into the room. The cool white pillowcase felt soothing against my cheek. As I yielded to an overwhelming wave of somnolence, I heard a solitary dog barking in the distance. I was blissfully happy without knowing what happiness was. I am reminded of that moment when I look at Giorgio de Chirico's surrealist painting *Mystery and Melancholy of a Street*, in which time stands still in a landscape of eternity.

Another early recollection involves a large festive gathering, probably on Christmas or New Year's Eve. Our house was full of guests conversing in loud voices. My mother and Lucine were busy and out of sight. I felt alone in the crowd. There was suddenly much commotion as my uncle Haroutune's family arrived with fanfare. Their teenage son, Antranig, trailed by his two younger brothers, was carrying on his head a huge brass tray loaded with assorted pastries, fruit, nuts, candy, and chocolate and surrounded by a circle of candles. It was a wondrous sight.

The first distressful memory of my early life comes from a year later. My mother had gone to Aleppo to buy clothes for Elvira's forthcoming wedding. It was the first time that I had been separated from my mother for any length of time. Lucine continued to take care of me as usual, so my mother's absence didn't disrupt the rhythm of my daily life. I never asked Lucine where my mother had gone and when she was coming back, but I must have missed her terribly. I took to sleeping with her gray cardigan pressed to my face. I was comforted by its soothing warm touch and heady scent. It was not the cologne she used liberally, but the natural aroma of her body (which years later became linked in my mind with the scent of the clothing of my young children). When my mother finally returned home, I was overcome by an intense longing for her and clung to her neck with sobs welling up in my chest. That evening, I came down with a fever.

This is perhaps why my mother took me along on her next visit to Aleppo. One afternoon, she and her two older sisters decided to go to a Turkish *hammam* for old times' sake. Since Lucine was not there to look after me, my mother wondered what they should do with me. My aunt Aroussiak said that they should take me with them since I was "too young to know." (Little did she know.) The prelude to our bathing was a feast of fruits, nuts, and watermelon in one of the warm anterooms (like the *tepidarium* of a Roman bath). We then got undressed, wrapped a bath towel (Turkish, *peshtimal*) around our waists and went into the bathing area. My mother took me into one of the alcoves and washed my hair with soap scented with bay leaves, pouring warm water over my head with a handheld brass bowl that had a small bulge in the center. A finger in the underside of the bulge helped one grasp the slippery wet bowl more firmly. (The Romans called it *omphalos*, navel). With my bath over, I was let go to roam around while my mother and aunts washed themselves at leisure.

I walked gingerly on the slippery marble floor into the large central area surmounted with a huge dome. The bottoms of green bottles had been stuck into the cement roof, letting in a diffuse light. The hall was dominated by an immense copper cauldron, which dispensed steaming hot water. The heat emanating from its red shiny surface kept me away at a cautious distance. Women carried hot water in wooden buckets to the stone basins in their alcoves, flip-flopping on wooden clogs—called *khap-khap* in Turkish—whose rhythmic sound echoed from the bare walls.

As my eyes got used to the sensuous translucent light filtering in from the skylights, I watched in wide-eyed fascination as the glistening mounds of moist female flesh moved slowly in and out of the mist. (I don't remember any faces.) Paradise had opened its doors for me. I had no inkling of sexual arousal. The feeling that washed over me was a sense of awe—the way worshippers must have felt in the presence of the great Neolithic Mother Goddess of Anatolia. Having seen that sight, my young life would not have been lived in vain had I died on the spot.

~

Elvira's wedding took place in the spring of 1938, when I was five. One festive gathering followed another. The groom was Mr. Khatchig Armadouni, a tall and dashing young man who was one of the few persons in Iskenderun to own a private car. His father had been dragoman—interpreter and guide—of the British consul, and he himself worked for a shipping company owned by an Italian called Catoni. (I remember my mother telling a friend that Catoni had an Armenian mistress whom he kept at a house in a working-class neighborhood; I didn't know what a mistress was, but sensed the disapproval in my mother's voice.)

Elvira, the vivacious and pretty eighteen-year-old "daughter" of the Khatchadourians, was quite a catch for Mr. Khatchig—and he knew it. Whereas people generally addressed my father as Aram *Effendi* (the Turkish honorific title for gentlemen), Uncle Khatchig always referred to him as Aram *Agha*, the form of address for Kurdish chieftains. I don't remember Elvira's wedding ceremony itself. But when the bride came to her new home from church, her mother-in-law, Zanazan Hanem, a woman of imposing presence, welcomed her by smashing a stack of dinner plates at her feet for good luck. It startled me out of my wits until the joyful ululations of their women guests from villages set my mind at ease.

This occasion was the setting for my first sexual experience of sorts. Since the wedding festivities were a drawn-out affair, a few of the younger children were entrusted to one of the maids, called Iskouhi, to be looked after. In a photograph of the wedding guests she looks like a mature teenager with dark hair and sultry eyes. Iskouhi took me and two other children to her small bedroom; one was a girl about my age, the other her somewhat older brother, called Garo. After we idled away some time, Iskouhi announced that we were going to play "wedding." As I watched impassively, she took off her white panties and tucked them under her pillow. She then lifted up her skirt and lay down on a long table set against the window with closed curtains. We must have been given some instructions, since the two other children dutifully lined up ahead of me and one after the other touched her between the legs. When it was my turn, I looked down and all I could see was a mound of black curly hair that looked like a little furry animal. I felt vaguely apprehensive. So instead of touching her directly, I put my fingers behind the curtain to scratch at her pubic hair, surprised at how coarse it felt. That was the end of the "wedding" game; if Iskouhi had higher expectations, we failed to live up to them.

Several days later, when Garo's parents were visiting us, he and I stood chatting in the street. There was no reference to the wedding game. Instead, Garo spoke enviously of how strong his older brother was, which he attributed to his "fucking" Iskouhi. I liked the idea of being strong but could not fathom the rest except for a vague feeling that it had something to do with Iskouhi's furry little animal. I also suspected that it entailed more than scratching at it through the curtain and had a sinking feeling that I did not have the wherewithal for doing whatever else was required.

~

Iskenderun turned oppressively hot in June. Winds from the arid interior howled through the Beylan Pass, making it worse. The saving grace was the delightfully cool air of the Amanus Mountains overlooking the city. We spent the summers at the mountain village of Sovouk Oluk. Lucine took me to play in the nearby pine forest. I very much enjoyed it despite being kept on a tight leash. The memory of that little pine forest became imbued with a deep feeling of nostalgia over the years. However, when several decades later on a trip to the region I took my wife and children to see that hilltop, the puny clump of bedraggled trees looked nothing like what I remembered.

One summer treat that would never fade into insignificance was the freedom from sitting on the potty. I hated its offensive smell. Now instead of the potty, I could use the open ground. To make it even more stench-free, I changed places after each small evacuation. A teenage boy from the village called Hagop who worked for us followed me patiently to scoop the poop. Despite this demeaning task, Hagop was fond of me, and years later in Beirut I would see him occasionally at

Elvira's house and helped him when he could no longer work at his job at a dairy because of the rheumatism in his hands.

I was also in the habit of picking my nose, which caused it to bleed. After the blood coagulated, I had an irresistible urge to pick at the clotted blood, causing a fresh nosebleed. I wonder if it was my way of getting attention, although attention was not something I lacked. One time, the nosebleed wouldn't stop and my mother ordered a car to take me to see the doctor in Iskenderun. When we got to the clinic, the place was in an uproar. My mother got caught up in the excitement and seemed to have forgotten me. The word that people kept repeating excitedly was *harb* (Turkish, "war"). It meant nothing to me, but as I sat there holding the bloody handkerchief to my nose, I realized that something momentous must have happened to take precedence over my nosebleed. It was September 3, 1939, and the British prime minister Neville Chamberlain had just declared war on Germany, with his speech being broadcast on the radio. ("This morning the British ambassador in Berlin handed the German government a final note stating that, unless we hear from them by 11 o'clock … a state of war would exist between us.") My nosebleed stopped, but the war raged on for the next six years, upending our lives.

I started kindergarten in the fall. It was my first foray into the world outside home and freedom from Lucine's constant hovering. I enjoyed mingling with other kids. Unfortunately, it was short lived. My father received a ransom note threatening that unless he deposited a substantial sum of money under a park bench by a certain date, he would never see his only son again. No one told me anything about this, so I couldn't understand why a strange man moved in with us and shadowed me when I went to school. I had no idea he was a police detective. I was startled when one day while I was returning from school an older boy chased me playfully and the man grabbed him by the neck until I told him he was my friend.

After some vacillation, my father agreed to have the police set a trap. The money was deposited under the specified bench and round-the-clock surveillance put in place. A day later, a couple came to sit on the bench, and when the woman bent down, the police pounced on them. It turned out she was merely pulling up her stockings. With the bluff exposed, my father took me out of school and dismissed the detective.

From then on, I was tutored at home by a schoolteacher. I have no recollection of what she taught me, but I do remember the overpowering odor of her body. I don't know whether it was her perfume or the perspiration from her unshaven armpits, but I tried to place myself as far away from her as possible. She was the only woman I had sat close to other than my mother and Lucine, and her effusively affectionate manner made me uneasy.

In my childhood photographs I appear as a poised and pensive child with a touch of gravitas. In none of them do I smile. Nor do I have any memory of getting angry or throwing a temper tantrum. This was largely because my needs and desires were anticipated and catered to. Besides, my dignity would not have allowed it. But I was a fussy eater. When giving me lunch, Lucine would make each morsel into an "airplane" that circled in a holding pattern until I deigned to open my mouth. All of this attention—not only at home but also in the ways that relatives and others treated me—made me feel special. My close resemblance to my mother made me easily identifiable, and everybody knew I was Mrs. Efronia's son. My mother's social standing was partly based on her being married to my father, but she also inspired respect ("demanded" would be closer to the truth) by the force of her personality. I always knew with absolute certainty that my mother loved me unconditionally. I was the most important person in her life as she was in mine. Freud noted that a mother's special love for an only son instills an unshakable sense of self-confidence in the boy. That was true for Freud as it was to be true for me.

One would think that all this attention would have spoiled me. However, I never felt like a spoilt child, nor did anyone suggest it. It would have been unthinkable for me to get upset with my mother or my mother with me. She never punished me; she didn't even raise her voice at me. I was punished only once by Lucine, when she was sorely provoked. An older boy had talked me into going to the nearby beach without telling anyone. When Lucine discovered my absence, panic ensued. She searched for me in vain and as a last resort ran to the seashore. There we were wading ankle-deep in the calm waters with our shoes and socks neatly placed on the shore. I called out to Lucine gleefully, but she was too undone to speak. She pulled me by the ear all the way home. I was puzzled by why she got so upset over nothing.

Figure 1.4 With my nanny Lucine.

Next to my mother, the person closest to me during my childhood was Lucine. I must have been even more central to her life, since I was essentially the only person she truly ever loved. She had no family. She never married. She had no lovers and she barely had any friends. I was all she had and she could not have loved me more even if I had been her own only child.

Lucine was a survivor of the Armenian genocide of 1915. She was an orphan who didn't know who she was, who her parents were, when or where she was born, or even her real name. She was given the name Lucine in the orphanage in Beirut, where she was brought as a child. She was among the thousands of Armenian children who were left behind when their parents were killed or succumbed to starvation and illness during the death marches that ended in the Syrian desert of Deir Zor. Many of these children were taken into their families by Bedouins or given to them by their mothers to save their lives.[5] Following the end of the war, American and European missionaries scoured the desert for these children and took them away, often buying them, from the Bedouins and placed them in orphanages in Aleppo and Beirut. These orphans received an elementary education and were taught various skills: the girls learned sewing and needlework; the boys, trades like carpentry. Some of these children were adopted by Armenian families and others taken in as maids. Some became family members while others were left in an ambiguous position, as was Lucine.

Most of Lucine's earliest memories were from the Danish orphanage near Beirut called the Birds' Nest (*Turchnots pouyn*). She was quite happy there, being well cared for and enjoying the companionship of other children. She remembered very little of her earlier life. The only visible residue of her life with the Bedouins was the tribal tattoo on her chin (which she later plucked out with her fingernails and scoured with ash and lemon juice). One thing she did remember was having a baby lamb that she was given as a pet. She cherished that memory because it made her feel that she must have been liked, perhaps even loved, by the Bedouin family. At the orphanage, she learned knitting and needlework. She excelled at both, and they became her lifelong passions.

Lucine was virtually thrust on my parents by the agent responsible for the placement of Armenian orphans. When the agent first approached my mother about taking in Lucine, she turned him down. She had all the help she needed, and she didn't want to adopt a child; she was hoping to have her own children. The agent told her, "Mrs. Khatchadourian, this is not a matter of your needing this child but her needing you. She is a sickly girl who is rather homely and doesn't have much of a future out in the world. You can afford to take her in and you have a duty to do so." My mother was shamed into saying she would discuss it with my father to see what they could do. My father was not averse to the idea. The cook and the housekeeper were older than my parents, and they felt awkward ordering them around. A teenage girl would be useful for minor chores, and when they had a child she could perhaps become the nanny. They also felt some responsibility for helping the survivors of the genocide. Besides, the girl would cost them next to nothing. Thus, a combination of compassion, social responsibility, and self-interest induced them to take Lucine into the family. It was the smartest thing they could have ever done.

A few weeks later, Lucine showed up with the orphanage agent with all of her earthly belongings in a bundle under her arm. My mother looked at her and sighed. A beautiful woman herself, my mother liked being surrounded by other attractive people, which Lucine was not. Short, with a thick head of coarse dark hair and a jutting jaw with protruding teeth, she looked like a peasant. Yet, there she was, and my mother felt a twinge of remorse for judging her on her God-given looks, sadly lacking as they were. Ironically, as Lucine grew old, people would comment on how "cute" she was. Her diminutive frame, white hair, and kindly twinkle in the eyes made her appealing even to strangers, who didn't know her saintly character.

After Lucine had settled in, my mother checked her drawer and was impressed with its neatness. The few pieces of needlework she had brought along were impressive. She had a meek and docile temperament that was hard to dislike. As my mother's heart softened, Lucine felt more at ease. She was now in a fine home, the likes of which she had not seen before. My parents were intimidating but kind to her, and her duties were not burdensome. Lucine knew she had to make this work. When she had built up enough courage, she asked my mother if she might call her Mayrig (diminutive for "mother"). My mother said she could. That started a relationship that lasted over six decades, longer than the time my mother lived with my father or with me. Toward the end of her life, my mother became totally dependent on Lucine's care. And after my mother died, Lucine became like a second grandmother to our children. Those were the happiest years of her life.

Lucine continued to think of herself as my nanny to the day she died in California at the age of eighty-six (when I was in my sixties). She could never fully trust me to look after myself. When I took her out in her wheelchair and the weather turned cold, she would look up and say, "Put on your sweater," and I did. As she was dying, she extracted a promise from my cousin Nora to cook for me the dishes I especially liked, after she was gone. It's not that Lucine didn't trust my wife, Stina, who was an accomplished cook, but there were some dishes that only an Armenian woman could be trusted to do right.

~

Then there was Elvira. When Lucine came to us, Elvira had already been living in our house for four years. She was the second child of my uncle, Yacoub. Her family lived close by and Elvira was a frequent visitor to our house. My uncle had been very eager to have a son and was disappointed that his first child was a girl. But little Laura was so adorable that he couldn't help growing fond of her. However, when Elvira, yet another girl, was born, he was crestfallen. The third time around, he was desperate. When escorting the midwife for the delivery, he showed her a gold coin and said, "You get me a boy and I'll give you this; if it's a girl, you get nothing." "Look here, Yacoub Effendi," the midwife answered back, "you're the one who planted the seed; I just pluck the fruit. Don't blame me for it."

The third child was indeed a boy. Yacoub named him Kevork, after his own father, who had been killed during the prelude to the genocide, when Yacoub was ten years old. My uncle was ecstatic, but Elvira was not. Between the adorable Laura and the little prince, she felt like a pariah. In desperation, she stuffed chickpeas into the baby's nostrils, nearly suffocating him and bringing down her mother's wrath on her head. It is no wonder that Elvira loved to visit her Aunt Efronia, who doted on her. As she poured out her anguished little heart over how unhappy she was at home, my mother would tell her she could always come to live with her. And that is exactly what Elvira did. At the age of four she showed up with her clothes in a small suitcase and declared to my mother that she was there to stay. In the evening, when her mother tried to take Elvira home, she kicked up such a fuss that they decided to let her stay, thinking she would get homesick in a day or two. She never did, and Elvira became a member of our family with her parents' acquiescence. She continued to call my parents Uncle and Aunt and maintained contact with her parents and siblings, but she didn't leave our house until she got married, at eighteen.

My mother's account of Elvira's coming into our family made it sound like it was all Elvira's doing, but she had aided and abetted it. My mother had been so desperate for a child that when her sister

Aznive became pregnant with her seventh child, my mother persuaded her to give her the baby; my aunt agreed, without her husband's prior consent. The baby was called Robere. Aznive nursed him initially, but my mother switched him to a wet nurse so that my aunt would not become attached to the infant and renege on their agreement. However, the baby got sick and died within a few months despite my father's bringing every physician in town to his bedside. I knew nothing about this until many years after my mother had died, when I heard about it from my cousin Alice (the ill-fated Robere's older sister).

I didn't have much to do with Elvira while we were in Iskenderun. She was too old to play with me and too young to help take care of me. Elvira must have felt secure enough not to resent my intrusion into the family. She was very fond of me and remained so all of her life. I occasionally went to her bedroom to watch her preening in front of the long mirror in the fine clothes my mother had bought for her from Paris. (No wonder the end of the Shirvan runner next to her mirror was noticeably worn out.)

There was a lot of friction between Elvira and Lucine when they were teenagers, even though one would not have suspected it from their affectionate relationship in their adult years. Elvira and Lucine were ostensibly "sisters." Yet their actual status in the house set them worlds apart. Elvira was my mother's niece, her own flesh and blood, whereas Lucine was a stranger out of nowhere. Elvira was pretty, vivacious, capricious, and demanding; Lucine was not. Her position was closer to that of a maid than a member of the family. It wasn't until my birth that she gained more status as my nanny. Consequently, she was fiercely attached to me not only because she loved me deeply but also because I was her anchor in the family.

Despite my mother's protestations that she loved Lucine no less than Elvira, everything from the clothes the two girls wore to their public face testified to the contrary. Lucine would be enraged when a tradesman came to the door and asked her to call the "Madame." That would never happen with Elvira; no one would ever mistake her for a maid. While Elvira was not responsible for these humiliations, she inflamed Lucine's resentments further in the way she treated her. My mother was exasperated feeling trapped between the two of them. However, when my mother reprimanded Elvira, she would threaten to

Figure 1.5 My mother and Elvira.

go back to her parents' house. Lucine couldn't blackmail my mother in the same way, for she had nowhere else to go.

My parents' claim that Lucine was treated like a true member of our family was further belied by the fact that her identity card, and later passport, referred to her as "Minassian," a name thought up by my father presumably to prevent Lucine from laying claim to my inheritance. When I would ask my mother why Lucine had a different surname, she would give me an evasive answer. I finally got this matter set right. Years later, when my mother and Lucine had fled from the

Lebanese civil war to come and live with us in California and were applying for American citizenship, I had Lucine formally adopted by my mother and had her name changed to Katchadourian. I told my mother we needed to do that so as to establish a legal relationship to me and let it go at that.

I wasn't the only one to be so attached to Lucine. She was everyone's *kouyrig* ("sister") or *tantig* ("aunt"). Even though I also called her Lucine *Kouyrig*, I never thought of her as my sister, nor did I confuse her with my mother; she occupied a unique place in my life. Lucine is the only person that I know who did more for everyone else than anyone did for her in return, including me. Yet, for most of her life she was like an extension of my mother and didn't come into her own until my mother died and she could come out of her shadow. Fiercely loyal to my parents, Lucine rarely complained to me about them. The only issue over which she was openly bitter was that my parents didn't send her to school when she came to our house when she was still a child. She had been a good student in the orphanage, especially good in math as she pointed out with pride, and had had a deep love of learning, yet she never had a chance to go beyond an elementary literacy.

I have said relatively little about my father so far because he was a more distant figure in my early life. I was aware of his importance and the deference with which people treated him, and I felt secure under his protection. However, it was more like the security provided by a distant monarch rather than a loving father. I have no memory of his ever playing with me or hugging or kissing me. There is only one photograph that shows him holding me in his arms affectionately. He no doubt loved and cherished me, but he was not able to express his love in ways that would be evident to a child. Or perhaps my heart was so full of my mother that there was no room left for my father.

Part of the reason for the distance between us was also cultural. Men of his generation and background didn't have much to do with their children in an intimate way. My father's character was also a factor. He was not one to give free rein to affectionate feelings. I never saw him hug or kiss my mother either. Nor did I ever see him cry, except years later at the funeral of his younger brother, Haroutune, with whom he had had a turbulent relationship. I was so taken aback to see my father in tears seated on a tombstone that I could only stare at him in silence. I was old enough to have gone and comforted him. How I wish I had done that, but the thought did not even cross my mind. One emotion my father had no problem expressing was anger. He raised his voice at the slightest provocation (one of the few traits that he passed on to me and that I have tried to overcome). However, his anger was mostly bluster that dissipated like a passing thunderstorm. Nonetheless, I was embarrassed when it occurred in my presence, even though his anger was rarely aimed at me.

There were two other important figures from my childhood who did not live with us: my cousin Nora, the daughter of my younger uncle Yervant, and our beloved grandmother Ovsanna, who lived with

Figure 1.6 With Nora and our grandmother Ovsanna.

them. After our family moved to Beirut and Nora's family to Latakia in Syria, I didn't see much of her although my mother continued to be in close touch with them. Years later, when Nora was going to college in Beirut, she lived with us. Since Nora was only a few years older than I, we could relate to each other easily and she became the closest of my cousins.

~

The region where I was born has a rich and complicated history. To make sense of our goings and comings easier and the impact of events on our lives, I will provide a short overview of its background. Most people in the area (including my parents) referred to Alexandretta as Iskenderun (Alexander is *Iskender* in local languages). However, the city did not officially change its name to Iskenderun until it became part of the Republic of Turkey in 1940. For over four hundred years, starting in the sixteenth century, the entire Near East (what we now call the Middle East), including Aintab and Iskenderun, was part of the Ottoman Empire. None of the countries in the region existed in their modern form as political entities until the twentieth century.[6]

The boundaries of modern Arab countries as political entities were created by the Western powers, in particular the British and the French following World War I; Israel was established by the United Nations following World War II. Arabia was a vast desert inhabited by Bedouin tribes. The custodians of the holy cities of Mecca and Medina were members of the Hashemite clan descended from the Prophet Muhammad. During World War I, T. E. Lawrence ("Lawrence of Arabia") incited Bedouin tribes to revolt against their Ottoman masters under the command of Faisal and Abdullah, the sons of the Sharif of Mecca (the great-grandfather of the present king Abdullah II of Jordan). Lawrence promised them that after the Allies won the war, Faisal would be made king of an expanded Arabia. Meanwhile, unbeknownst to him, a secret agreement had been reached on May 16, 1916, between the representatives of the French and British governments, François Georges-Picot and Sir Mark Sykes, to partition the Near Eastern provinces of the Ottoman Empire into two respective spheres of influence. This was accomplished with the assent of Tsarist Russia, which had its own designs on Istanbul. Originally intended as a trade agreement, it evolved into Britain and France's exercising political and economic control over the region. In 1918, Faisal was declared king of Greater Syria. He was based in Damascus and lasted only two years, until deposed by the French, who also disbanded the indigenous Arab government.

The final plan, ratified at the Conference of San Remo in 1920 and approved by the League of Nations (the precursor of the United Nations), created a set of Mandates under which Great Britain and France would foster the establishment of national governments, ostensibly deriving their authority from the initiative and free choice of the indigenous population. France became the mandatory power over the Republics of Syria and of Lebanon, whose boundaries were gerrymandered to create a Christian majority. Great Britain set up two monarchies: Faisal became king of the new country of Iraq and his brother Abdullah king of the desert kingdom of Transjordan, with Palestine to the west under direct British rule.

These arrangements were supposed to last until the people in these countries were ready for self-rule. Yet, the French and the British did not leave until the end of World War II, when they began to dismantle their colonies. King Faisal of Iraq (not to be confused with Faisal Ibn Saud, the founder of Saudi Arabia) died in 1933 and was succeeded by his son, and then his grandson, Faisal II. In 1958, the young king and his family were killed during a military coup. After that, Iraq was ruled by a succession of strongmen, culminating in Saddam Hussein's coming to power in 1979. King Abdullah of Transjordan was assassinated on July 20, 1951, in the Al Aqsa Mosque in Jerusalem, in the presence of his grandson Hussein (who was saved by a medal on his chest deflecting a bullet). The assassin was a Palestinian who believed that Abdullah had colluded with Israel in dismembering Palestine. Abdullah was succeeded briefly by his son, who became incapacitated by illness, and then by his grandson, King Hussein, who ruled Jordan for forty-six years, followed by his son Abdullah II.

The British Mandate of Palestine, formally established in 1922, was administered through a high commissioner until 1948, when Palestine was partitioned by the United Nations into Jewish and Arab states. The Palestinians and neighboring Arab countries rejected the partition plan, and in the ensuing war Israel expanded its boundaries and took part of Jerusalem while Transjordan took the rest of Palestine, constituting mainly what is now known as the West Bank. Since Transjordan now occupied both sides of the Jordan River, its name was changed to Jordan. The final redrawing of the map came with the occupation of the West Bank and Gaza by Israel in the Six-Day War of 1967, leading to the further aggravation of the intractable conflict.

The United States emerged from World War II as a superpower and stepped into the shoes of the British and the French as the dominant Western power in the Middle East. Three key interests led into its increasing engagement in the region. The first was Arab oil, on which America became increasingly dependent. The second was the Cold War rivalry with the Soviet Union for influence over the region, with Egypt and Syria siding with the Soviets and countries in the Persian Gulf and Lebanon with the United States. The third was the unequivocal support for Israel by successive American governments. The first Iraq war was the initial major military engagement of the United States in an Arab country since World War II. The euphoria of its successful outcome became dissipated and wasted during the second Iraq war. More than anything else, the horrendous attack on the World Trade Center on September 9, 2001, and the rise of Al-Qaeda radically changed American attitudes and perceptions of a radical Muslim threat. Americans were stunned as much as they were horrified. Why would anyone do that to them? ("Why do they hate us?") From a dim presence in a distant world, Muslims became very real overnight. There was suddenly so much interest about Islam, and so few people who could speak about it responsibly and in ways that Americans could understand, that I took it upon myself to go around talking about Islam to Stanford alumni groups and others like them, even though I was neither a Muslim nor an Islamic scholar. I had come a long way from my childhood perceptions of Turks/Muslims as the murderers of my grandfather to my view of Muslims as people who needed to be understood and respected.[7]

~

My parents and Armenians more generally were caught in the shifting political sands that followed World War I. They could have moved out of the region, as some did, but others stayed as long as they had some ability to control their lives. When my parents settled in Iskenderun following their marriage in 1919, it had become part of the French Mandate of Syria. Except for its harbor the town did not have much going for it. It was surrounded by swamps infested with mosquitoes and endemic malaria. I remember sleeping under a mosquito net enjoying the feeling of being in a little space of my own. I watched the mosquitoes land on the netting and occasionally managed to squeeze one or two of them between my fingers, staining the net with their blood, much to my satisfaction. It was small compensation for the itch I suffered when they managed to get through the netting and bite me. The use of quinine to treat malaria had become so habitual that after we moved to Beirut, where there was no malaria, my mother would never take aspirin without also taking quinine; it was always "aspirin and solfato" (quinine sulfate).

I must have had malaria myself, although I don't remember suffering its bone-shaking chills. I do remember, however, the suction cups that were applied to my back for colds and coughs. I have faint scars on my back that testify to my being bled into these cups for more serious conditions. The cupping was not painful, but hardly enjoyable, as was Lucine's rubbing my back with warm olive oil and then covering it with newspapers to protect my pajamas from getting soiled.[8]

Iskenderun's main asset was its location on the Mediterranean. Next to Beirut to the south, it was the most important port on its eastern coast. Its harbor, sheltered in the bosom of the Bay of Iskenderun, had been a trade entrepôt for many years with shipping lanes to Beirut to the south and to Turkish ports on the southern coast of Anatolia to the west. When my parents settled in Iskenderun, the first few years were a miserable time for my mother both for deeply personal reasons and for the deplorable condition of the town. In addition to its sickly climate, it had no social life to speak of. Compared with the Aintab of my mother's youth, the place was a cultural backwater. My father was absorbed in his business and offered little respite and distraction. The fact that my mother could not bear children rankled as people whispered about who would inherit the Khatchadourian fortune if there was no heir.

Gradually life got better. The French drained the swamps. Expanded trade opportunities brought in European expatriates. The little town took on a cosmopolitan and polyglot character. Its largest communities consisted of Turks, Armenians, and Arabs. There were smaller groups of Greeks, Italians, and the French as well as a few British families. One of my mother's closest friends was the English Mrs. Watt, who spoke fluent Turkish. The French administrators and their wives added much flair to the city's elite. They called Iskenderun Little Paris. My mother had the status and the social graces to fit right in. She bought her clothes from Parisian catalogues, and my father indulged her wishes, including having a piano hauled in from Germany. My mother's patience and perseverance in turn made my father more amenable to getting involved in social circles. The photographs from the balls held on special occasions show the women dressed in elegant clothes with bare backs and plunging necklines and men in formal suits.

The problems that plagued my mother at this point were mainly caused by our relatives. When the families of her two sisters were struggling financially, they gravitated to Iskenderun, where my father could help them, as he generously did. My father's brother, Haroutune, who worked for him, got into trouble with some customs documents. Although my father was exonerated of complicity, he was deeply embarrassed. He threw his brother out of the business but ended up taking him back at the pleadings of his wife. Uncle Haroutune was a troubled man, but it could not have been easy to be my father's younger brother and to work for him. I remember him years later in Beirut when he suffered from excruciating sciatic pains and became addicted to morphine, which was his only source of temporary relief. He was the first person I knew to be so utterly crushed by pain and suffering—a gentle man whom life had treated harshly. He was always kind and affectionate to me when I visited him, trying to smile while wincing with pain. I felt deeply sorry for him. He died in the hospital while undergoing detoxification. The doctors may have bungled the procedure, but it was a merciful end for him and his family, and no one made a fuss.

Some of my mother's other woes were caused by her widowed sister-in-law (and cousin) Marie, who lived with my parents with her three children. The contrast between the lives of the two cousins was striking: they had married two brothers but ended up with very different fates. Marie was a widow with no husband, no house, no money. Efronia had it all. No wonder Marie felt envious and resentful. However, with the passage of years and my mother's capacity to bury the hatchet, all was forgiven. Marie's son, Uncle Khatchig, became my godfather. (I called him Khatchig *Ammo*, while Elvira's husband was Khatchig *Enishde*, in Turkish.) After we came to Beirut, I dutifully visited Marie Nanoug and Uncle Khatchig every Easter. They were both very fond of me and always received me joyfully, showering me with colored eggs (which I didn't care for) and chocolate and candy (which I liked). Uncle Khatchig gave me the most useful Christmas presents, including a large Webster's dictionary when I was in high school, which I used to the end of my years in medical school.

~

The French Mandate in Syria, including the *sanjak* (district) of Iskenderun, was imposed on Turkey by the Allies following World War I. Mustafa Kemal Pasha—the charismatic Ataturk—came to power and established the modern Turkish republic in 1923. Soon after, he laid claim to the *sanjak* of Iskenderun, which he called *Hatay*, on the grounds that the majority of its population was Turkish. The French tried to mollify Ataturk, but they would not relinquish control over the region.[9]

The French position changed as the clouds of World War II gathered over the horizon. The prospect of Turkey joining Germany (as it had done in World War I) was deeply unsettling. Had that happened, the course of World War II may well have been different. The Germans would have had access to the oil fields of Iraq and Rommel would not have run out of fuel in North Africa and may have chased Montgomery out of Egypt. On the eve of the war, the French signed a Franco–Turkish friendship treaty and agreed to hold a referendum that would determine whether the *sanjak* of Iskenderun should become part of Turkey or remain attached to Syria. The referendum came out in favor of the Turks, and in 1939 the region officially became a province of Turkey.

Some historians have claimed that the referendum was rigged (Syrian maps continued to show the region as part of Syria).[10] As the referendum was in progress, the senior French administrator had contacted my father and told him that they were going to leave irrespective of the outcome of the referendum; the Armenians would be better off if Turkey took over the region peacefully rather than by marching in with its troops. My father sent word to stuff the ballot boxes in favor of the Turks. When it was over, the Turks moved in with great fanfare. I was part of the welcoming committee. There is a photograph showing me standing next to my father, dressed in my fine black velvet suit and struggling to hold on to a large bouquet of flowers towering over my head.

Prior to the Turkish takeover, the French authorities declared that those who wished to leave the region of Iskenderun would be given Syrian or Lebanese citizenship and could move to either of those countries. Some 50,000 people chose to leave, including most of the Armenians in the region. They included the villagers of Musa Dagh, who during World War I had fought the Turks to a standstill until rescued by French warships. Hence, they had special reason to fear retaliation at the hands of the Turkish government. The French resettled them in what is now the town of Anjar, in the Beka'a Valley of Lebanon.[11]

My parents were in a dilemma. Given their tragic experiences in Ottoman Turkey during World War I, they felt a strong inclination to leave. On the other hand, my father had too much to lose. Besides, the Turks who took over were civilized and gracious people. The family of a senior Turkish officer rented the ground floor of our house, while we lived upstairs. I fondly remember his wife, a petite and vivacious woman, who was always delighted to see me. When I was sick with a cold, she brought me a toy machine gun that made a huge racket throwing out sparks. It was the most amazing toy I ever had. My parents had a cordial relationship with the Turks. They spoke fluent Turkish with them and culturally had much in common. It looked as if we would have no problem living with them. There was an occasional sour note, such as when my grandmother told a Turkish lady that their Ottoman ancestors acted like "savages" in the way they treated the Armenians during World War I. My mortified mother tried to contain the damage by explaining the circumstances (including the murder of my grandfather) that led her to say what she did; she meant no disrespect to the lady personally. My grandmother was hardly apologetic.

Our Turkish friends, suspecting that we may be contemplating leaving Iskenderun, pleaded with my parents to stay, assuring them that no harm would come to us in modern Turkey. My father said he had no intention of leaving, when in reality that prospect loomed large in his mind. Long before these developments, my mother had urged my father to move out some of his assets to Beirut and to open a branch of his business there. His nephew, my godfather Khatchig, was already working in Beirut and could have easily taken charge of such an enterprise. But my father would not yield to my mother's pleadings; he listened to nobody when it came to running his business (or for that matter, to taking advice about anything else).

In the fall of 1940, as the war began to heat up, my father decided to take us on a visit to Aleppo and on to Beirut to reassess the situation from there. I vividly recall the day we got into a car and I was told that we were going on vacation. Why then, I wondered, were my mother and Lucine sobbing so inconsolably? I must have suspected that there was more to what was happening than I was being told. Shortly before we drove away, Lucine found me in our garden, plucking peaches from a tree and flinging them to the ground. She asked me why I was doing that and I told her, "I don't want the Turks to get our peaches."

As the car started on its way, we sat in a gloomy silence. I had no clear sense of where we were going and when we would get there. I must have dozed off soon after but awakened when the car stopped at the border crossing into Syria. It was chaotic. Buses and cars jockeyed for position as their drivers argued and yelled at each other. I was seated between my mother and Lucine, while my father stood outside talking with some agitated people. A five-gallon can of olive oil had fallen off the rack of a bus, splattering oil all over the tarmac. Someone was trying to smuggle it across the border to avoid paying duty on it. The commotion made me apprehensive and compounded the discomfort of being stuck in the car for hours.

Even though it is the tedium of travel that I remember most, leaving Iskenderun was far more traumatic for all us. Our family passport photograph betrays our mood. My father looks somber, my mother's eyes are brimming with worry, and I look lost. We stare at the camera with expectant, almost plaintive, expressions. We were not refugees in tattered rags but on our way to exile nonetheless.

~

Forty-five years later, in 1985, we went on a visit to Aintab and on to Iskenderun with my wife Stina and our good friend Sanford Gifford, a great travel companion with an insatiable curiosity and a wealth of knowledge. Stina was working on her book based on my mother's memoirs, and we wanted to see some of the places that had played a central role in her life. We were able to find my mother's old house in Aintab through fortuitous circumstances and could easily match the original building with my mother's descriptions down to the wood paneling of the living room.

Like so much else in my life, serendipity had a hand in this discovery. One of my mother's cousins, a dentist in Aleppo, had visited Aintab fairly recently and could give me a general description of where my mother's old house was located. The fact that it was next to the house of the Mufti (a Muslim legal scholar and community leader) was going to be of help. Earlier in our trip, we stopped in Ankara. On a friend's recommendation, we met a young American woman who worked at the U.S. embassy. She took us to a restaurant, whose owner turned out to be a woman from a prominent Turkish family in Aintab. Her parents lived in the vicinity of my mother's house, and there was a street named after them. Her brother was in Aintab, and he would help us find the Mufti's and my mother's house. That is what he did, and he made it possible for us to visit both houses.

We then went on to Iskenderun but were less fortunate in finding our own house. I remembered very little of where we lived. Moreover, during the following four decades the town had changed a great deal. I had a photograph of our house, but as I walked around the neighborhood, I could find nothing that remotely matched it. I knew that our house was located close to the sea, so I walked into a shipping agency along the shore and showed the photograph to a cordial, well-dressed man, who instantly recognized it. He knew the people who used to live there, but the house had been torn down thirteen years earlier and replaced by a four-story apartment building. He kindly took me there, and the location fit perfectly with the background mountains in my photograph. I went into the grocery store on the ground floor of the building and was led to the backyard, which was a small part of what remained of our garden. I was reluctant to ask too many questions, since I could already see suspicion in the shopkeeper's eyes. (Who was I, and what was I looking for?) Unfortunately, I could not contact the couple who had last owned our house and presumably had lived in it. I was told that they were old and unwell. I did not quite believe it but decided to let it go; I now wish I had been more persistent. (Yet, I shudder to think what my reaction would have been had I seen my parents' brass beds there.)

~

Every life is unpredictable, but mine has been perhaps more unforeseen than most. It is tempting, if ultimately futile, to speculate on the various paths my life could have possibly taken instead of following the meandering course it did take. When I was born in Iskenderun there was a ready-made script with a predictable scenario for my life. I would have attended local Armenian schools and then probably gone to the American University of Beirut (which I did anyway). I would have probably studied business or civil engineering but certainly not medicine (which is what I did).

I would have then gone into my father's business and eventually taken over its management and inherited it. I would now be a portly businessman in Iskenderun owning a good slice of the town and probably would have become an even bigger fish than my father in a somewhat bigger pond. I would have married a good-looking young woman picked by my mother from a local Armenian family, or perhaps a prominent one in Beirut, and I would have had a large family with many children and grandchildren. During our visit to Iskenderun, I actually met a man who fit this model quite well. He was a successful businessman of Italian extraction. Unlike us, his family had stayed in Iskenderun when the Turks took over. Our fathers had been contemporaries and served on the same town council. I asked him if he had been happy living in Iskenderun. He sighed. Yes, he said. He had had a good life but had always felt like a second-class citizen, even after serving as a captain in the Turkish army. His greatest concern was that his eighteen-year-old son who was about to leave for college in the United States would never come back.

An alternative, and less likely, scenario is that I would I have chafed under my father's control and become embroiled in a contest of wills with him over running the business. Consequently, I may have taken off for Beirut or elsewhere, to make my own way. That would have been hard on my father but he would have not given in. My mother would have been caught in the middle, but she would have probably managed to work out a compromise; in any case, she would have never abandoned me.

Then there is a third possible outcome. Had my father taken my mother's advice and opened a branch in Beirut, it would have created a whole set new of prospects. There is a good model for that version, too. One of my father's friends who was in the same type of business left Iskenderun in good time and restarted his business in Beirut. He and his sons (the oldest of whom was my own age) were immensely successful in the booming war economy and eventually became the wealthiest Armenians in Beirut.

When we left Iskenderun under the circumstances that we did, all bets were off. There was no longer a script for my life. The circuitous path my life took then on with its unpredictable twists and turns led me to where I am today. Would I have been more or less happy had I stayed in Iskenderun? Would I have missed what I didn't know existed? Certain events in our lives may feel like a disaster at the time. Our family treated the loss of Iskenderun as a disaster, and I fully accepted that view myself. But now it looks as if it might have been a blessing in disguise that led to a more exciting, rewarding, and fuller life for me and to who I am today.

2

The Khatchadourians and Nazarians

WHAT DOES IT MEAN to say that my parents were Armenian? There is a country in the Caucasus called Armenia, but my parents had never been there. They were born and lived in Ottoman Turkey, but they were not Turks. After they married, they settled in the French Mandate of Syria, but they were neither Syrian nor French. They moved to Lebanon and became Lebanese citizens, but they were never "really" Lebanese. Finally, my mother emigrated to the United States and died in California, without "really" becoming American. Yet my parents always thought of themselves as "Armenian," more specifically, *Aintabtsi*—Armenians from Aintab. And they would say so within five minutes of meeting another Armenian.

The ambiguity of who I "really" am has bedeviled me more than it did my parents. I can hardly ride in a taxi from an airport (especially JFK) without the driver asking me, "Where are you from?" If say I am from San Francisco the next question is, "Yes, but where are you *really* from." If I say I'm Armenian, then, "Where is Armenia?" If I say I'm actually not *from* Armenia, that raises a host of other questions. If I say I'm from Lebanon, it's no better. If I say I'm from Iskenderun, then I'm in for a really long ride. Why should the driver care? Because he senses that I'm not "really" American even if I may look like one. Often nor is he, and that creates a kinship between us. He actually hopes that we are both from the same country, but there is little chance of that being the case. Only once was the taxi driver Armenian. No wonder the ride turned into a nightmare, because during the entire trip through heavy traffic he hardly took his eyes off the rearview mirror, all the while talking to me trying to establish if we were related.

No matter how Armenian my parents felt they were, there was also a lot of the Turkish in them. They spoke Turkish to each other and to relatives and *Aintabtsi* friends. They spoke in Armenian with me, but I spoke in Turkish with my grandmother, who didn't speak Armenian. The heavily Armenian provinces of Anatolia were in the east. The Armenians of Aintab, farther to the west, were a minority in a largely Turkish-speaking region. Over time, they lost their facility in Armenian because of the lack of use and through compulsion. (There was a time when the Ottoman Sultan's Janissaries would cut off the tongue of anyone who dared to speak Armenian.) Yet, at some level, they clung to their mother tongue. Apostolic Church services continued to be in classical Armenian. My grandmother's Bible was in Turkish but written in the Armenian script. It was not until the nineteenth century that Armenian began to be taught again in schools as part of the Armenian cultural renaissance. My parents belonged to generation in Aintab that became bilingual.

The food my parents ate at home was the distinctive cuisine of Aintab, which was more Turkish than Armenian. Dishes originating in the various regions of Anatolia were specified as such, like *Adana kebabeh* or *Amasia bamiaseh*. Armenians in Turkey generally dressed like Turks. Differences were mainly regional, between peasants and city folk, the rich and the poor, rather than between ethnic groups. One exception was that Armenian women (and other Christians such as Greeks) did not veil themselves, although they might wear a headscarf in public. More important differences set Armenians apart from their Turkish neighbors. Armenians were exclusively Christian and Turks Muslim. They had a sense of kinship with Armenian history and culture, they were more Westernized through their contact with American missionaries (particularly the Protestants), their social interactions were largely restricted to their own community, and they almost always married other Armenians. (One had to convert to Islam in order to marry a Turk, which happened very rarely.)

Armenians, like other groups, have distinctive names. Their surnames typically end with "ian," in reference to a forebear, a trade, or a geographic location. It essentially means "from" or "related to."

My surname comes from my great-grandfather Khatchadour—hence Khatchadourian (*Khatch* is "cross" and *dour* "given"). Names based on trades and places were often derived from Turkish. For instance, the Turkish word for blacksmith, *demirji*, becomes the Armenian surname Demirjian. A person from Istanbul would be Istambulian. My mother's surname, Nazarian, is derived from the Arabic word for "gaze" (*nazar*). Occasionally Armenians would have Turkish first names, such as Yacoub (for Hagop or Jacob), but no Armenian would be given an Islamic name such as Ahmed or Aisha. Currently, Western names have become much more prevalent among Armenians. In our extended family we have Laura, Nora, Lena, Sona, Tania, and Nina, as well as Edward, Paul, and Kai.

The spelling and pronunciation of my name was to plague me all of my adult life. As soon as I give my name, I am asked, "How do you spell it?" It becomes more challenging when earnest people insist on pronouncing my name "correctly." To avoid this, I sometimes use a simpler name, for instance "Kay," when making restaurant reservations, but it doesn't always work. Once I said my name was Smith to a dry cleaner and I was still asked to spell it ("Oh, you mean, Smith"). Finally, I discovered one name that never requires spelling: *Tom*. The problem is compounded by my accent. It's hard to place it because it was shaped by the four languages I spoke before I learned English (Armenian, Turkish, Arabic, and French). When asked, "What kind of accent is that?" I tell people to guess, and they come up with astonishing suggestions. Hungarian is the most popular choice, for reasons I cannot fathom.

I learned the danger of using false names some years ago when I signed up for a ski class in Vermont. I told the instructor my name was "Henry," but then forgot about it. As the lesson was under way, I noticed that the instructor was making comments to everyone else but not to me. I knew I wasn't skiing faultlessly, so I asked the instructor why she had nothing to say to me. She looked miffed. "What's the use?" she said. "For the last twenty minutes I've been yelling, 'Henry, bend your knees,' but you act as if I'm not talking to you."

My daughter Nina, who is a conceptual artist, made a video piece called *Accent Elimination* (2005). It was inspired both by the posters she noticed around New York City advertising coaching for "accent reduction" and by her own long-standing frustration that she had never been able to accurately imitate her mother's or my accents. Nina's video piece required that we work with a renowned professional speech coach, Sam Chwat, to help Stina and me acquire a standard American accent and for Nina to learn to speak with our accents. Stina and I each wrote a one-page script that described how it came to be that we speak with the accents we have (which also ends up being a condensed version of our life stories). The coach worked hard with the three of us for nearly three weeks, often imitating us and making us sound hilarious, to help us learn the scripts without our accents. I hadn't realized that it is not only how one pronounces words but the cadence of one's speech that shapes one's accent. I was told, for example, that I pronounce each word in a sentence (doesn't everyone?) and had to learn to slur my words. For instance, "I also" had to sound more like *Ialso* as if a thick sauce had been poured over the words. The thirteen-minute video that resulted showed our struggles and also our occasional success.[1]

~

The origin of Armenians is lost in antiquity. According to the biblical account of the Flood, Noah's ark came to rest on Mt. Ararat in the heartland of ancient Armenia. Haig, the legendary eponymous ancestor of Armenians, is claimed to have been a great-great-grandson of Noah. He defeated the Babylonian king Bel in single combat and established his nation in the region of Mt. Ararat. Hence, Armenians call themselves *Hay*, after Haig, and their country *Hayasdan*. The historical origins of the Armenians go back to the seventh century BCE, when Armenia emerged as a nation in the region of Lake Van in eastern Anatolia through the intermingling of Indo-European invaders with the native population of the Kingdom of Urartu (the Assyrian name for Ararat). The first king of Urartu was Aramu (880–844 BCE), and hence his people came to be known in the region as *Armenians*, and their country *Armenia*.

Urartu was absorbed into the vast Persian Empire at the end of the sixth century BCE. From then on, Armenia came under the sway of one great power after another. For a brief period, a Greater Armenia established in 189 BCE reached its peak under Tigran the

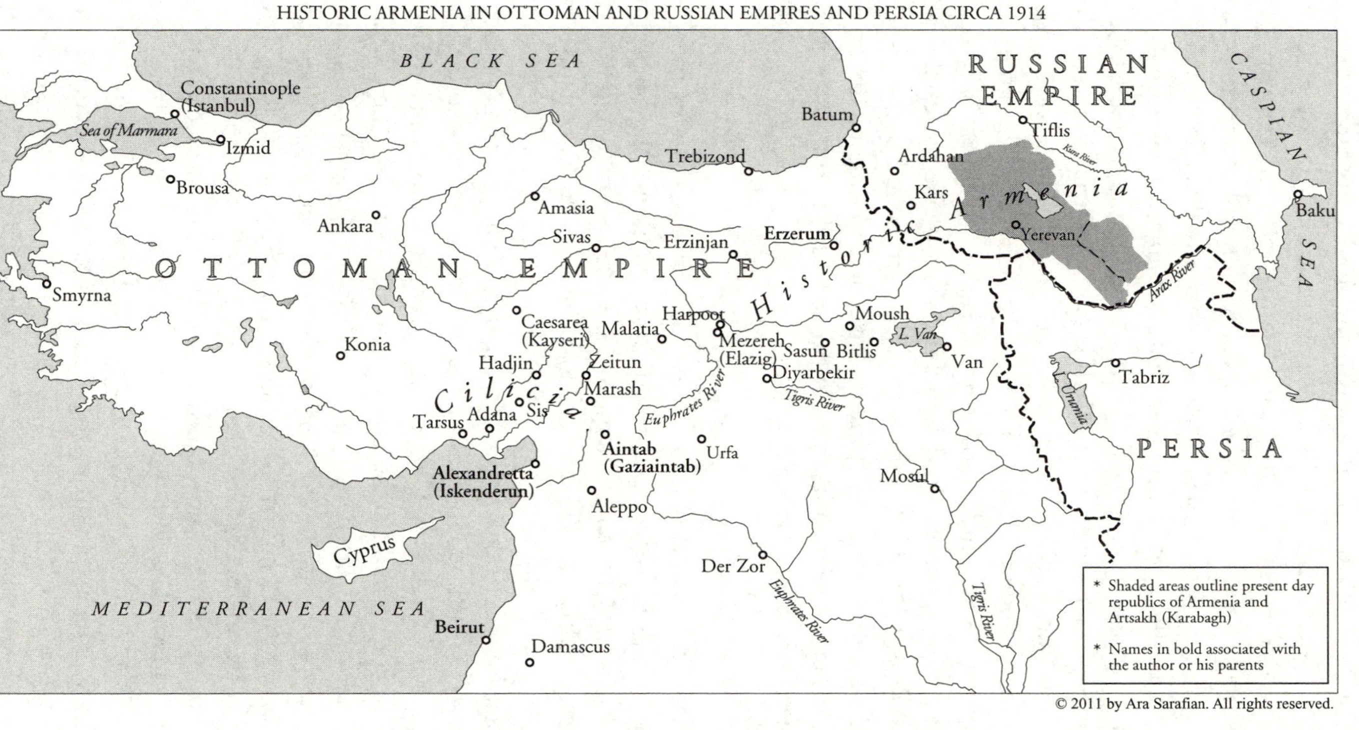

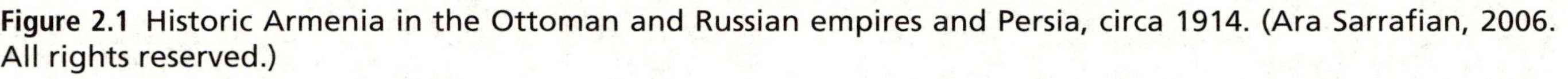
Figure 2.1 Historic Armenia in the Ottoman and Russian empires and Persia, circa 1914. (Ara Sarrafian, 2006. All rights reserved.)

Great, but then fell to the Romans a hundred years later. The last independent Armenian principalities were in Cilicia, in south-central Anatolia. Armenians came to Cilicia in the eleventh century from eastern Anatolia, fleeing from the Seljuk invasion and occupation of the Armenian capital of Ani. In Cilicia, they established medieval fiefdoms that became allied with Crusader states from the eleventh to the thirteenth centuries and served as a foothold for their establishment. (The last Armenian king, who was half French, is buried in the Basilica of St. Denis outside Paris.) Both Iskenderun and Aintab were within the historical region of Cilicia. Eventually, the region became part of the Ottoman Empire and is now part of Turkey. At the end of World War I, a small and short-lived Republic of Armenia was established in the Caucasus, until overtaken by Communists and becoming the Armenian Soviet Socialist Republic. After the breakup of the Soviet Union, Armenia regained its independence in 1991 and now exists as the modern Republic of Armenia.

~

The great Armenian historical claim to fame is its being the first nation to accept Christianity as its state religion, in 301—before the Roman Emperor Constantine legitimized Christianity by the Edict of Tolerance of Milan in 313 and Theodosius I made Christianity the official religion of Rome in 380. The majority (around 85 percent) of Armenians belong to the original "mother church," which is variously referred to as Orthodox, Apostolic, and Gregorian. The others are Protestants and Catholics. Christianity was brought to Armenia by the apostles Thaddeus and Bartholomew, two of the seventy who followed the twelve original disciples, which is why the Armenian Church is called Apostolic (*Arakelagan*). Christianity, however, did not take root until Gregory the Illuminator—Krikor Lusavorich—cured King Trdat (Tiridates) III of the delusion that he was a wolf (a condition known as lycanthropy). In gratitude, the king and his court became Christian; hence the church is commonly known as Gregorian (*Lusavorchagan*). The Armenian Church is also part of the Orthodox, or Eastern/Greek, Christianity, which split from the Roman Catholic (Western/Latin) Church in the great schism of 1054. Armenian Protestantism started as a reform movement within the Apostolic Church in Turkey in the nineteenth century and became the autonomous Armenian Evangelical Church in 1846. The Armenian Catholic Church dates back to the Crusades, and Pope Benedict XIV formally recognized it in 1742.

In principle, all Armenians are Christian. During centuries of oppression and persecution, it is the church that preserved the Armenian national identity, and Armenians have paid dearly for their faith. Consequently, even those who are secular feel a cultural kinship to the church. An Armenian who converts to another religion is typically seen as no longer Armenian, or at best "less Armenian." My father belonged to the Apostolic Church, in which I was baptized and to which I remain culturally attached. However, my mother was Protestant, and I was brought up mainly in that tradition. I went to an Apostolic elementary school and an Armenian Evangelical high school and became far more intimately engaged with the Evangelical Church during my adolescence.

~

The diaspora has been critically important throughout Armenian history.[2] Over the centuries, Armenians have emigrated to every corner of the world. Consequently, there are almost three times as many Armenians living in the diaspora than the three million citizens of the Republic of Armenia. The largest Armenian communities in the diaspora are in Russia, numbering over a million. (The Russian MIG fighter plane was designed by Artem Mikoyan, and his brother Anastas was part of Stalin's inner circle.) There are about 400,000 Armenian Americans, concentrated mainly in and around New York, Boston, Fresno, and Los Angeles. There are also important communities in Europe (especially France) and in the Middle East (especially Lebanon).

Despite their relatively small numbers, Armenians have survived and thrived by adapting to their host countries and becoming part of their fabric. Part of my education as an Armenian boy was to learn about the important contributions that Armenians had made to world culture. For example, several Byzantine emperors and empresses were of Armenian origin. When the roof of the Hagia Sofia collapsed in the eleventh century, the Byzantines swallowed their pride and brought in the architect Trdat from the Armenian capital of Ani to rebuild it (it still stands). In Turkey, the greatest Ottoman architect, Sinan, who lived in the sixteenth century, was a Janissary and most probably recruited as a boy from an Armenian village (to which he made donations). In the nineteenth century, the Balian family produced nine architects, who were responsible for many major public buildings in Istanbul, including some of its Baroque palaces and mosques. Armenian rug weavers made the exquisite *Kum Kapu* carpets for the sultan. The use of red clay, which defines the third phase of Iznik pottery in the sixteenth century, was an Armenian innovation. My father was especially proud of the fact that the Armenian Church was the custodian of a third of the Church of the Holy Sepulcher in Jerusalem. He said it made him feel like Armenians counted for something in the world.

My own family illustrates how widely Armenians are dispersed in the diaspora. I have an archeologist cousin in Yerevan (whose son was the Armenian ambassador to Argentina and to Lebanon). I have a few cousins left in Lebanon, a cousin in Syria (a retired general), and others who lived in Jerusalem, Amman, and Baghdad. In Europe, I have cousins who are in the diamond business in Antwerp, and others who live in Brussels, Paris, and London. I even had a cousin living in Cork who was married to an Irish officer whom she met in Cyprus when he was with the UN peacekeeping forces. I was surprised to receive a letter from her daughter, who had seen one of my books at the Cork city library and recognized my surname. In North America, I have cousins in Montreal, New York, Princeton, Boston, San Francisco, and Los Angeles. During the Montreal wedding of the daughter of my cousin Christine there were six hundred guests at a spectacular sit-down dinner in fairyland surroundings. One of the bands with an Armenian vocalist had been flown in from Los Angeles.

~

My father's family originated in Erzurum, one of the heavily Armenian provinces in eastern Anatolia. We can trace the family only to my great-grandfather Khatchadour, who was born around 1800 and moved to Aintab, where he died in the 1850s.[3] He was a wealthy copper merchant, nicknamed *Gazandji* (Turkish, "cauldron"), and owned several villages with their farmlands, making him an Anatolian version of landed gentry. Since it took ninety donkeys to transport the produce from his villages to Aintab, the family was nicknamed *Doghsan Eshekli* (Turkish, "Ninety Donkeys").

My grandfather Adour (1846–1908) was Khatchadour's second son. In the few photographs we have of him, he looks like an imposing and stern man with an air of supreme self-confidence. He was widowed four times and fathered twelve children. At the funeral of one of his wives, a young woman watching the funeral procession told her companion, "One woman's misfortune is another woman's good fortune." She became Adour's next wife. In a subsequent version, Adour saw a woman watching the funeral cortege of his third wife go by. He smiled at her, and she smiled back and forty days later became his last wife and my grandmother. (Otherwise, I know nothing about her background.) There are many stories attesting to the severity of my grandfather's character. He bought shoes for his boys all at the same time. Those who wore them out early went barefoot until the rest of the brothers had worn out their own shoes. (They wore *yemenis*, which are like leather slippers.) My father would take off his shoes in less-frequented places and go barefoot to save his shoes. My aunt Khatoun was one of Adour's older daughters, and as he became successively widowed, she took care of her younger siblings. She was an unusually intelligent and competent woman. After she married a lawyer, her husband's clients would often seek her advice as well.[4]

Of the persons identified in the following photograph, I knew my uncle Haroutune and my aunt Hripsimeh, as well as Meroum Nanoug and Marie Nanoug. There is a striking contrast between the traditional clothing of Adour and the Western-style suits of his sons. The three sisters-in-law also look quite similar in their more elaborate clothing and hairdo and different from Adour's daughters. The person I identify with most strongly is my grandfather Adour, perhaps because he is so obviously the center of gravity.

My father, Aram, was born on January 24, 1886, the oldest of my grandfather's three children by his last wife. He attended the Vartanian high school in Aintab from which he graduated in 1903 at age nineteen. Otherwise, much to my regret, I know very little about his upbringing and early life. He never talked to me about it (perhaps because I never showed any interest). Most of the information I have about both sides of my family comes from my mother. There is a picture of my father on

Figure 2.2 My father and his family in 1906. My grandfather Adour is in the center. On his far left is his son Sarkis with his wife, Meroum, behind. On his right is Yeghia with his wife, Marie. Next is Haroutune. Standing behind is my aunt Khatoun, and kneeling in front on the left is Hripsimeh.

a bicycle taken on Cyprus, which intrigued me, but I have no idea what he was doing there. When he was in his twenties, he and his two older brothers, Sarkis and Yeghia, established their construction business in Iskenderun. My father was a man of intimidating intelligence, highly competent and self-assured—a man of his word and scrupulous in his business dealings. The one ethical principle he inculcated in me was that the Khatchadourians always pay their debts (which I have extended to honoring my obligations).

My father was tough in his business dealings but scrupulously honest and fair minded. A neighbor had entrusted him with his family jewels for safekeeping in his office safe. Thieves managed to crack the safe and took the money, but they had to leave in a hurry and left behind the box containing the neighbor's valuables. My father could have appropriated them by claiming that they too were stolen—but he did not. This may seem like the natural thing to do, but it was not so perceived at that time. In another instance, a bank clerk counted out to him an extra gold sovereign. My father returned it, and the grateful clerk told him the mistake would have cost him his monthly salary. I never heard my father lie, and I cannot imagine him ever doing so. I too have never told a bold-faced lie, even though I have not always told the whole truth and have occasionally embroidered it.

One episode in particular conveys my father's entrepreneurial skill and daring. At the outset of World War I, the Turkish army was having difficulty provisioning its troops in mountainous regions in the area since the horse-drawn carts kept breaking down on the rough roads. My father approached the regional Turkish commander in Aleppo with a proposal. For an agreed sum and forty exemption certificates from the army, ostensibly for the people who would be working for him, he would undertake to ferry the goods to the troops with no questions asked. The commander, with his back to the wall, agreed, with the understanding that if the plan failed my father's life would hang in the balance. My father got rid of the clumsy carts and bought a herd of camels instead. Camels do not break down on rocky terrain, and it only takes a few men to lead a caravan. He gave a few of the exemption certificates to his close relatives and sold the rest at prices worth a man's life. He gave some of his profits to his two brothers as his business partners and walked away with 600 gold sovereigns—a substantial fortune at the time. (I still have my father's red kilim traveling belt with concealed pockets for carrying gold coins, but most of the gold is gone.)

~

The family of my mother's father were originally from Persia. The Nazarians have been traced back to the sixteenth century, but the founders of the family branch in Aintab came from Persia around 1700. They were three brothers on their way to Jerusalem on a pilgrimage. On a Sunday morning, they went to church and were dismayed by its decrepit condition. The community leaders regretfully told them that they could not afford to do better for fear of incurring higher taxes. The Nazarian brothers offered to pay for the renovation of the church. The youngest brother stayed behind to oversee the project and settled in Aintab.[5]

My mother's paternal grandfather, Hagop Nazarian, lived in the nearby town of Kilis (hence he was called *Kilisli Agop*). After losing his first wife, he moved to Aintab and married a widow. In 1860, Hagop switched from the Apostolic to the Protestant Church because the Protestant minister had helped him to win a lawsuit whereas the Apostolic priest had not lifted a finger. Various stories circulated in Aintab about why people left the mother church to benefit financially and to curry favor with the American missionaries. After a Protestant convert reverted to the Apostolic Church, his explanation was, "When the flour ended, so did the faith" (Turkish, *Oun bidti, din bidti*). When American missionaries provided money to be given to the poorest of the poor (the "naked poor"), the church elders went to a Turkish bath, took off their clothes, stood around, and said, "Well, we are all naked," and pocketed the money. Protestants attributed such scurrilous tales to envy since they tended to be more educated and better off through their association with missionary schools.

Kilisli Agop was a goldsmith. His second son, Kevork—my mother's father—owned a shoe store in the Arasa bazaar of Aintab and was a respected member of the Protestant Church council. My maternal grandmother, Ovsanna, was the pampered only child of Hagop and Trvanda Matossian. (Both of my mother's grandfathers were called

Hagop.) Trvanda was a renowned beauty with green eyes. When she reached puberty at thirteen, suitors flocked to ask for her hand. The candidates were narrowed down to two eligible young men, but choosing either meant rejecting the other and offending his family. So the family went to the local judge (*kadi*) to have him settle the matter. As the two suitors stood side by side in court, neither of whom Trvanda had seen before, the judge said to her, "My dear girl. Both of these two young men want to marry you. Which of them do you want to be your husband?" Trvanda was puzzled. "What does that mean?" she asked. The judge sighed and told her, "It means you will go and live with his family and then you will find out the rest." That was good enough. Trvanda looked over the two men, pointed to my mother's grandfather and said, "This one."

The child bride was brought up by her mother-in-law. Trvanda's friends continued to visit her, and after she had her first child they could play with a real baby instead of playing with dolls. Trvanda grew into a formidable woman. In his older years, her husband lived in fear of her. One day, Trvanda gave him some money to buy groceries. He did that, but he also bought a pack of cigarettes and a watermelon. He hid the cigarettes and cheerfully presented the watermelon to his wife. Trvanda took the watermelon, smashed it against the courtyard wall, and told him to go eat it himself. Hagop died a painful death when he mistook a bottle of lye for raki and drank it.

When my mother's father, Kevork, turned nineteen, his parents asked for the hand of Ovsanna. Kevork was not allowed to accompany his parents on this ceremonial visit, but he followed them surreptitiously to get a glimpse of the girl and was greatly pleased by what he saw. After their engagement, he could visit her in the presence of her mother. One rainy day, as Kevork was leaving her house, he ran into Ovsanna coming down the steps. He pulled her under his umbrella and kissed her. Kevork's kindly and cheerful temperament endeared him to everyone. There are many stories attesting to his playful character. In the family photograph taken before my mother was born, Kevork's clothes appear to be in slight disarray. When my grandfather saw the picture, he told Ovsanna, "That photographer was so busy looking at you that the rascal had no time to tidy me up." When my grandmother went to the roof to hang the laundry, the men in the neighborhood would go to

Figure 2.3 My mother and her family (c. 1890). Aroussiak and Yacoub stand in the back. Aznive is in front. Yervant is on his father's lap. (My mother was not born yet.)

their windows to look at her. One day, Kevork went up to the roof with her, exposed his buttocks, and shouted, "Here is something for you to look at." One of their infants kept my grandmother awake at night. Kevork told her that he would attend to the child on the following night so she could sleep undisturbed. The child started crying in the middle of the night, but my grandfather continued to sleep. When his wife nudged him, he drowsily said to her, "Don't worry. You go on sleeping. The child can cry during my watch."

During nine years of marriage, Ovsanna gave birth to seven children, two of whom died in infancy. My mother was her last child. Except for my grandfather, I knew all of the members of my mother's family, along with their children and grandchildren. I was generally closer to my mother's than to my father's side mainly because I was closer to my mother than to my father (a theme I will return to repeatedly) but also because the Nazarians were more gregarious.

My mother was born on September 12, 1894. She was a month old when her father went to the Arasa bazaar on a Saturday morning to buy pickling peppers for his widowed sister-in-law and never came back. He was killed during the massacres ordered by Sultan Abdul Hamid that took place between 1894 and 1896 and claimed hundreds of thousands of Armenian lives—a harbinger of the genocide of 1915. My grandfather Kevork need not have fallen victim to these events. His family lived in a predominantly Turkish section of town, and his next-door neighbor was the Mufti (whose house we visited during our trip to Aintab, as related in the last chapter). On that fateful Saturday, in an attempt to save his Armenian friend's life, the Mufti sent word with his son to have my grandfather stop by to see him before going into town. My grandfather muttered that the Mufti no doubt wanted to order another pair of shoes when he had yet to pay for the last two pairs. Despite repeated urgings, he went on his way, telling the boy he would see the Mufti on his return. When he was shopping in the Arasa bazaar, he was set upon by the mob. Several men pinned him down and his own butcher slit his throat; then they took his keys and looted his store. His body was never found. The Mufti was stricken with remorse. When the infant Efronia cried next door, he would plead, "For God's sake, can't someone stop that child from crying?"

During our visit to Aintab when we found the Mufti's house next to my mother's old home, we were invited in by the women. I told them my grandfather's story, and one of the women produced a charcoal portrait of the old Mufti. Her husband, who must have been the grandson, was not home, but word must have reached him about our visit, and he caught up with us in the marketplace with several companions. He was a stocky man in his late forties. He greeted me in a polite yet guarded fashion, and then the conversation became tense. He had learned that I was looking for my mother's old house and wanted me to know that *no Armenians* had ever lived in that part of town at any time. I had just seen my mother's house, but many pairs of eyes were looking at me intently, and I didn't want to get into a confrontation. I thanked him for the information. He invited us for coffee, but I declined. (That was a mistake. It would have been interesting to talk to him about his family.) I then walked into a nearby shop and found a small copper plate with an Armenian name inscribed on its back. It must have belonged to another of the "nonexistent" Armenians in the area.[6]

At the age of twenty-nine my grandmother was left widowed and destitute with five young children. She managed to raise her family on a meager allowance from the church working as a "Bible woman," reading the Scriptures to illiterate families. She never asked for help but was grateful for whatever assistance she could get. Despite their strained circumstances, my mother grew up in a remarkably joyful family. Their neighbors marveled at the sound of pealing laughter emanating from their house. Nonetheless, my mother missed having a father. Whenever she saw a man she liked, she would tell her playmates that the man was her father. My father, who was not lavish with praise, called my grandmother Ovsanna a heroine.

Before these tragic events, particularly the deportations and the genocide of 1915, Aintab had a vibrant Armenian community. Its churches, schools, and cultural institutions had earned it the name of the "Athens of Cilicia." In addition to the large Armenian Apostolic church (later converted to a stable and then a mosque), and an imposing Catholic church, there were three Protestant churches. In 1874, the American Board of Commissioners for Foreign Missions established the Central Turkey College in Aintab, and American Congregationalist missionaries founded a secondary school for girls (the "Seminary").

Missionaries were not allowed to proselytize among Muslims; hence they interacted mainly with Armenians, mostly Protestants. Consequently, most of the college and seminary students were Armenian, as was the faculty. Professor Alexan Bezjian had a master's degree in physics from Yale. My two uncles attended the American Central Turkey College and my mother the Seminary. My mother's cousin Dr. Hovsep Bezjian was the chief assistant to the renowned Dr. Fred Shepard at the American hospital. My mother's family was active in the Protestant church, and she and her brother Yervant sang in the choir. She had a deep melodious alto voice, which she kept to the end of her life. After my mother graduated from the Seminary she became a schoolteacher. Both of my uncles went to study at the American University of Beirut to become pharmacists. (Five decades later I was to study there and then teach in the medical school.)[7]

~

The transformative event that has defined the ethnic consciousness of modern Armenians was the genocide of 1915, which led to the virtual extermination of the Armenian population of Anatolia. Whatever the historical disagreements over these tragic events, there is one stark and undeniable fact: prior to World War I, about a million and a half Armenians lived in Anatolia. Today, there are no Armenian communities left in Turkey outside of Istanbul. What happened to them? There are two diametrically opposed views on this question: one claims that the extermination of the Armenians was an act of genocide, and the other denies it.

The official position of the Turkish government and the version of history taught in Turkish schools is that Armenians in the Ottoman Empire were a law-abiding and trustworthy community living in peace and prosperity for centuries with their Turkish neighbors. Nefarious Western influences that fueled revolutionary ambitions for independence turned them into a rebellious and seditious people. Given the exigencies of the war, the Turkish state under the Young Turks (who ruled the country between the fall of the Ottoman Empire and the establishment of the Republic of Turkey by Ataturk) had to resort to stern measures to deal with this internal threat by deporting Armenians to distant regions for security reasons. In this process over several hundred thousand Armenians died from starvation and illness in the chaos of war (as did a great many Turks). There was, however, no genocide, and these events were neither planned nor orchestrated centrally by the government.

This position is challenged by Armenian and most Western scholars, as well as world opinion more generally. There has been official recognition of the Armenian genocide by nineteen nations including France, Canada, Russia, and Lebanon, as well as the European Union.[8] The majority of state legislatures in the United States have made similar declarations. However, attempts to get a U.S. congressional resolution have failed repeatedly because of the efforts of lobbying groups and the U.S. government's reluctance to offend Turkey, which is an important ally.

There is extensive evidence consisting of eyewitness accounts of survivors, missionaries, and diplomats—in particular the memoirs of the American ambassador to Turkey Henry Morgenthau, as well as a considerable body of work by Western, Armenian, and a few Turkish scholars—that supports the claim that there was a centrally planned and orchestrated program for the extermination of Armenians of Anatolia.[9] The analyses of the motives and causes of this tragedy range from simplistic explanations on both sides to more nuanced views of the interplay of historical, political, economic, and social factors whose confluence, like a perfect storm, made such a tragedy possible.[10]

Lucine's own recollection of these events was virtually lost either because she was too young to remember or because she repressed these painful memories. Ironically, when Lucine was in her eighties and suffered from dementia, some long-buried childhood recollections began to resurface. She had a hazy memory of a man yanking her out of bed, pulling her by her hair, and yelling at her. She recalled marching down a dirt road with bodies strewn on the sides. It is hard to tell if she had actually witnessed these events or read about them and seen pictures in books. I tried to elicit more of these memories, but they proved too painful for her to deal with. Numerous accounts from other sources describe what was most likely the experience of Lucine's family as well. The deportations would start with an announcement by the town crier that certain designated families, or those in a given section of town, should be to ready leave, sometimes in less than a few days, for an undisclosed destination. They were told to lock their houses and rest assured that they could reclaim their property when circumstances made their return possible. This process of deportations was an efficient way of exterminating large numbers of individuals. Had people been killed outright in their homes, some of their Turkish neighbors and friends might have been outraged. Moreover, getting rid of hundreds of thousands of bodies would have been a daunting task. The higher technology of Nazi concentration camps was not yet in place.

On the designated day, the convoy of a few hundred persons began to march, accompanied by a few soldiers who were ostensibly there to guard them. For a day or two, the march went on in a more or less orderly fashion. Those who were better off traveled in carts, or on mules and donkeys, while others walked, carrying what they could on their backs. In a few days, people began to run out of food and drink. There were no provisions or shelter provided. Even when crossing a stream, the deportees might not be allowed to quench their thirst.

As these groups left more populated areas, the guards and an assortment of predators began to take away their money and belongings, down to their clothing. The few remaining men were killed and the women and children left defenseless to be abducted, raped, murdered, or left to die of exhaustion and starvation. The more horrendous atrocities, one would hope, were perpetrated by criminals who had been let out of jail and unleashed on the defenseless victims. Those who were marched toward Aleppo were more likely to survive and be sheltered by the local population; those who ended up in the Syrian desert of Deir Zor did not. In this living hell, mothers taught their starving children the Armenian alphabet writing with their emaciated fingers in the sand of the desert.

The graphic descriptions and images of these atrocities are not only heartrending but also numbing in their intensity. I can deal more easily with accounts that are less dramatic but no less compelling. One such account that has made a deep impression on me is about a mother carrying a child and urging her young son to walk along with her. At some point, the exhausted boy could no longer go on and resolutely sat down on the road. A cart carrying the narrator and two other men approached the woman. As they passed her, she looked at the men pleadingly, but they moved on as one of men said, "She is not our responsibility." The narrator looked back, and the expression on the mother's face haunted him for the rest of his life.

My mother witnessed the pitiful sight of convoys of refugees passing through Aintab, where they were temporarily held in a large quarry at the edge of town. Many of her friends and members of her church were led away, never to be seen again, including the beloved conductor of the church choir. My mother's family was spared through the intervention of her cousin Dr. Bezjian, who was the personal physician of the commanding officer in the region. The officer reluctantly agreed to spare my mother's family. It took great courage to do so since Turkish officials, including regional governors, who were lenient toward the Armenians, were summarily dismissed.

The fact that Armenians as a people have been able to prevail against all odds is a core element in their national consciousness.[11] One evening, when my mother was cooking in our kitchen at Stanford, she told me to get her some parsley from the garden. When I brought her a few sprigs she looked up and asked, "Is that all?" I told her it was a very small plant. "Don't be afraid. Go get some more," she said to me. "Don't you know that the more you cut parsley, the more it grows—just like Armenians?" I was horrified but did as I was told.

My mother was a woman of strong faith but she could not understand how God could have tolerated the plight of the Armenians. When my son Kai was a young boy, he used to sleep over occasionally in his grandmother's room as a special treat. One evening, after my mother had turned off the lights, Kai heard her mumbling in Armenian. "Medzmama" (grandma) he asked, "who are you talking to? I'm the only one in the room and I don't know Armenian." "I am praying," answered my mother. "But Medzmama," said Kai, "in America God doesn't understand Armenian. You have to pray in English." The next day, when my mother told me this, she added ruefully, "I have long suspected that God does not understand Armenian."

~

It was in the crucible of centuries of persecution that the Armenian character was forged. The historian H. F. B. Lynch considers *grit*—toughness and endurance—the distinguishing characteristic to which Armenians owe their existence. Armenians have what the Chinese admiringly call the ability to "eat bitterness" (*chi ku*). In the realm of work, the ultimate Armenian virtue is being successful at whatever one does—be it a surgeon or a butcher. Being successful requires a good education, hard work, diligence, and perseverance. Entrepreneurship, ingenuity, shrewdness, and being clever (*jarbig*) are highly valued; laziness and helplessness are looked down upon. In the worst case, it is better to be a thief than a beggar. There is a materialistic undercurrent to these virtues. Some professions are more honorable than others, but the ultimate measure of success is how much money you make. I don't know if these values would be applicable to Armenians generally, but they were the values that I grew up with in Lebanon fifty years ago. In the personal realm, devotion and loyalty to family are paramount, followed by responsibility to friends, who are typically Armenian, and ultimately to Armenians in general. There is less emphasis on helping outsiders or saving the world. You are loyal to the country in which you live in and you do what is expected of its citizens, but you do not lose your identity by becoming (or marrying) one of "them." In this respect, Armenian Americans tend to lump together all non-Armenians as "foreigners" (*odar*); Armenians in Lebanon make finer distinctions. There is an almost fanatical dedication among "good Armenians" to maintaining their ethnic identity. The danger and prospect of being assimilated varies. The Armenians in the Middle East had little desire or fear of becoming assimilated. In the United States and Europe, where ethnic identities are more readily soluble in the local brew, the prospect of getting dissolved is far greater.

Why should one's ethnic identity matter? Americans are aware of their ethnic roots and are proud of them, but it constitutes only a small part of their overall sense of identity—the answer to the question "Who am I?" There are also Americans whose ethnic identity is less diluted and is a crucial part in their historical experience, often by way of discrimination and persecution. Being an Armenian in the waning years of Ottoman Turkey was likewise not a cultural nicety but a serious matter. In 1915 it would have amounted to a death sentence.

My own sense of Armenian identity went through a gradual transformation. My parents took their Armenian identity for granted and expected me to do the same. But they were not fanatical about it, and they didn't press it on me. My father was determined that I go through the Armenian school system with the expectation that I would also obtain a good education. Otherwise, I doubt if he would have compromised my career prospects by sending me to an Armenian school. My primary exposure to Armenian culture was in the parochial elementary school of the Apostolic Church. The high school of the Armenian Evangelical Church was more focused on one's being a good Christian than being a good Armenian (although for some they amounted to the same thing). As I got older and went to college, I continued to uphold the Armenian virtues that had been inculcated in me. However, I rejected some of the social and political presumptions that went with them. I came to especially distrust the divisive intolerance of Armenian political parties with their with-us-or-against-us mentality.

As a boy I was enthralled by the exploits of nineteenth-century Armenian revolutionaries in Ottoman Turkey, even when these amounted to nothing more than acts of desperation against overwhelming odds. My mother was more critical. Decades later, she still bridled at the Armenian revolutionaries in Aintab who had buried their guns in a mock funeral to hide them from the Turks. They should have instead "killed and been killed," as my mother would say, and not be "led to the slaughter like sheep." I was also fascinated by turn-of-the-century novelists who chronicled the life of the Armenian community of Istanbul when Üsküdar was an Armenian district. I particularly liked the writings of Hagop Baronian, who satirized the pretensions of the Armenian bourgeoisie. His books were the first works of literature that I truly enjoyed.

My increasing involvement with Armenian Evangelical youth groups gradually weaned me away from these more nationalistic interests. When I entered college, and learned more about the rest of the world, I came to appreciate and admire other peoples and cultures—including the Islamic cultures of Arabs and Turks. I continued to mourn the persecution of Armenians in the Ottoman Empire and bridled at the continuing denial of the genocide. But I became disenchanted with my sense of ethnic identity—not the Armenian part as such but with the idea of having an ethnic identity at all. I liked cultural diversity, but ethnicity seemed to have a way of getting tinged with bigotry and intolerance. Be that as it may, I felt that if I must have an ethnic identity, I might just as well be Armenian. Moreover, I became adept at taking on the coloration of my surroundings—whatever country or culture I happened to be in. Ultimately, I wanted to be judged as an individual rather than as a member of one or another ethnic or national group. If that meant being a perpetual foreigner, and out of place, so be it.

~

The events that took place over the next several decades after we moved from Iskenderun to Beirut, and eventually my mother's coming to the United States, will be part of the next six chapters. However, the key events discussed in her memoir—written at age eighty-seven, when she and Lucine were living in Palo Alto, close to us at Stanford—chronologically belong in this chapter. The idea of having my mother write her recollections came from Stina. It was mainly intended to provide my mother with something interesting to do, as well as producing a legacy of her life for us and her grandchildren. At first my mother was reluctant to undertake the task and was unsure about how to do it. I suggested that she write about whatever she wished but be as specific as possible and not omit painful events. She said she would try. A year and a half later she astonished me by handing me a 450-page, neatly handwritten manuscript. I took the manuscript home to read in bed that evening. The first few chapters were about my mother's family background and the travails of her early life, which were fairly familiar to me. A few years earlier, when my mother was visiting us at our summer house in Finland, I had had lengthy conversations with her about her life. After the rest of the family went to bed, we sat together by the light of a candle and she would tell me about the various suitors who had pursued her relentlessly at the end of World War I and how she had rejected them all. There was a certain urgency and passion that animated these accounts, and her face would turn radiant in the soft glow of the candlelight. I had a vague feeling that there was more to these stories than she was telling me, but I had no idea what that might be. (I tape-recorded six hours of these conversations, but three decades later, I still cannot bring myself to listen to them. I don't know if I ever will.)

As I went on reading, I came to a chapter titled "The First and Last Time I Was in Love." I knew right away that my mother was not referring to my father. My parents had a good marriage, even though they were quite different in their characters. My father was a devoted husband who had made it possible for my mother to live in the lap of luxury during their years in Iskenderun. However, he was a cerebral man, not given to the expression of tender feelings. He was the kind of man women love to marry but don't fall in love with (what Nietzsche described as an Apollonian, in contrast to a Dionysian, character). I was stunned by this sudden revelation. Who was this first and only love? Could he have been my real father? The implications were bewildering.

When I went to see my mother the next day, I was feeling quite apprehensive. How were we going to broach this subject? Would she be concerned over how I might have reacted to these revelations? I decided to bide my time. I congratulated my mother for her fine writing but said nothing about her lost love. We both felt relieved, and a few days later we could start talking about it. Soon after, my mother was taken ill and had surgery for a blocked bile duct. It turned out to be pancreatic cancer. She was given six months to live. I tried to convey to her these grim prospects, but she was not eager to hear about them. I decided I would tell her only if, and when, she wanted to know. She never did—even when she knew she was dying.

I did promise my mother that I would set aside all I could and translate her manuscript to get it published. That pleased her greatly. We had long conversations that opened new vistas into her life and allowed me to look more closely into my own life. I translated my mother's memoirs from the Armenian into English by dictating into a tape recorder whenever I had a moment to spare, whatever the circumstances. I recall sitting in my car at the beach at Waddell Creek, watching my son Kai windsurf in the Pacific. I marveled at the contrast between my mother's tragic experiences in her youth and the carefree life of her grandson as he and his friends flitted over the waves with their colorful sails like butterflies.

I hoped to have my mother's manuscript published while she was still alive, but it was not to be. Some publishers wanted to limit the book to its historical material and others to its more personal side. No one wanted to do both, and I would not settle for less. Instead, a few years after my mother's death, Stina wrote a compelling and touching book based on my mother's memoir titled *Efronia: An Armenian Love Story*, centered on the account of my mother's long-lost love.[12] Stina was close to my mother. She had learned to speak Armenian (which greatly endeared her to my relatives) and learned about Armenian culture. Moreover, from her vantage point as an outsider she could provide an objective and insightful historical and personal context to help readers understand the book more easily. Since her book is readily available, I will focus mostly on events that have a more direct bearing on my own life.

~

My mother graduated from the Aintab Seminary in 1913. The following summer, she accompanied her mother and grandmother to Iskenderun, where her oldest brother, Yacoub, ran a pharmacy owned by the Khatchadourian brothers. The ten months Efronia spent in Iskenderun started as a miserable time. The weather was stiflingly hot and humid. Most people had left to pass the summer in the mountains. With nothing to keep her busy, and no one to talk to outside of her family, her life was stultifying. To make matters worse, two of Yacoub's close friends hounded her with insistent marriage proposals. Efronia had faced the same problem in Aintab and didn't have the slightest intention of marrying any of these men. Yet, being stranded in Iskenderun, she found it hard to avoid them.

By the end of the summer, the situation became more bearable. People returned from the mountain resorts, and the pace of life picked up. With much effort, Efronia convinced her mother to let her teach school. It would get her out of the house, and she would earn some badly needed cash. Soon after, Efronia found a wonderful friend in a young and cultured Turkish woman called Nouriyeh Hanem, who was married to a respectable businessman, which made the friendship acceptable to my uncle. Efronia and Nouriyeh Hanem became inseparable. Late one afternoon, while having tea at a seaside café, she introduced Efronia to her cousin, a young man called Ramzi. At the end of the evening, when Ramzi escorted Efronia home, they had fallen in love. Ramzi was all that Efronia dreamed of. He was handsome, kind, gracious, with refined manners, and from a prominent Persian family. There is no photograph of Ramzi. In a picture taken of Efronia at about that time she appears as a lovely young woman with fine features and a wistful longing in her warm brown eyes. The fact that Ramzi was Muslim did not trouble Efronia, but it would have greatly troubled her mother and family. Ramzi was a Persian, not a Turk, but a Muslim nonetheless. They were her father's murderers—how could she even think of marrying one of them? During the heady days of their romance, Efronia didn't dwell on this problem, but it lurked at the back of her mind. As they walked in the cool of the evening on the pier jutting into the Mediterranean, she was enchanted with Ramzi's sweet words and tender affection. He was going to leave for London to continue his studies, and he wanted to marry her and take her with him; he loved her and wanted to spend the rest of his life with her. His parents liked Efronia. Her being Armenian did not trouble them; others in the family had also married foreigners. Ramzi's father feared that should war break out, the Armenians in Turkey would be in a precarious position. He offered to pay for Efronia's going to London with Ramzi to attend school even if she didn't feel ready to marry him. Efronia was torn. She couldn't imagine parting from Ramzi, nor could she contemplate abandoning her mother, to whom she was deeply attached. How was she going to bridge this chasm?

Ramzi's mother, Farouz Hanem, wanted Ramzi to get engaged to Efronia prior to his departure to England. They exchanged rings and had their photograph taken. Ramzi left for London with the greatest reluctance, and Efronia headed back to Aintab with a heavy heart. Fortunately, Ramzi had a cousin in Aintab, a married woman called Munever Hanem, through whom they could keep in touch. She lived close to Efronia's married sister Aroussiak, which made their contacts easier. Efronia confided in Aroussiak, and she was to become a crucial source of support.

In June 1914, Efronia was overjoyed to learn that Ramzi was coming to see her in Aintab. With the clouds of war massing on the horizon, Ramzi's father urged him again to get Efronia out of Turkey. Ramzi made one last desperate attempt to persuade Efronia to leave with him for London. He kneeled in front of her and offered to become a Christian if that would make it possible for her to marry him. Efronia was deeply moved but would not hear of his renouncing his religion. Finally, she decided to confront her mother. Her mother listened to her gravely and fell silent. Efronia knew what that meant. The second separation from Ramzi was even more heartrending, and Efronia's life became caught up in the maelstrom of the war. The one glimmer of light was her contacts with Ramzi through Munever Hanem, who conveyed their news and letters to each other. When she asked Efronia to tutor her young son, it became easier for them to stay in touch. Three agonizing years passed. In the spring of 1917, an ominous development plunged Efronia into despair. The letters from Ramzi stopped. Munever Hanem no longer sent her son for lessons and severed all contact with Efronia. She was bewildered and sick with worry. What could possibly be wrong? Finally, at Aroussiak's urgent pleas Munever Hanem consented to meet with Efronia. Accompanied with Aroussiak, she went to see her. When the door opened Efronia had a horrifying premonition: the woman was dressed all in black. She came up to Efronia, held her in her arms, and cried softly. Ramzi had been killed in a car accident in London. My mother collapsed to the floor. Her life was shattered. She spent the night at her sister's house. The next day they told their mother what had happened. My grandmother was grieved to see the depth of her beloved daughter's sorrow and full of regret and remorse.[13]

~

Following the Allied victory, British troops moved into Aintab on December 18, 1918. By the terms of the armistice, Turkey had to accept Allied occupation of certain territories that were considered of strategic importance. The region of Aintab was one of them. The Armenians could now breathe more easily. The survivors of the deportations began to trickle back in. The Turks were forced to vacate the Armenian churches, schools, and homes and return them to their owners. As the chaos of the times gradually abated, life began to take on a semblance of normalcy. The community had been decimated and everyone had lost someone, but life had to go on. Like salmon swimming to their spawning grounds, young men who had been living and studying in Europe and America came back to Aintab looking for wives. This led to a swarm of suitors for Efronia, but my mother had vowed not to marry anyone and offered no explanation for it. Her determination puzzled and alarmed her family. Some of the suitors were relatives or came from prominent families. In one case, my grandmother had pledged Efronia to marry the son of a cousin when the woman was on her deathbed; her son now wanted to redeem the pledge. If Efronia persisted in rejecting all suitors, what was going to happen to her? Where was she going to live? Her mother was going to Beirut to be with her son Yervant, who was studying at the university; there was no room for her there. Life in her brother Yacoub's household would have been intolerable. His wife, Yester, was a difficult and contentious woman who had been a thorn in Efronia's flesh since they were in kindergarten together. Efronia thought of becoming a nurse, but she had no money to pay for the tuition. She tried to enter a convent, but they wouldn't have her. The Mother Superior wanted to know why a young and beautiful Protestant woman wanted to be a Catholic nun. It didn't sound like she was responding to a divine call. Besides, the last time the convent had taken in a beautiful novitiate there had been no end of trouble as young men scaled the walls to get to her.

This is when my father came on the scene. My mother's cousin Marie was married to his brother Yeghia, so Aram and Efronia knew each other. When Aram requested to see Efronia in private, she had no choice but to meet with him. He was a highly eligible man. His wealth and status would have solved all of Efronia's and her family's problems. That was not, however, what turned Efronia around. Whereas the other suitors had thrown themselves at her feet with ardent declarations of love—thus competing with Ramzi's living memory—Aram made Efronia a rational, businesslike proposal. He told her he had long admired her competence, intelligence, and fine personal qualities. They could have a good marriage and a happy life together. This appeal to her reason rather than to her heart swayed her.

In the fall of 1922 my parents went to Aleppo for their wedding, on November 7. It was a grand affair with music provided by the Nalbandian brothers' orchestra. (The younger brother was to become my first violin teacher, and the older brother went to Ethiopia as the director of the palace band of Emperor Haile Selassie.) The only car in Iskenderun was a Red Cross ambulance. My father paid a pretty

Figure 2.4 My parents in 1923, a year after their wedding and ten years before my birth.

penny to have it take them to Aleppo. Going through the ceremony with a cheerful face could not have been easy for my mother. When she was getting into her wedding gown, she lost one of her earrings and became very upset that this might augur bad luck. My father's assurances that he would buy her two pairs of the same earrings didn't allay her fears. The lost earring was finally found in the folds of her dress.

~

In 1919 British troops in Aintab were replaced by French forces. Armenians welcomed them and hoped that Cilicia might become an autonomous region under the protection of Western powers.

However, the rising tide of Turkish nationalism under the charismatic Ataturk made the presence of foreign troops unacceptable. Growing tensions led to the outbreak of hostilities. In one Cilician town after another, small contingents of French troops assisted by the Armenian population fought pitched battles with the Turks. Having learned their lesson, the Armenians barricaded themselves in their quarters and this time fought back. The battle of Aintab was celebrated by a French author in *Le Verdun de L'Anatolie*. It included the account of a solitary French soldier who ran from one post to another firing at the Turks to give the impression that a whole contingent was fighting them. The Armenians threw all they had into the battle. Master metalworkers improvised guns and munitions, including a "cannon" called *Vrej* ("Revenge"), which only generated an explosive sound, but was enough to terrify the Turkish rabble massed for the plunder. The war-weary French, however, had no interest in becoming bogged down in yet another conflict in the region. They had secured the League of Nations Mandate over Lebanon and Syria, including the district of Alexandretta, and they could do without the rest of Cilicia. France signed the Treaty of Ankara with Turkey, thus ending the conflict. The French troops began to withdraw, with the dejected Armenians, fearing the worst, following in their wake—this time never to return.

With the loss of Aintab, my parents had to decide where to settle down. My mother had made it a precondition for marrying my father that they were not to live in Iskenderun. She claimed that she hated it when she had lived there earlier. It was of course the prospect of returning to the scene of her romance with Ramzi that made living there inconceivable. My father knew nothing about it and agreed. With the assets he had on hand they could live wherever they wished. Actually, he was thinking of settling down in Beirut or Istanbul, both of which had greater business prospects than Iskenderun. My mother suggested emigrating to the United States to make a clean break with the troubled region, but my father would not agree; America was too far, too foreign, and the pace of life too hectic. Had they emigrated to the United States, what would their lives have been like? And had I been born in the United States, what would my life have been like? Would that have made me a "real" American?

As my parents mulled over these choices, an urgent appeal came from my uncles in Iskenderun—their business was facing a crisis and they needed Aram's help. My father reassured my mother that this was to be a temporary arrangement, at most for a few months. My mother objected, but not too forcefully so as not to raise suspicions, and finally acquiesced. They boarded ship in Beirut. The trip was agonizing as my mother fought to keep her composure. The ship docked at Iskenderun at the very pier that she used to walk on in the evenings with Ramzi. When she set foot on the ground, she fainted. The people around her ascribed her reaction to the humid weather, but she said nothing. As months followed each other, Iskenderun began to look more like a permanent arrangement. Efronia decided to confront Aram in his office. This was the breach of a solemn promise and not a marital squabble to be settled at home. In direct and forceful words, she reminded her husband of their agreement. He listened to her intently and fell silent; then he walked out without a word. My mother remained in her chair and faced a critical choice: she would either have to leave her husband or never bring up this issue again.

~

Once my mother made up her mind, she set out to make the best of a difficult situation. Locked in her heart was the aching memory of Ramzi. It is testimony to the strength of her character that she could fashion a new life for herself while remaining true to the lost love of her life. She tried to contact her dear friend Nouriyeh Hanem, but she was in Ankara. She was alarmed to learn that Ramzi's parents were still living in town and tried to avoid meeting Ramzi's mother Farouz Hanem. However, after Nouriyeh Hanem returned to Iskenderun there was no way for Efronia to keep her distance from Farouz Hanem. Soon enough, Nouriyeh Hanem suggested that they pay a visit to Ramzi's mother. It had been almost ten years since Efronia's last visit to his home. The women embraced silently and sat down to drink their tea in a somber mood. Nothing was said about Ramzi. Finally, Farouz Hanem offered to take my mother on a tour of the house to show the changes they had made over the years. At the end of a corridor, she stopped in front of a room and opened the door. It was Ramzi's bedroom. My mother looked in and saw Ramzi's silk pajamas folded neatly on his bed with his slippers resting at its foot. On the opposite wall was a life-size enlargement of their engagement photograph with a Persian wedding wreath attached to her head. She slumped to the floor. When she reopened her eyes in the living room, my father and a doctor were peering at her anxiously. My mother said that she had felt dizzy but was now fine, and they went home. This was the second time that Efronia—a vigorous and healthy woman—had fainted for no apparent reason after coming to Iskenderun. Did my father have any suspicions about why this was happening?

~

After I read Ramzi's story and spent many hours talking about him with my mother, it became clear to me that although my mother had come to care deeply for my father and was a loyal and devoted wife, she loved Ramzi at age ninety as much as she did when she was nineteen. Moreover, a number of puzzles from my own life began to sort themselves out. After I had gained an awareness of my parents as individuals, I had a nagging feeling that they were somehow mismatched. When my mother talked about her past, her mood became clouded over. One day when she was expressing to me some vague disaffection with my father, he walked in unexpectedly. Lucine reacted with alarm, but my father took no notice of it, or at least pretended not to. My mother periodically suffered from excruciating migraines. She would shut herself up in her darkened bedroom with a scarf bound tightly around her head and refused to talk to anyone, not even me, until the headache abated. Even if Ramzi was not the cause of any of this, it is hard to imagine how his loss would not have cast a shadow on her life and my mother's relationship with her husband.

I also wondered about how my mother's relationship with Ramzi may have affected me. I had been baffled by the way I had so little in common with my father. He loved me and was proud of me, just as I loved and respected him. Yet, there was no real intimacy between us, and we had little in common. Had he not been my father and I his son, we would have had little to say to each other. In contrast, I became aware of how much I seemed to have in common with Ramzi in my mother's descriptions of him. Had my mother raised me as a substitute for Ramzi? If she had done that, I doubt if she would have been conscious of it. But if in fact I had been a substitute for Ramzi, how would that have influenced my own relationships with women? The implications of that are too obvious to need spelling out and too close for comfort to speculate on. Moreover, I had reached a point in my life where there would have been little point in ruminating over these imponderables. I was over fifty years old, happily married with two teenage children I adored, and more concerned with being a good father and husband than dredging up my relationship with my long-dead father and ninety-year-old mother. Yet, I continued to be burdened by how the loss of Ramzi might have affected my mother's own life. As I read and reread her account of losing Ramzi, I grieved for her. Given a choice, I would have gladly forfeited my life and never been born if my mother could have had another son with Ramzi instead.

3

Coming of Age

THE INTERMINABLE JOURNEY from Iskenderun to Beirut finally ended. We stayed with Marie Nanoug and godfather Khatchig while trying to decide what to do. My parents faced an agonizing choice: should we go back to Iskenderun to become Turkish citizens and hold on to our possessions? Or should we stay in Beirut to become Lebanese citizens and risk losing everything? It had to be one or the other; we could not be dual citizens. My father was eager to go back and take charge of his business, but the political situation was volatile. If Turkey entered the war on the side of Germany, as it did in World War I, then who could tell what would happen? The memories of World War I were fresh in everyone's minds. Our relatives and friends urged my father not to go back. They said the Turks would take all he had from him anyway. Why risk an uncertain future for his only son? My father was not easily swayed, but this time he yielded to their advice. He had given power of attorney to a Turkish associate for such an eventuality and now asked him to liquidate his assets. The man sold everything but reported only a fraction of the proceeds and pocketed the rest of the money. Even that small residue that was due to my father was sequestered by the government and never reached our hands. The fruits of a highly successful business spanning a whole generation were taken from us in one fell swoop.

It was not until we left Marie Nanoug's house and moved into a place of our own that I realized we were not going back to Iskenderun. Our lives had changed irrevocably. Instead of a big house, we now lived in an apartment. I was painfully aware that something important had been taken away from us. I was too young to realize the full implications of the loss of Iskenderun, but it induced in me a lifelong yearning for a lost world. It also fueled in me a relentless ambition and a drive to excel at whatever I did to make up for the loss. Even though making money never became the primary objective of my life, money became a significant consideration in whatever I did. This influenced not only my career choices but my more personal decisions as well, with some unfortunate consequences.

Another residue of Iskenderun was a feeling of entitlement. Since I came from a family that had been affluent and highly regarded, I was determined never to be poor or insignificant. My sense of *noblesse oblige* would not allow me to do anything mean or despicable to disgrace myself or besmirch my family name. I don't know whether these attitudes were inculcated in me by my parents or I formed them on my own. This sense of self-importance may have been misplaced. Nabokov said he never met a Russian émigré who had not been part of the Russian aristocracy. Likewise, maybe our position in Iskenderun was not as grand as I construed it to be. Like the awe a child has for a kindergarten teacher, our status may have loomed large in my eyes because I was so little.

My father must have salvaged some of his assets because we were never poor, nor did we feel like refugees in Beirut. It was more a matter of what we *could* have had and who we *could* have been that troubled me. Beirut had a large and vibrant Armenian community that we became a part of. The Lebanese treated us well. There was no reason for me to feel discriminated against or disadvantaged in any way. Nonetheless, especially after I came out of the narrower confines of the Armenian community, I felt like an outsider (what Edward Said called being "out of place").[1] Even when the circumstances of my life became vastly different after I moved to the United States, the sense of being out of place lingered on. Perhaps that was an inevitable part of the immigrant experience. However, unlike typical immigrants who have an "old country" to relate to, there was no such place for me. I would be out of place in Armenia, Turkey, Lebanon, Europe, America, or anywhere else in the world. Hence my sense of identity became that of a "foreigner" no matter where I lived. This didn't mean feeling alienated or oppressed but being "different"—a theme that would run throughout my life.

My father never talked to me about the loss of Iskenderun, and that made it worse. I blamed him for letting it happen and for not reversing its effects. Had he tried to explain to me how and why it happened, it would have helped me to come to terms with it. Had he expressed regret for his mistakes and miscalculations, I could have empathized with his loss and shared his burden. His silence suggested an unwillingness to take responsibility for his actions and to shut me out of his life. This may have contributed to the distance that kept us emotionally apart until he became old and I became more forgiving.

In contrast, my mother talked at length about Iskenderun but never in my father's presence. She was careful not to be openly critical of him, but it was clear that she too blamed him for gross misjudgment and for refusing to take her advice. "Your father is a very intelligent man," she would say. "He rarely makes a mistake, but when he makes a mistake, it's a huge one." Proud of her greater foresight reinforced by the benefit of hindsight, my mother never tired of rehashing my father's miscalculations. Since he was the sort of man whose "left hand didn't trust his right hand," how could he place his entire fortune in the hands of a Turk he could hardly have called his friend? My mother further blamed my father for not starting a branch of his business in Beirut before the war and then for losing his nerve and not starting a business after we came to Beirut. Had he invested the money he had salvaged in some new enterprise in Beirut, no matter what, he would have easily recouped his losses in the war-fueled economy. It would have made him "forget Iskenderun."

I now realize that my assessment of my father's responsibility for all this was overly harsh. I had basically adopted my mother's views on these matters, as I was to do with so many other things. Actually, I don't know what the real reasons were that led to my father's business decisions. He was caught up in the turmoil of the times, over which he had no control. Entrusting his fortune to a Turk who was hardly a friend was a mistake, but what other choice did he have? There were hardly any Armenians left in Iskenderun that he could turn to. And simply because the man was a Turk did not mean that he would be dishonest; an Armenian could have done the same thing. As for my father's sharing his experience of loss with me, what could he tell a six-year-old boy that would have made sense to him?

My mother finally got some satisfaction. In anticipation of the likelihood that we might not return to Iskenderun, she had arranged for our rugs, silver, and other easily movable items to be shipped to Beirut soon after we left. She did this without my father's knowledge, and it took a lot of nerve to do it. After my father had decided we were not going back, he told my mother that they needed to rent an apartment and furnish it. "But," he added sheepishly, "I don't know what we are going to do about the rugs." My mother looked at him in the eye and said, "The rugs are here." She was to save these rugs once again three decades later. In 1975, at the outset of the Lebanese civil war, she had them shipped to us at Stanford, without asking or telling me beforehand. The huge crate that was deposited in our driveway took us totally by surprise. We didn't know what to do with the dozens of fabulous kilims and antique rugs, none of which interested me at the time. The next day I called my mother to report the arrival of the rugs and to thank her. I also suggested very mildly that she should have perhaps alerted me before shipping them over. She was silent for a few seconds and then said, "You sound just like your father."

My father never discussed our family finances with me, but my mother did. She told me candidly how much money we had and what our yearly expenses were. It was reassuring to know that we had enough to live on for at least five years. Moreover, I was greatly relieved that my father regained his initiative and joined with three younger engineers to start a construction company, of which he became the president. The company did well for a number of years. (I particularly enjoyed riding occasionally in the company car, a Lincoln Zephyr, with its distinctive V-shaped sweeping end.) But the company faced stiff competition from larger corporations with greater political patronage in securing government contracts. However, what actually broke up the partnership was the deterioration of my father's relationship with his colleagues. I think they got tired of arguing with him. I happened to be in the waiting room of their offices when they were discussing this issue. I couldn't make out what the others were saying, but I recall hearing my father's voice (which was always louder) telling them that business partners should be able to discuss and argue over issues freely.

However, he said this with an uncharacteristic undertone of pleading in his voice. It was met with a stony silence. I felt a mixture of sympathy and anger for my father for yet another setback. I could understand why his partners would no longer want to put up with him, yet I was also disappointed that they closed ranks against him when he meant well. (Twenty-five years later, when I had become a psychiatrist, one of these partners came to consult me about problems he was having with his much younger wife. I hadn't seen him since I was a teenager, and it felt strange that I should be advising him over such a personal matter.)

Now out of a job, my father spent most of his time with his friend (and my mother's cousin), who with his sons had a booming business importing used clothing from the United States. My father acted mostly as an unpaid advisor, but at least it got him out of the house. And when they set out to build a five-story rental apartment my father agreed to oversee its construction. He was paid for his work, but he must have saved them many times over the money they gave him. This was followed by his entering into a partnership with them manufacturing men's clothing in the free zone at the port. As usual, my father poured his heart and mind into the business. The burden also fell on my mother and Lucine since the head tailor came to dinner at our house several days of the week. The business was quite successful, but it too didn't last. In this case, it wasn't my father's fault. His partners wanted to reduce my father's share of the profits by paying him a salary as an employee. My father felt betrayed, but it would have been beneath him to complain. However, it was not beneath my mother to write a blistering letter to her cousin, who lamely blamed his sons for the injustice.

It was especially galling for my mother to watch her brilliant husband toiling for the benefit of others, and she never tired of berating him for it. It wasn't quite fair, since for my father to gamble his limited resources on an independent venture would have been too risky. Finally, as my father got older, his nephew Jirair, who was an engineer and a successful entrepreneur, provided him with better opportunities to work with him on construction projects—a partnership my father would have dearly loved to have had with me. It was hard work, and my father came home exhausted. It hurt to see him in his seventies soaking his aching feet in warm water while I still toiled in medical school without being able to help him. It was not until my father was in his eighties and I was on the medical school faculty that I was able to provide the money for him to construct his own apartment building, which he then rented to support his family.

~

The Armenian community of Beirut that we were a part of was quite large. Armenians had lived in the region since the Middle Ages, but most of the more recent arrivals had come as refugees following the genocide of 1915. The Armenians from the region of Iskenderun who came at the onset of World War II constituted the last large influx. And as neighboring Arab countries became increasingly nationalistic, more of their Armenian population gravitated to Beirut, swelling the community to over 200,000 individuals, who by the 1960s constituted close to 20 percent of Beirut's population. That made Armenians the largest non-Arab community in Lebanon. Many of the poorer Armenians settled in the shantytown of Bourj Hamoud, at the northern edge of Beirut, which then evolved into a tidy working-class community and became the Armenian quarter. More affluent Armenians lived in mostly Christian neighborhoods, as we did.

The self-segregated neighborhoods in Beirut reflected the political organization of Lebanon as a whole. The system of religion-based *millets* of the Ottoman Empire was preserved under the French, after independence, and persists to this day. The National Pact of 1943 stipulated that the president of Lebanon would be a Maronite Christian, the prime minister a Sunni Muslim, the speaker of Parliament a Shiite, the defense minister a Druze, and the finance minister a member of the Greek Orthodox Church. These allocations ostensibly reflected the relative proportions of these groups in the general population (Maronites presumably constituting the majority). However, since there had been no census in Lebanon for decades, these assumptions became increasingly untenable. Yet, they were maintained and no new census was conducted so as not to disturb the balance of power and keep the peace. Parliamentary elections continued to be denominationally based whereby Apostolic Armenians would vote only for several Apostolic Armenian candidates who would be part of a larger political slate. Catholic Armenians voted for their own candidate. Protestant

Armenians and Arabs were lumped together because there were not enough of them to constitute separate *millets*. Since Armenian Protestants were more numerous, the Protestant representative in parliament would usually be an Armenian. Most Armenian deputies at the time could not have given a speech in Arabic. There was a joke about an Armenian deputy who never said a word, until he surprised everyone by asking for permission to speak. As heads turned to him in expectation, he said, in halting Arabic, "Mr. Speaker, would it be possible to close the window in the back? It's freezing in here."

The *millet* system provided considerable autonomy to faith-based communities in running their affairs, including the administration of civil law, which dealt with matters like marriage and inheritance. This meant that getting a divorce was relatively simple for Muslims but very difficult for Catholics. Under this system, Armenians did well in Lebanon. They had their own churches, schools, and cultural organizations. This led to a certain degree of insularity, but that didn't bother them. It still enabled Armenians to be fully engaged in the economic and social fabric of Lebanon. Many of them became successful in business, in the professions, and as craftsmen—as they had been traditionally. Occasionally they attained important political positions as well, including serving in the cabinet. (A former Lebanese president, Emile Lahoud, had an Armenian mother and an Armenian wife.) Perhaps one of reasons that the Lebanese were so accepting and supportive of the Armenian diaspora is that they too had a large diaspora and knew what it was like to live among strangers. When I was growing up in Beirut, there were three times more Lebanese living outside of Lebanon than within Lebanon itself. A European economist who studied the Lebanese economy concluded that the country would be bankrupt without the remittances that emigrants sent home each year.

Our apartment was located in a pleasant part of town close to the public gardens on a street parallel to the residence of President Camille Chamoun. We lived on the third floor of a six-story building. Like other structures of the period, it had no elevator, no central heating, no running hot water, and only an oriental-style squat toilet. Nonetheless, its dated style and architectural flourishes endowed it with a certain charm in contrast to the bland modern structures that came to surround it over the years. The apartment was quite spacious. Its high ceilings and large windows gave it an airy feeling. My parents and I had separate bedrooms. Lucine slept on a sofa bed in the dining room. Coming into the house, one walked into a large living room furnished with Morris chairs and a sofa covered with kilims. A large Tabriz medallion rug covered almost the entire floor. Next to it was the salon, which was used only for entertaining formal guests. Its beautiful gossamer curtains were one of the few things saved from our Iskenderun salon, of which it was a pale shadow. Despite its limitations, my mother did wonders furnishing our home with the lovely old rugs and kilims, brass and copper braziers, and whatever else she had managed to salvage from Iskenderun to create a tastefully decorated, warm, and welcoming home. It had far more charm and character than the ostentatious homes of our wealthier friends. It was meticulously maintained by Lucine, who now shouldered the entire burden of housekeeping. I still marvel how she could carry the huge oriental rugs to the roof to clean them every spring and why no one helped her.

A small balcony jutted off my parents' bedroom overlooking Rue Spears, the wide street running in front of our apartment building.[2] I used to sit there and daydream. When I developed a ravenous appetite in adolescence, Lucine would have a huge plate of French fries and a large carton of milk ready for me when I came from school. I would sit on the balcony and watch people in the street and hope to get a glimpse of women undressing in the opposite apartment building. A larger veranda provided a cool place for the family to sit in the afternoons and have our dinners when the weather turned warm. Lucine used it to grill kebabs and hang the laundry to dry. When cherries and strawberries were in season, my mother would soak them in syrup and let them stew in the sun before making jam. I would eat them avidly, plucking out the fruit one by one. As the layers floating in syrup got thinned out, Lucine would patiently replenish them. That gave her so much pleasure that I felt as if I was doing her a favor.

The only outside help was a Druze woman who came once a week to do the family laundry by hand. My father hated disorder, so the washing had to be finished in a rush before he came home. Since we no longer had a cook, Lucine took over the preparation of the daily meals, while my mother cooked for guests. She was an outstanding cook, as was Lucine. The kitchen was one place where my father and I were not welcome—not that we had any wish to be there. Perhaps that's why the kitchen left so much to be desired. It had little counter space, some crude shelves, a primitive sink, and a solitary bare bulb hanging from the ceiling that cast an anemic light. I marvel at the spectacular dishes that were prepared there nonetheless. Over the years, my father made various improvements to the bathroom and had a Western-style toilet installed, but the kitchen stayed as it was. Years later, I could have had the kitchen remodeled myself and decent lighting installed over the sink, making life easier for Lucine. To my regret, I never did so, even though it would have cost only a little more than a few skiing trips.

Food has been important beyond sustenance for Mediterranean people going back to the Romans. The cuisine of Aintab was famous (there are restaurants that serve it in Istanbul). The Nazarians, more than the Khatchadourians, were fond of good food, which is the tradition I grew up with. As gourmets, they were said to have "open gullets" (Turkish, *boghazi atchek*). I was too young to remember the exploits of our renowned cook in Iskenderun, but I heard about the dinners he prepared for the Catholicos Sahag II of Cilicia (supreme head of the Apostolic Church in the diaspora) and his retinue. During a pastoral visit to the region, the Catholicos was our houseguest and slept in my parents' bedroom. At one dinner, the *pièce de resistance* was a whole lamb stuffed with rice, chunks of meat, and almonds, and baked in a crust. Even my mother was awed by it. One of the old priests in the retinue, who was from Aintab, had no use for such exotic fare. He took my mother aside and asked her if, for God's sake, he could have some *chi koefte* (a delicious mix of bulgur and raw meat pounded into a pulp) served to him on the side. My mother assured him that she would prepare it for him with her own hands. The Catholicos noticed it at the table and glared at the old priest, suspecting that he must have asked for it.

The formal dinners in Beirut were on a lower scale but still quite spectacular. On special occasions, the table would be set with a stack of several plates in front of each guest. A dozen appetizers (*meze*) would be on the table to nibble at and wash down with raki (Arabic, *arak*). I liked these small dishes more than the main courses, but I

Figure 3.1 Childhood.

wasn't allowed to make a meal out of them; it wouldn't have counted as dinner. One elaborate course would then follow the other until the final dish was left untouched and taken back to the kitchen—proof that no one could have possibly eaten more. In addition to the fish, fowl, and meat dishes, there were Aintab specialties, like *koefteli yaghene*, a yogurt-based soup with small balls of *koefte* filled with walnuts and sheep fat and a thick layer of hot butter with mint floating on the surface, with chunks of chicken and lamb added at the table. It was a defiant declaration of calories and cholesterol be damned. To please the hostess at Armenian dinners, one had to rise from the table feeling not only full but slightly sick. There were various ways in which the hostess would cajole and shame the guests to eat more.

If, for instance, you said, "I'm sorry, but I absolutely have no room left for dessert," you would be told, "Don't you know that no matter how crowded the church is, when the bishop comes in people always manage to make room for him to get through." The crowning blow came when a guest staggering away from the table would be told by the hostess, "What's wrong with you? You ate nothing" (*Pan mu chi gerar*). "Are you not feeling well?"

Some of my fondest memories are from more intimate winter evenings when we sat around the magnificent brass brazier (made by the renowned Kalemkerian brothers in Aleppo). I would listen to my parents conversing with their friends, talking about other friends. My father had the habit of fiddling with the live coals and arranging and rearranging them into neat piles, much to my mother's annoyance ("Aram, don't play with the fire.") The coffee simmered in the *ibrik*, sending off its intoxicating aroma. As I leaned drowsily against my mother, I felt contented and happy.

Every fall, a man would come carrying what looked like a huge bow strung taut to tease out the wool in our quilts and then re-sew them. We always had a bath on these nights before going to bed under the fluffed-up comforters. Our bathroom, with its boiler heated with wood, was like a fabulous steam room. When it got replaced by a gas heater, it became simpler to use but lost its charm. We bathed once a week, usually on Fridays. Years later, after I got used to showering every day, I couldn't imagine how grimy I must have felt between baths, even though I didn't feel any less clean at the time than I do now.

When I was a young boy, my mother would bathe herself first and then wash me. I loved the comforting feeling of leaning against her legs, but I hated the soapsuds that got into my eyes as she washed my hair over and over. I carefully avoided looking at her nude body, which she shielded with upraised hands like Botticelli's Venus rising from the sea. Lucine never bathed me after I was no longer a baby. When I was much older, I heard a piercing scream one evening when Lucine was taking a bath. I rushed into the bathroom, setting aside all considerations of modesty. She had distractedly poured hot water directly from the boiler on herself. I drenched her with cold water, but she still ended up with second-degree burns. It broke my heart to hear her whimpering in pain.

While I cherish now the happy memories of my younger years in Beirut, the charms of our old apartment and our style of life were lost on me when I was growing up. I felt especially embarrassed by the squat toilet when we had guests. I envied the bathrooms of our friends, with their modern toilet bowls, bathtubs, and showers. It wasn't that these modern conveniences worked better, but they reflected a more European, or "modern," way of life.

~

Several of the other tenants in our apartment building were particularly interesting. The family that lived on the ground floor kept to themselves. I learned later that the man worked as a government inspector in the red-light district. His wife, now past her prime, must have been an attractive woman and still had a provocative allure when lightly dressed for housework, with her dark eyes enhanced by eyeliner. I wondered if she had met her husband when working at a brothel. Her older and careworn maid may well have come from a similar background. The couple had a son who was about my age. I would occasionally see him with boxing gloves at one of the public swimming beaches (even though he hardly looked like a boxer). There was a short man, supposedly his trainer, who would massage him in one of the changing rooms. The whole thing had a strange feel about it. This was reinforced by the trainer's association with a couple of men with bleached hair and in tight shorts. I suspected their relationship had something to do with sex, but I didn't quite know what that was and didn't want to know. One day, I was horrified when a burly police detective took a blushing and frightened young man off the tramway and slapped him mercilessly on the sidewalk and then took him away in handcuffs. No one intervened, and everyone seemed to know what was going on. This encounter too became connected with my mind with a dim realization that men (but not women) could have sex with each other.

Even more fascinating was the couple on the second floor. He was an elderly physician with a Turkish name, but I learned that he was actually an Armenian who had converted to Islam in order to marry the daughter of one of the Ottoman pashas in Istanbul. They had met when he had made a house call to see her. She was now a frail, odd-looking old woman with straggly hair and disheveled clothes. I was fascinated by her excruciatingly refined manners and the exquisitely ornate Turkish she spoke with my mother. I never saw her venture beyond the landing in front of their door. I would occasionally run into the old doctor on the stairs. He was always dressed in a fading three-piece dark suit that had seen better days. Beneath his air of dignified formality, I detected a deep sadness. He looked forlorn like a loose black balloon. I would nod at him in a furtive greeting while looking away, but we never spoke. I wondered about his life in Istanbul. What was it about that woman that could have induced an Armenian physician to make such a drastic sacrifice by becoming a Turk? I could conceive of doing something like that for another neighbor's sexy daughter who was a grade school teacher. Maybe I would not go as far as becoming a Muslim, but I could have seriously considered becoming a Catholic for her sake. (Years later, some of my closest relatives would be Catholics. But at the time, Catholicism felt remote and forbidding.)

Although my parents and the Turkish couple never visited each other (I never asked why), Lucine occasionally helped them with their grocery shopping and had a key to their front door. When she hadn't heard from them for several weeks, Lucy went in and found them lying dead, sprawled on the living room floor in a pool of vomit, urine, and feces. Despite the overpowering stench, Lucy had gone into their bedroom and found next to their bed an open suitcase full of cash. She hurried home and told my father. Had they taken the money, no one would have known about it. Instead, my father called the police, who carted away the bodies and gave him a receipt for the money and then, I assume, pocketed it.

~

At the age of six, I entered the parochial school of the Armenian Apostolic Church of St. Nishan. Since I had not actually attended school for any length of time, I was placed in kindergarten. During my first day, the teacher wrote a few elementary additions and subtractions on the blackboard and asked us to solve them. I quickly raised my hand and gave the answers. That led to my being promptly moved up to the first grade. I don't remember much from that year except that I enjoyed being in school and found it easy.

There was one traumatic experience that year that left its mark on me. On the eve of Thursday before Easter, Lucine took me to the Apostolic church service in the evening to watch the ritual enacted by a priest of Jesus washing the feet of his disciples. The courtyard was full of people milling around in the darkness holding candles in paper cones, creating an eerie atmosphere. Somehow, I became separated from Lucine and panicked. I ran frantically from one end of the courtyard to the other until I found her. She hadn't been aware of my absence, so I couldn't have been gone for long. But it was long enough for me to hold on tightly to her hand for the rest of the evening. I trace back to that experience my lifelong fear and recurrent nightmares of getting lost. In a typical dream, I am in a foreign city and can't remember the name or the location of my hotel. In another, I am to give a talk and can't find the auditorium. Then there are examination dreams that become tangled up with the idea of getting lost. I run into a classmate while blithely walking around and find out that we have an important exam the next day that I didn't know about. Or, I am in front of a class and my mind goes blank. This actually happened to me when I was a medical student. As we stood around a patient's bed during hospital rounds, I realized that the next patient was mine. But to my horror, I could remember absolutely nothing about her case. I feigned a coughing fit and left for the nursing station. As I picked up the patient's chart by its bed number, it suddenly all came back to me. It's hard to imagine why that single experience in the churchyard should have left such a lasting residue. Perhaps it fed on my deeper fears resulting from our leaving Iskenderun for an unknown destination. Or, maybe it triggered deeper unconscious fears of being abandoned by my parents, something that I obviously had no reason to fear.

Toward the end of the summer of 1941, when I was about to start second grade, I came down with a severe case of rheumatic fever. Before the era of antibiotics this was a life-threatening illness that could damage the heart valves. It started one afternoon when my neck began to feel stiff as I was walking home at the mountain resort of Mrouj. That night, I could hardly move my head, and I developed a high fever. I ended up spending most of that year in bed with migratory joint pains but no heart symptoms. I was hospitalized several times, during the last of which my tonsils were removed while I was running a fever. It was a risky move intended to eliminate the source of the streptococcal infection thought to be causing my illness (actually, it is the autoimmune reaction to the infection that causes it).

During the nine months that I was bedridden, my mother never left my side. When I was in the hospital, she slept in a lounge chair next to my bed and only occasionally went home to bathe, and Lucine took over until she came back. The prospect of my dying must have been so terrifying to my mother that by physically holding on to me she must have thought she could keep me alive. Sick as I was, the treatment felt worse than the disease. I was placed on sulfa drugs, which were the only antibiotics available at the time, and on sodium salicylate for its anti-inflammatory effects. In its liquid form it had a disgusting taste and smell. Even after the critical period of the illness had passed, I was kept on it for months, and it made my life miserable. As the recovery lingered on, my father arranged, with some difficulty, for me to be seen by a famous French physician called Dr. Calmette.[3] He was retired and lived in the enormous mansion of the Druze Jumblatt family. The doctor was a short and imposing old man with white hair and a beard—the first man I saw with a beard who was not a priest. He listened to the history of my illness told by my father yet again in excruciating detail and then examined me. There followed an interminable discussion in French, which I did not understand. Finally, Dr. Calmette wrote out a prescription and gave it to my father, who folded it and put in his pocket. The fee was a gold sovereign, as stipulated, in advance. As we walked down the curving driveway, my father took out the prescription and read it shaking his head. I snatched the paper and looked at it—it was for sodium salicylate, the bane of my life. I sat down on the sidewalk and cried. I had borne my illness with fortitude and reconciled myself to the prospect of death, but this was too much to bear.

Except for the intermittent crises during which I felt acutely sick, I suffered most from the boredom of being confined to bed. There were no children's books to be had, so my mother got for me what she could by way of adult novels. My favorite was *The Count of Monte Cristo*, by Alexandre Dumas. I read it many times, as I did a collection of short stories that I virtually memorized.[4] The rest of the time, I lay in my bed looking at the ceiling and letting my mind wander. When my grandmother Ovsanna came on an occasional visit, she would sit with me for hours telling and retelling the few stories she knew or quietly wait for me to doze off. Even after I no longer needed to be in bed, I was kept at home for a few months to prevent a recurrence. Since I was no longer sick, I felt like a caged animal and made life difficult for my mother by refusing to take my drugs. On one occasion, she became so exasperated that she began to cry softly. When I saw her tears, I stopped whining and drank the sickly syrup without saying another word. I was surprised and happy when one of my classmates called Miko showed up one day after school to play with me. He came from a poor family that could not afford to buy candy. Miko used to put granulated sugar in his pocket, dip in his wet finger, and lick it off. We played a game of marbles, and despite his being much better at it, he let me win. As he was leaving, I noticed that my mother slipped something into his hand and Miko thanked her. I realized that she had actually *paid* him for playing with me. I felt so humiliated that I told my mother I didn't want to see Miko again. She knew why and said nothing.

I missed the entire second grade, but when I went back to school in the fall, I was promoted to the third grade. My teachers thought I was smart enough to handle it. While I had actually learned more during my illness by reading books for grownups than I would have in school, I still wished I had had the textbooks of the second grade to study and perhaps a tutor to help me. It would have facilitated my catching up with my class, especially in math. It didn't occur to me to ask them for the textbooks, and my parents must not have thought about it. I remember the excruciating efforts in relearning how to write with an old-fashioned pen, getting ink all over my fingers and struggling with the simplest math assignments. I did catch up in a month or so, but I always feared that the gaps in my learning would trip me up sooner or later.

The most profound impact of my illness involved the hours I spent reflecting on matters of life and death. I knew that I might die—it was written all over my parents' anguished faces—and I came to terms with it as well as a child could. From then on, death held no terror for me. And if I was not afraid of death, what else was there to fear? I developed an absolute and unshakable faith and a deeply personal bond with God that was to sustain me for the rest of my life. I did not associate my bond with God with the church. It was a purely personal conviction: I just *knew* that God existed as surely as I did. I loved him and he loved me and would always take care of me in my hour of need. No one instilled these ideas in me. I felt that it was God who reached out and touched me.

One evening, my wrists and hands began to swell with the effusion of fluid into the joints. I watched it happen with a sense of calm resignation and the certainty that God would not abandon me. I told my anguished mother, sitting by my side, that I had prayed to God and that he was going to make me well. She turned away to hide her tears. That image of the boy with the swollen hands has been my refuge during times of adversity for all of my life. It was the defining moment that shaped the core of my identity: seventy years later, I am essentially the same person as that nine-year-old boy. That experience also left a melancholy cast on my character and a sensitivity to the tragic element in life. Thus, my favorite season is the fall, I like foggy and rainy weather, my preferred color is gray, I play the viola with its mournful tone, and I prefer music in minor keys. And I feel fine about it.

After I was allowed to leave the house, my mother and Lucine continued to be overprotective, going to ridiculous lengths to ensure that I stayed well. I was taken to see another famous physician—this time the legendary Doctor Asadur Altounian of Aintab, who now lived in Aleppo. By then he was a very old man married to his much younger English nurse. I remember her, dressed in shorts, coming from the tennis court and my staring at her legs. As my mother peeled off layer after layer of my clothing for the examination, Doctor Altounian said to her, "Mrs. Khatchadourian, if you want this boy to be healthy, you have to raise him like the *Mutwellies* [impoverished peasants from Shiite villages] and let him run around barefoot and half naked." Of course, that's not what happened. Instead, Lucine would run after me in the middle of a soccer game on the neighboring playing field to stuff a towel into my back so I wouldn't catch cold. The game had to stop while this was being done. The reason my friends put up with it was because I owned the football.The last phase of my recovery was at the village of Kessab, the only Armenian village in Syria. We spent the summer there with my uncle Yervant, his daughter Nora, my grandmother, and my aunts Aznive and Aroussiak. I was fussed over and pampered by everyone. My mother finally felt secure enough to let me wander about on my own. My most vivid memory is from one evening when we had some visitors. I was in bed at a far corner of the room getting drowsy. In the dim light of oil lamps, one of the guests told a story about a man who was "married" to a female *jinni* (a spirit that could take human forms). I wasn't sure what that meant, and some of the other guests sounded skeptical. The storyteller persisted and claimed that the malevolent spirits, who were the relatives of the "wife," had finally come to claim the husband. No one else could actually see them, but the man kept pleading that they were taking him away. He died of fright, even though the doctors claimed it was a heart attack. The story haunted me for a long time.

~

The third grade in elementary school was a joyful time. I was happy to be back in school after being absent for a whole year. My favorite teacher was Mr. Vicken Sassouni. He came from a distinguished Armenian family. His father had been mayor of Erevan during the brief period of Armenian independence following World War I; his mother was a good friend and poker partner of my mother and head of the Armenian Red Cross. Mr. Sassouni was younger than the other teachers and had been teaching for only a year. (He went on to become a dentist in Philadelphia.) He taught us French and was our homeroom teacher. At the end of each month, we stood in front of the class and he gave each of us an evaluation of our performance. When it was my turn, I went up with some trepidation because I had gotten a mediocre grade in French. However, Mr. Sassouni gave a glowing report of my class performance, alluding only briefly to the French class and asking me gently if for some reason I found French difficult. I was astounded that he could be so fair and objective in assessing my class performance without being influenced by how I had done in his own class.

My favorite subject was Armenian history. I did less well in math and Arabic (two subjects my father excelled in). I had a knack for telling stories. When our teacher, Mr. Bedrossian, called on me to give an account of the battle of Avarayr of 451—when Armenians fought and died for their faith—I began with a moving description of the glorious sunrise with the rays of light reflected on the shields of the heroic Armenian soldiers. I then described the battle under the command of the Armenian hero (and saint) Vartan Mamigonian against the massed Sassanid Persian troops with their trumpeting elephants. The fact that I had never seen an elephant didn't stop me from providing compelling details. My recitation brought tears to the eyes of our patriotic teacher, and I became his favorite pupil.

By the time I graduated from high school, I felt confident that I had mastered the Armenian language. I reveled in my ability to use the right word with precision and grace to express whatever thoughts and feelings I wished to convey. It was like having perfect pitch. The presence of an audience was important but not essential. Over the years, I lost that sense of mastery and never fully regained it in the five other languages I learned to speak. Even though I acquired a far larger vocabulary and fluency in English, I still feel that I am using someone else's language when I speak or write in English.

Armenian schools were tough. Virtually all subjects, other than math and science, were taught in triplicate: the history of Armenia, the history of Lebanon, the history of the world; the geography of Armenia, the geography of Lebanon, the geography of the world; and so on. We also spent as much time making trouble and getting punished for it as we did in learning anything. The boys, but not the girls, were yelled at, slapped, and caned by our teachers at the slightest provocation. We tormented them in turn in ingenious ways. One was to insert a double-edged razor blade into the wooden desk. While the teacher was at the blackboard, one of the boys in the back would pluck the blade to produce a tinny, *tingting* sound to general laughter. As the teacher went to investigate, there would be another *tingting*, now from the front. So it went, until we were ordered to stand up and open our palms to be struck by a ruler. During one recess, we collected all of the canes from the classrooms and dumped them in the back of a truck parked in the street. That bought respite for a week, until the canes were replaced.

One boy in particular could get away with murder. Garbis was short and chubby, with bulging eyes, which made him look like Peter Lorre in *Casablanca*. When a teacher had barely slapped him, Garbis would fall to the floor, writhing convulsively and rolling his eyes. Or he would stagger and crash into the free-standing blackboard, scattering the erasers and pieces of chalk all over the floor, to our great merriment. The teacher and the rest of us knew it was all contrived, but Garbis's performance was so convincing that it was impossible not to wonder if *this* time it wasn't real. Nonetheless, we never felt mistreated or abused in school. There were limits, however, to how hard our teachers could hit us. If a teacher acted capriciously or vindictively, we resented it. It was also important that we not be humiliated by making us cry in front of the class.

In only two instances did I became upset over being punished. The first was in music class. Our music lessons were boring, as we sang in unison one song after another. To relieve the tedium, I tried to harmonize, but the teacher standing behind me thought I was singing out of tune and slapped me on the neck. What an idiot, I thought—he is teaching music and can't recognize harmony. In the second case, we hatched an elaborate plan where we rigged the benches with firecrackers that would explode when we sat down hard on them. We picked Mr. Bedrossian's history class since he was the most excitable teacher and would be the most rattled. As he came in, the class stood up and then we dropped on the benches to trigger the charges. All went according to plan, except that the firecracker under my bench wouldn't go off. So I got up and kicked the bench. Mr. Bedrossian saw me, came at me in a rage, and slapped me hard. That was fine, but when I saw the *et tu Brute* look in his eyes, I was filled with remorse. I was, after all, his favorite pupil.

Although I was at a parochial school, there was little by way of religious instruction. We were taken to the church once a week, but the service was in classical Armenian, which we could not understand except for a phrase here and there. I enjoyed the music and incense, but the ceremony went on forever, with the priest droning on endlessly. We had to stand up during certain times during the service and had to follow the lead of one of the two teachers. However, they were not always in step, so we broke into two factions based on which teacher we imitated. None of all this had anything to do with my personal belief in God. Yet, at some level, I would always feel part of the Apostolic Church.

The most traumatic experience during my elementary school years was my father's imprisonment. A lawyer working for his construction company had a sack of flour delivered as a bribe to a government official to facilitate getting a contract. The lawyer was arrested along with my father, who was held accountable as the president of the company. He was held in a minimum-security jail opposite the public gardens close to where we lived. Every day, Lucine would take him lunch to spare him from eating the prison food. On Saturdays, I accompanied her to visit my father. It felt strange and scary to see him sitting on a mattress on the floor among tough-looking strangers. In exchange for a pack of cigarettes a week, he could keep his mattress in a more desirable location. My father would hug me, exchange a few words, and then eat his lunch in silence. I worried and felt sorry for him.

There were lengthy discussions between my mother and influential relatives on how to get the case dismissed, or at least obtain a lenient verdict at my father's trial. Mr. Azad, who was married to one of my cousins, was the editor an Armenian newspaper and politically well connected. He was also the cousin of the lawyer working for my father who was also in jail. Mr. Azad obtained an audience for my mother to meet with Prime Minister Riad El Solh (the first to serve in that position after independence). My mother came back from the audience greatly encouraged. Since the prime minister spoke Turkish, my mother had been able to make a compelling impression in pleading her husband's case. Finally, the case came to trial, and my father was sentenced to six additional months in prison. The news hit us like a bolt of lightning. Our relatives gathered at our house to comfort my distraught mother. It felt like a funeral. I sat sulking in a corner and refused the offer from two older cousins to take me to the movies as a consolation.

There was no dishonor attached to my father's imprisonment. Everyone blamed the company lawyer for selling him out. To make matters worse, the lawyer was set free on the grounds that he was merely carrying out my father's instructions and was unaware that he was breaking the law. My mother was incensed. She was absolutely sure that when Mr. Azad took her to see the prime minister, he had contrived to make it look as if my mother was the wife of his cousin the lawyer. Consequently, the lawyer was freed because of Mr. Solh's intervention as a favor to the lady who had made such a great impression on him in eloquently pleading for her husband. No one dared to contradict my mother's conspiratorial theory, and Lebanon being what it was, it might just as well have been true.

~

My social life through elementary and high school was narrowly confined to the Armenian community. I didn't have a single Arab friend until I went to college; nor did my family socialize with our Arab neighbors, even though they maintained a cordial relationship with them. Lebanese Arabs were more amused than annoyed with the way many Armenians spoke a broken Arabic with a distinctive accent. One television comedian became very popular by his imitation of Armenians, but some of them took umbrage, beat him up, and had the show canceled. My aunt Aznive, who hardly spoke ten words of Arabic, used to say, "What's wrong with these Arabs? We've been here for twenty years and they haven't learned Armenian." When she called the grocery store, she would simply order "One kilo fresh tomatoes" and hang up without identifying herself. Since no one else did that, the grocer knew where to send the tomatoes. After the Lebanese baccalaureate became a requirement for entering college, subsequent generations of Armenians became fluent in Arabic. It would have been a nightmare for me if this rule had been passed when I was still in high school.

Despite my father's urgings, I resisted learning proper Arabic. His admonishment "You have to learn the language of the country" fell on deaf ears. Arabic tutors during the summers didn't help. One of them spent the time cleaning his nails as we both waited for the hour to end. After I left Lebanon, I came to regret that I never learned classical Arabic. But at the time, I had no interest in other cultures. It was enough to be Armenian. I was close to becoming fluent in French (my first Western language). And after I married Stina, I made some attempts to learn Swedish during our summers in Finland. But instead of just speaking it as well as I could, I tried to learn the conjugation of verbs, which didn't get me very far. Learning Finnish was beyond my reach since it belonged to a linguistic family that was totally different from that of any of the other languages I spoke.

I used to walk through the Jewish quarter of Beirut on Saturdays on my way to violin lessons. I realized these people were not Arabs, even though they spoke fluent Arabic and I couldn't tell them apart. I noticed the Hebrew writing on the school building and knew that their football team was called the Maccabees. However, I didn't associate the Jews of Beirut, about whom I knew very little, with the Jews of the Bible, about whom I knew a good deal more. One Saturday, a woman called me into her kitchen to light her stove. I was happy to oblige but rather puzzled since she seemed quite capable of lighting the stove herself. I had no idea that it was because of the Sabbath. I think she sensed my puzzlement and smiled at my naïveté.

One morning, our Lebanese teacher of Arabic provided by the government came to class highly distraught. She was waving an Arabic newspaper with screaming headlines and photographs of bodies strewn on the ground. I could make little sense of her tearful account beyond the fact that some sort of massacre had taken place involving Palestinians. The teacher expected us as Armenians to sympathize with the plight of the Palestinians. However, we failed to see the parallel, and our seeming indifference must have filled her with despair.[5]

After France fell to the German onslaught, British and Free French troops invaded Syria and Lebanon from Palestine, meeting little resistance from the forces of the Vichy government, which still had mandatory authority over Lebanon. However, no German forces ever set foot in Lebanon. We had blackouts and painted our windowpanes deep blue, but only once did I see an airplane drop bombs, and that was on an oil depot. Allied troops brought with them a variety of soldiers. I was astonished to see African soldiers with tribal scars from the French colonies, particularly Senegal. I had seen dark-skinned people from Egypt, but no one who was that dark. Especially interesting were the Australians, with their distinctive hats. When drunk on beer, they played childish pranks such as peering under the veils of Muslim women or picking up short men and depositing them in the large baskets on the backs of porters. Fights would break out, but the Australians were so good-natured that it was hard to stay angry with them for long. They also learned how to outsmart Lebanese drivers in the chaotic Beirut traffic. When cars entered a one-way street from both sides the Australian truck driver would simply take out a newspaper and sit calmly reading it. The Lebanese driver facing him would yell and

Figure 3.2 Adolescence.

scream but finally surrender and back up. Australians would sell and barter their rations of corned beef, which I hated, and chocolate, which I loved. They also had a peculiar and pleasant smell, which must have been due to the soap they were issued.

Armenians were generally more interested in the war on the eastern front than in Europe. This was because Soviet Armenian troops fought in the Red Army in the battle of Stalingrad in 1942 under commanders like Marshal Ivan Bagramian, who served with Chief of Staff Marshal Semion Timoshenko. (One of our friends named his son Timo.) There were sixty Armenian generals in the Soviet forces, and over 200,000 Armenian soldiers lost their lives. An Armenian division made it into Berlin. Armenians were not partial. They also claimed that the German general Heinz Guderian was

Armenian because of the ending of his name, even though he was nothing of the sort. (Armenians would claim the devil as their own if his name ended with "ian" and was famous enough.) I was too young to care about any of this and had at best a hazy idea of the war raging in the world. What bothered me most was that before every movie, we had to suffer through interminable propaganda newsreels of bombings and marching troops to the grating prattle of a disembodied voice. Despite occasional shortages, my family hardly suffered from food shortages during the war, although I know that the families of some of my classmates did.

~

My mother pressed me into taking violin lessons when I was quite young and no child prodigy. I hated it. When Lucine took me to my lessons, I would bang the violin case against the wall as we walked along, hoping to break it. When I got older, it got worse. I had an ill-tempered teacher (actually a dentist) who struck at my knuckles with his bow if I played a false note. These lessons also interfered with going to the movies on my free afternoons. The teacher was often late or the lesson dragged on, so by the time I got to the theater the show was already in progress. It was not until high school that I became proficient enough to enjoy playing the violin. It also provided opportunities to get close to girls who played the piano and who would otherwise be out of bounds. Unfortunately, as soon as the music stopped, the girls' mothers would walk into the room. And since the violin keeps both hands occupied, there was not much that could be done with the music in progress.

Movies were the primary source of entertainment in Beirut. I tried to see at least one movie a week no matter what was playing. In my younger years, I mostly went to Italian serials like *Diavolo*, where each episode ended with the hero in the standard hopeless situation, tied to train tracks in the path of an onrushing locomotive; then in the next episode, he would be back in the street with no explanation of how he survived. The movie theater was a fleabag crowded with screaming kids. When the theater ran out of seats, latecomers were herded on the stage, next to the gigantic figures moving about on the screen. At moments of peak excitement, fistfights would break out, emulating the action on the screen. Ushers would then rush and restore order by kicking the belligerents into calming down.

Tarzan was another great favorite, and the lead actor Johnny Weissmuller became *Jon-ny Musmuller.* Zorro provided a better opportunity for imitation. We draped our black school uniforms around our shoulders like a cape, covered our faces with a handkerchief folded into a mask, and swung into action. There was an apartment house nearby with two separate staircases that gave access to the roof. This allowed Zorro for the day to engage in a fierce combat on the roof and then escape down one staircase when his enemies blocked the other. The racket drove the people living in the apartments below to distraction. One of the tenants finally stood at the exit of the escape staircase and as Zorro emerged flushed with victory, he delivered a resounding slap, sending the masked hero sprawling on the sidewalk in a state of utter confusion, not knowing what hit him.

Westerns were also great hits, as were romantic movies in which there would be kissing and glimpses of nudity, such as in *The Outlaw*, starring Jane Russell. What fired my imagination was not that movie itself, but the poster showing the heroin in a barn with slightly exposed, incredible breasts—the most exciting sight imaginable to my fevered teenage mind. Nor was I the only person to be so aroused by the poster that became an iconic image. In a typically Lebanese show of prudishness, state censors wouldn't allow a serious French film with a discreet scene in a brothel to be shown at the Cinema Roxy, located across the street from the red-light district operating under government auspices.

In my last year of elementary school, I became obsessed with the idea of becoming a boy scout. Since it involved going camping, my mother wouldn't hear of it. Ultimately, we struck a bargain: I could join the scouts but not go camping. I reveled in wearing the uniform and proved quite adept at all the activities that didn't involve camping, such as playing the drums in the band and playing on the gymnastics team. I soon won my first stripe and then a second stripe. I could now line up the other scouts and order them around. However, whenever a camping trip came up, I always had an excuse. It got so embarrassing that my mother finally relented. I went to my first camping trip and had a miserable time. I couldn't sleep on the ground with just a blanket under me, and the soup, "cooked" by a few of the boys, was disgusting. My happiest memory as a scout was going to a jamboree. We were driven there packed in the back of an open truck. There was a very cute girl scout behind me. We were standing back to back, and on sharp curves, when Lebanese drivers always accelerated, our buttocks were pressed together. It felt so wonderful that we kept doing it even when there was no sharp curve. I was ready to marry that girl on the spot but did not even manage to ask her where she went to school.

Summers meant living in a different world. Some families owned houses in the mountains. Most others, like us, rented houses in various resorts to get away from the oppressive heat in Beirut that made your shirt stick to your back even when standing still. Renters usually made these arrangements in the spring or early summer. My father insisted on holding out until the period of peak demand had passed and houses that were left over could be had more cheaply. This irritated me to no end. It meant suffering the heat in Beirut for several more weeks, albeit partly compensated by spending more time at the beach. Even worse, it was embarrassing to show up with a truck full of belongings at a resort when everyone else had already long settled in.

This also reduced our choices of where we could spend the summer. The preferred locations such as Dhour el Shweir or the grandly named Bois de Boulogne would often have no houses left. So we ended up in villages like Dwar or Mrouj on their outskirts. In these more rural settings, I spent a lot of time walking among the village shops. The butcher provided the most fascinating, if gruesome, spectacle. He slaughtered the sheep by cutting their throat and draining the blood into a basin. He then made a small cut in the skin at the ankle and inflated the body with a hand pump, separating the skin from the underlying flesh and making it easier to skin the animal. He finally opened the abdomen, took out the organs, and hung them from meat hooks. On Saturday mornings, a few men would gather around and the butcher would offer them thin slices of fresh liver with small glasses of arak (reminiscent of the ancient Greeks who "tasted" the entrails of sacrificial animals). Sheep organs were cheaper than flesh, but we rarely ate them. I decided I hated liver, spleen, and kidneys, without ever having tasted them, but I liked sheep's brain and testicles, which my mother prepared by breading and frying them. My father would ask me how I knew I didn't like liver when I had never tasted it, and I responded that since I already knew that I didn't like liver, why would I need to taste it?

One year we couldn't find a house even in Mrouj, and we ended up in the remote village of El Mtein in the middle of nowhere. We were the only family renting a house, and I resigned myself to a lonely summer. However, the *mukhtar* who owned the house and lived on the first floor had three sons who took me under their wing and I went hunting with them. I wasn't a good shot, felt sorry for the little birds, and did not particularly enjoy eating them, but hunting is what boys did and that is what I had to reluctantly do also. There were all sorts of weapons in their house, including grenades in a kitchen drawer. I also befriended the *natour*, a strapping young man who guarded the vineyards. I would accompany him on his early morning rounds and gorge myself on the luscious grapes with the dew of the morning still on them. They were wonderful people, and I became like a member of the family. Even the women of the household did not mind my going in and out of their house.

Mtein turned out to be the hotbed of a political party (which I recall as the *Hezb al Khowme*). At a massive political gathering that drew people from the whole region, I was featured to play the party anthem on my violin. I remember walking on stage and looking with some apprehension at the massed crowd standing still in the darkness. My grandmother was delighted when my name was announced on the loudspeaker ("It's Hrant!"). What startled me most was the amplified sound of my violin that boomed out of the loudspeakers as I played the first note. That was the only political gathering I was ever to attend in my life. It made me a celebrity in the village.

The following summer brought my hunting days to an abrupt end. We were living in Mrouj, and I went hunting with another boy. As we walked through the vineyards, we helped ourselves to the grapes. Then, while we were sitting in the shade of a fig tree looking for birds called *khoury* (or "priest," because of their black head), a *natour* barged into the clearing yelling at us for eating the grapes. As I stood still, he came and took my shotgun. He then went to my friend, who resisted, and the guard slapped him until he let go of it and he took both guns to the police station. My gun was not registered and I had no hunting license. My father and I went to recover the gun and he tried to talk his way out of it by raising his voice, but the police chief yelled back at him and threatened to arrest him. Scared witless, I begged my father to just leave—I no longer wanted the gun.

~

I don't recall when I entered puberty, but I do remember vividly my first encounter with its early signs. One day when urinating, I looked down and was astonished to see several long and coarse black hairs curling out of my scrotum. Suddenly, I was flooded with a feeling of joyful pride. I didn't know why. I had never seen a man's pubic hair, yet these few odd strands looked like the harbinger of how my body was going to change. No one, of course, ever spoke to me about these matters. Lucine used to sew my underpants out of cotton cloth, until my mother took me to a store to buy readymade shorts. When I saw the slits in front, I didn't know what they were for and said I didn't want them. The saleswoman looked baffled, and my mother turned her attention to something else so as not to embarrass me. She must have then bought them and put them on my bed. When I tried one on, it became clear what the slits were for. A replay of sorts occurred years later when I was having a suit made by the bespoke tailors Anderson and Sheppard on London's Savile Row. After taking detailed measurements, the trousers tailor asked, "Sir, on which side do you wear it?" Wear what? There was an awkward silence until the matter was clarified by furtive gestures. We agreed that the extra space in the crotch should be on the left. (I didn't know it at the time that the left testicle usually hangs lower than the right, a fact known to Aristotle and classical sculptors.)

I didn't associate these bodily changes with anything sexual. Rather, it meant that I was growing up to become big and strong. My year-long illness had left me feeling like a weakling, and I was desperate to change that. When in high school, I found old copies of bodybuilding magazines in a shop and was astounded to see the pictures of men with enormous muscles exercising with barbells and dumbbells. That was the way to do it—but where would I find the equipment? I asked my

Figure 3.3 With my parents and Lucine (who is showing off her new watch).

father, and he sent me to a friend who owned a foundry. The man was very kind and intrigued with the idea, so he had a forty-kilo (88 lb.) barbell custom-made for me. I figured out how to exercise with it and eventually could lift it with one arm. By the time I graduated from high school, I was the strongest boy in class. I also took up boxing and became quite good at it. This led to a lifelong commitment to physical fitness. I can still wear some of the clothes I had fifty years ago. And I continue to ski. When I was told that as a "super-senior" I no longer had to pay for lift tickets, I thanked the cashier for their generosity. He asked how many people my age did I think still skied—it didn't cost them very much.

~

The transition from elementary to high school meant moving into a quite different social setting. My father decided that I should continue my education at an Armenian high school rather than attend International College, the preparatory school for the American University of Beirut. He was convinced that I would still make it into the university. I enrolled in the Armenian Evangelical High School. My mother was happy with the decision even if she wasn't closely involved with the church that sponsored the school. Another alternative would have been to attend the Collège Armènien (*Djemaran*), the other Armenian high school in our district. It had a faculty of distinguished Armenian intellectuals but also a strong affiliation with the Tashnag Party (short for the socialist Armenian Revolutionary Federation). Besides, since its primary foreign language was French, it would have further complicated entering AUB. (One of my freshman classmates who was a graduate of that school pronounced Miller Highlife beer as *Mil-ler Ikhliff.*)

Good students from my elementary school could skip seventh grade and start at the eighth grade of high school, which is what I did. This made up for my late start in elementary school. Skipping seventh grade posed no problems with most courses, except for math (a replay of my skipping second grade). Since I didn't like the subject to start with, it added an additional burden. Perhaps that's why I struggled with math throughout high school and in my freshman year in college, after which I wanted to have nothing more to do with it.

Over the next four years, I obtained a good but not an outstanding education. The curriculum was geared for preparation for college. Moreover, the inculcation of Evangelical beliefs took precedence over the fostering of Armenian culture, which reshaped my values and ideological loyalties. My academic interests continued along the lines established in elementary school. I preferred, and was far better at, subjects like history and literature than math and science. Words came to me more easily than numbers (and still do).

I became close to only one of my teachers. Mr. Guiragos taught our religion class in the eight grade, after which we became friends. He was in his early thirties, with a balding head and horn-rimmed glasses. He stood out by his effeminate mannerisms (students called him "Miss Guiragos," and he knew it). He was a kind and idealistic person who had a high regard for me but who also encouraged intellectual pursuits over my more materialistic attitudes. (He would chide me, saying, "You think like a merchant.") Mr. Guiragos was a graduate of the Near East School of Theology, but he was not an ordained minister. He had highly liberal views. I was shocked to hear him say (but never in public) that the Virgin Birth was not an essential Christian doctrine.

Why would it matter if God had chosen to come into the world as the illegitimate son of Mary and Joseph or even being fathered by a Roman soldier? Who were we to judge?

Mr. Guiragos was desperately in love with a stern woman pianist. Since the piano was his true passion too, they had a common bond—but that's all they had. She gave no indication of being interested in him otherwise. He spoke to me about his hopeless love, which he couldn't imagine ever revealing to her. He described how when playing a duet together their hands had lightly touched and how that had sent chills down his spine; he visibly shivered as he told me this. She ended up marrying a physician. I have no idea of what became of Mr. Guiragos.

My relationship with my mother and Lucine did not significantly change as I went through high school. They adapted themselves to my becoming a more independent and self-assertive adolescent and continued to indulge me. Lucine tried to teach me to be orderly and not to drop my pants on the floor and walk away, and I have never done that since. Otherwise, they made few demands on me. I was a good student, respectful toward my elders and trustworthy. My mother may have wondered if awakening sexual urges were in danger of leading me astray, but said nothing about it. Only on one occasion, when we were in the kitchen, she made some obscure and indirect comments on what, I assume, were the dangers of masturbation. I didn't take the bait, and she dropped it. Otherwise, I was essentially left on my own to pick my way through the process of becoming an adult.

My relationship with my father during adolescence became more contentious. I thought he reacted ambivalently to my emerging manhood. When I grew a light fuzz on my upper lip, I shaved it off and told my father proudly about it. I thought he would be pleased, but he was not. Instead, he chided me because shaving my beard early would make it grow coarser and harder to shave in later years. That made no sense, and I wondered if for some reason he didn't want me to become a man. My father and I now talked more mostly about mundane topics but never delved into more personal matters. He was happy about my performance in school and let it go at that. One day while I was waiting for my father at his construction company office, a pathetic-looking man walked in, leaning on an improvised cane and wincing with every step. His trousers had been cut off above the knee to expose a horribly swollen leg covered with pustules with angry red borders and necrotic patches of black skin. I assumed he had been injured when blasting rocks with dynamite at a construction site. The man walked through the open office door of one of my father's partners. He was plaintively begging for money, but was curtly told to go away. He left looking utterly dejected. I was appalled, but when my father appeared, I said nothing, assuming he would side with his partner. I was to be troubled by this encounter for years and regretted not giving the man whatever little money I had on me.

In a subsequent discussion with my father, I raised the issue of ethical responsibility, with no reference to the above incident. My father maintained that he would be morally responsible only for problems he had caused himself. In rebuttal, I posed a hypothetical situation: Suppose he came across a man dying of thirst in the desert. Would he be ethically obligated to help him? My father said if he had not caused the man's plight, then he would have had no moral obligation to help him. I think he was trying to be logically consistent, since he was not a heartless man. I couldn't argue the case any further, but I must have appeared to be perturbed enough for him to have second thoughts about it.

At various periods during my adolescence I nursed a smoldering anger at my father. An ostensible source of resentment was the paltry allowance he continued to give me on the misguided premise that money spoiled children. This had mattered little when I was younger, but now it put a damper on everything I wanted to do. The five Lebanese pounds a week I received would buy a movie ticket and a couple of pastries, but nothing more. I recall standing at the pastry shop with one of my richer friends (and most of my friends were from rich families), watching them gorge themselves on mille-feuilles, while I took little bites out of the one piece of pastry I could afford. On my birthdays, I would get an extra five pounds. But on one birthday as I was leaving for school, I overheard my mother asking my father if he was going to give me the extra money. He gestured that he was not, and did it dismissively. I was enraged.

These experiences brought back the ghost of Iskenderun to haunt me. The legacy of our family name made it even more difficult to accept our reduced circumstances: What good was the family name without the money to back it? I recall telling my father about a new friend and pointing out that his family was not rich, but belonged to the upper class. My father angrily muttered something about "upper class" and was visibly upset. I was surprised, but I wonder if I was not actually taunting him. During these times of alienation, I hardly spoke during meals. I leaned back on my chair, precariously balancing on its two back legs, and ate in sullen silence. I was never openly disrespectful, nor did I ever raise my voice at my father, but my simmering resentment must have been obvious to him. I must have been insufferable, yet my father didn't react to my provocations; probably he was restrained by my mother, who counseled patience while I was growing up.

~

My social life picked up considerably in high school, especially during the summers, when I had different friends than during the school year. We spent many afternoons at the two cafés in Dhour el Shweir—Hawie's and Nasr's—where we would watch couples dancing in the evening or just listen to the band. There were no girls for us to dance with (besides, I didn't know how to dance), so I mostly imagined myself in the shoes of the man dancing with the most attractive woman, whom I chose after careful consideration. The waiters were accommodating and took their time before asking us to order. I usually asked for *baba au rum*, which I could nurse for a long time even if I disliked its sickly sweet taste. Following my freshman year in college, when my cousin Nora lived with us, I could go dancing with her at Hawie's and came to feel more like a grownup.

There was one particular couple that attracted my special attention. He was a handsome Armenian medical student and a gifted violinist; she was a lovely Lebanese Arab girl. I could see how he would be attracted to her, but I was puzzled by his seriously courting a non-Armenian and equally surprised that the girl's brothers would allow him to do it. I didn't quite know what was wrong with such a relationship, except that it just was not done. (I would have been astonished if someone had told me that years later I would be married to a non-Armenian myself.) The couple eventually got married and moved to the United States. He became a psychiatrist, and many years later I met him again at a clinic in Arlington County, where he was on the staff. I said nothing

Figure 3.4 Nora at her graduation from college.

about my earlier reservations about his courtship. It would have been too silly for words.

Autumn was my favorite season. As a thick fog blanketed the pine forests, I walked through them with a wistful feeling about the ending of summer and the reopening of school. The high point of September was the feast of *Mar Takla* at the village of Mrouj, a harvest festival celebrating the picking of grapes (*vendange*).[6] People flocked to the village square packed with stalls selling countless varieties of fruits, vegetables, pickled goods, as well as coarse woolen socks, sweaters, and other handiwork of village women. The heady aroma of freshly baked bread permeated everywhere. With coins clutched in their hands, children ran about looking for treats and excitement. The village brass band, consisting of the baker, the policeman, the shoemaker, the grocer and his son, blared marches and national anthems with their tarnished and dented instruments. Embellished by the lilting rhythms of Arabic music, whatever they played sounded more or less the same.

"God Save the Queen" was orientalized to the point where the Queen herself would have hardly recognized it.

There was the old man with his panorama box, ornately decorated with tiny mirrors, tassels, bits and pieces of metal strips, cloth, and shells. A circle of portholes with large magnifying glasses served as viewers. You peered into one of them and saw a succession of enlarged postcards of mountains, waterfalls, cathedrals, pyramids, and other great sights from around the world. As the old man slowly rolled the images around, he gave a singsong recitation of their marvels embellished with florid expressions of wonder: "And-here-is Jebel el *Avarest*, the *biggest* mountain made by God the Almighty. Its peak is covered with ice *ten meters thick*, over which storms howl day and night. No man has ever set foot on it—even *angels* dare not fly over it." What amazed me most was the disembodied-looking head of a girl resting on a mirror inside a large wooden box. People looked at it, muttering the equivalent of "I'll be damned" or "That's crap." Children tugged excitedly at their parents, asking, "Where is the girl? Where is the girl?" I knew, as did everyone else, that it was some sort of optical illusion created by mirrors but it still bothered me not to know how it was done. (To this day, I get more annoyed than amused by how a magician does the seemingly impossible.)

An important rite of passage for teenage boys was the ringing of the church bell. It wasn't hard to make the bell ring single tones (*ding, ding, ding*). The challenge was to make the bell sing in a continuous rhythm (*ding, diding, ding, diding*). For that, you had to hold the thick rope down, close to the ground as the bell swung back and forth, then you let go of it until the next pull. If you weren't heavy or strong enough, the rope would lift you up into the air, leaving you dangling helplessly as the bell struck a solitary and mournful *ding*. It was humiliating and took me a couple of years until I could make the bell sing, triumphantly announcing to the world that I was now a man.

I was involved in two accidents while in high school, either of which could have killed me. The first was in the ninth grade during a school bus trip to the snow in the mountains. When we started back, it was already dark and drizzling. At an unprotected railroad crossing, our bus collided with an oncoming train. I had dozed off and didn't see it happen. When I woke up with a startle, the bus was in a pitch black chaos with a faint odor of gasoline in the air. I managed to crawl out of the wreck and somehow made my way to a dimly lit house on the mountainside despite line fractures in both legs. The young daughter of one of the teachers was killed and a classmate suffered compound fractures that left him lame. At the hospital, I asked that they wait until my parents came and saw me before taking me to surgery. I was bedridden for a month, and when the casts were removed my legs were so weak that I had to crawl before being able to walk again. No one seemed to have heard of physiotherapy.

The second accident was my own doing. One of my classmates in the senior year had a Norton motorcycle with a sidecar, like those used by the police. He asked me if I would like to drive it. I said sure, even though I had never been on a large motorcycle before. On a Saturday afternoon, my friend took me to a road leading out of the city. It was a wide avenue with a deep drainage ditch with cement walls running on its side. After some perfunctory instructions, I began to drive. As the sidecar pulled the motorcycle to the right, I overcorrected by veering sharply to the left, then to the right again, until I lost control and landed in the ditch. I was thrown over the handlebars and badly scraped the side of my right leg against the cement. My friend was unhurt. I limped to a nearby pharmacy, where they cleaned and bandaged the wound without charging me. I took the tram home and changed into my pajamas. My father was lying on the bed reading the paper. I stretched out next to him on my mother's bed. I told him about the accident, playing it down. He asked no questions. When my mother returned home, she immediately noticed the smell of iodine and after she found out that I was hurt she took me to see a doctor. I was given anti-tetanus horse serum, to which I had a horrific allergic reaction that made me forget the accident. My classmate who owned the motorcycle said nothing about paying for its repairs. However, a week or two later my father told me that someone had come to him for that purpose. I told my father that I would take care of it. I don't know whether my father paid for fixing the motorcycle or not, but I heard nothing further about it. Meanwhile, I began to save what money I could setting aside virtually all of my allowance. A month later I had accumulated enough to make at least a partial payment. My classmate took the money, looking surprised. Coming on top of my childhood illness with rheumatic fever, these accidents convinced me that I was destined to suffer some serious mishap every year. I waited for it to happen and may have even provoked it in order to be done with it. This morbid frame of mind was probably linked to larger psychological problems that plagued me during my high school years.

~

The faith in God established in my childhood persisted into my adolescence. During my last year of elementary school I had began to attend the Apostolic church sporadically. I only stayed for part of the Sunday service to light two wax candles and pray briefly, asking God's help and blessings on my life. It was more of a private ritual than a participation in the religious ceremony. However, I did it fervently, kneeling down and touching my forehead to the floor in full public view. The deacon must have been impressed enough to compliment Lucine for my piety. When she told me about it, I felt embarrassed. I did not realize I was being observed. When I entered the Armenian Evangelical High School, I shifted to attending the Protestant Sunday service. I continued to think of God as a powerful patron with whom I enjoyed a certain level of intimacy, but I was mainly interested in currying his favor.

Soon after entering high school, I joined the Christian Endeavor Society (*Tchanasiratz*), the youth organization of the Protestant Church.[7] Initially, my interest was mainly social. (I was to meet my first girlfriend there.) However, as I continued to be associated with the organization throughout high school and college, it came to play a much more important role in my life. The Bible study groups helped me gain greater spiritual depth and added more substance to the beliefs of my childhood, without fundamentally changing them. I also enjoyed being part of a group of young people who liked and admired me. As I gradually took on a greater leadership role, I learned how to speak in public and gained the self-confidence to do it well. The summer conferences at a mountain camp were particularly important in making the Christian Endeavor Society a central part of my life.

I never formally switched from the Apostolic to the Evangelical Church (and was never pressured to do so). However, I became an integral part of it through my association with its youth organizations.

Despite the very positive consequences of these associations, there was also a negative aspect to this experience. The Armenian Evangelical Church espoused a dour puritanical morality that stunted my emotional growth, particularly with regard to sexual matters. It created a schism between my public image as a self-confident young man and my private self as a troubled adolescent with an inhibited and constricted emotional life. It is hard to tell whether the Evangelical Church was responsible for creating my psychological problems or whether those problems distorted my religious beliefs. A theologian colleague pointed out that my religious experiences as an adolescent were hardly unique. Rather, they were quite similar to those of others brought up in an Evangelical tradition, except for the fact that I could not point to a specific conversion experience, which is typically the case.[8] Yet this religious tradition alone could not explain my obsessional preoccupations at the time. Decades later, when I was teaching an undergraduate seminar on guilt, one of my students wrote a compelling paper detailing her struggles with guilt, which had a similar mix of Evangelical religious beliefs and psychological problems—in her case, a severe depressive illness going back to her childhood. She ended her paper with "rare is the person who wouldn't hate the Christianity I created for myself." The parallel between our experiences was striking.[9] In both cases there was a mix of psychological and religious elements that came together in a perfect storm. I am therefore hard-pressed to say which of these came first or was preeminent. Did the "Sturm und Drang" of my adolescence bring to the surface psychological conflicts which were then given expression through religious imagery and ideation? Or, did the teachings of the Evangelical Church turn the normative turmoil of adolescent into a morbid experience?

A hundred years ago, William James distinguished between *healthy-mindedness* and a *sick soul* in psychological and associated religious terms. The healthy-minded temperament is fundamentally optimistic. It expects and sees what is good in the world. It has no tolerance for prolonged suffering and sees no merit in it. The religious belief that goes with it does not focus on evil or dwell on remorse and repentance. One needs to avoid sin, not wallow in it. The sick soul is the opposite. It has a pessimistic outlook that expects and sees what is bad in the world. The religious belief that goes with it accepts suffering and evil as integral parts of life and makes a virtue out of remorse and repentance.[10] My childhood was characterized by healthy-mindedness. God was a benevolent and loving presence that I could count on to help me out of difficulties and make me happy. That helped me to survive my childhood illness and come out of it with greater strength, hope, and optimism. Then something changed during my adolescence that turned me into a sick soul. I became burdened with an excessively sensitive and punitive conscience. Life became a minefield of temptation and sin that would lead me astray from God, who now had turned into a forbidding and frightening presence. Even though I was able to gradually free myself of this sick soul and regain some my healthy-mindedness, the residues of psychological conflicts and religious morbidity were to plague me for the rest of my life. But they also endowed me with a higher moral sensibility and deeper appreciation of the tragic nature of life than the more superficial healthy-minded view that turns a blind eye to all that is wrong with the world.

The psychological conflicts of my sick soul started insidiously as I entered adolescence and then grew into full-blown obsessive-compulsive symptoms that permeated my young life, causing a lot of anxiety and distress. A central element was an increasingly oppressive sense of guilt. The lifelong struggle with a hyperactive and punitive conscience eventually led to my teaching a college course and writing an extensive book on guilt. However, it took years before I could gain a measure of control over *when* I would feel guilty, *why* I would feel guilty, and *how long* I would feel guilty after a real or imagined transgression. And while the ever-present prospect of guilt may have helped keep my life on the straight and narrow moral path, it also robbed me of the legitimate pleasures of growing up into manhood.

The fear of doing wrong and thereby offending God spilled over into compulsive behaviors to ward off that danger. For instance, I developed the ritual of offering a short prayer for the souls of people who had died. It was customary in Beirut to paste black-bordered death notices on the walls of buildings. If there were a few of them, I could deal with them but if there were many, I became overwhelmed. Another compulsive ritual pertained to school exams, which were even more problematic. Thus, before I could respond to a question, I would have to offer a short prayer asking for God's help. This had to be done

Figure 3.5 Senior class in high school with our class teacher. (I am third from the left in the back row.)

in a particular way and could not be rushed. Otherwise, it would have to be repeated. As the minutes ticked away, I sat paralyzed while my classmates worked on their exams, adding enormously to the pressure I was under. To get around it, I devised a system where the prayer was represented by a symbolic dot. But where would I put the dot—on the first page of the exam, on each page, on each question? The dot had to be visible but not conspicuous so as not to attract the teacher's attention; that meant erasing dots repeatedly and redoing it until I got it right.

I told no one about these problems, although some of my classmates became aware of the surreptitious rituals and would make fun of me by mimicking me. I didn't seek help because I didn't know where to turn. Had I been physically sick, I would have asked my mother to take me to a physician. But this was different. It involved my thoughts and feelings. I was baffled and embarrassed and struggled on my own for several years until the symptoms gradually abated. It's remarkable that despite these burdens and distractions, I was able to function as well as I did. Their residue persisted in the form of compulsive features in my character. In some ways, these proved to be assets in making me a highly organized, meticulous, and tenacious person, but they also took their toll.[11]

The summer following my graduation from high school was full of reflection and self-assessment. We had rented a house on the edge of the village of Mrouj, with no other houses around. I spent many solitary hours in the evenings in a reclining chair under a blanket, gazing at the stars in the crisp and clear mountain air. I stared long at the Big Dipper, which looked like an oasis of order in a bewilderingly chaotic universe. What did its pattern signify? Could it reveal anything about my future? I didn't have the slightest clue at the time that decades later I would be staring at the same constellation from the porch of our summer house on an island in Finland and marveling at the course my life had taken: where I started and where I ended up; who I was and who I had become. Yet, despite all these changes, my sense of who I was hadn't changed at its core any more than had the Big Dipper.

In the four years I spent in high school, I had worked hard and had quite a bit to show for it. I graduated second in my class and would probably have been first had I not been weighed down with my psychological problems. Moreover, I was the most well-rounded student. My name appeared on the graduation program five times: graduation with honors, commendation for character, athletics pin, declamation prize in French, and playing the violin during the ceremony. However, my life hadn't been easy, and I was relieved that the ordeal of my high school years with its attendant turmoil appeared to be coming to an end. I was now headed for the next phase of my life, to college and beyond, a prospect full of promise. I could now look to the future with self-confidence and the certainty that God would never let me fall on my face.

4

That They May Have Life…

WHEN I APPLIED to the American University of Beirut in the spring of 1950, I was virtually certain that I would get admitted. However, I fretted until I actually received the acceptance letter, which I read several times. I was keenly aware that going to college was going to mean a far more significant change in my life than the transition from elementary to high school—even though, since I would continue to live at home, the everyday circumstances of my life wouldn't change that much. I knew I would now be facing greater challenges, but it was far from clear what courses I would take, what career direction I would choose, and in what ways my social life would be different. This uncertainty added to the exhilaration as well as the anxiety over what the future held for me.

I would be leaving the parochial and insular world of the Armenian Evangelical High School for the larger, secular, and far more diverse world of the American University of Beirut (AUB), with its 2,500 students from several dozen countries. Its motto, "That they may have life and have it more abundantly" (*Ut vitam habeant abundantius habeant*), held much promise. But I did not quite realize that I was embarking on a lifelong association with AUB: four years as an undergraduate, four years as a medical student, four years as a member of the faculty, and finally eight years as a member of the board of trustees until my retirement in 2008, when I became an emeritus trustee.

My undergraduate years provided the fundamental building blocks of my higher education, while my medical training prepared me for my career. In this decade-long process, I developed from a gifted but troubled adolescent with limited aspirations to a more self-assured adult with a wider worldview. The years on the medical faculty launched my academic career, which was to span close to fifty years, most of it at Stanford. My time on the board of trustees brought me into a group of accomplished and successful men and women with a rich legacy of service to an extraordinary institution that over one and a half centuries had left its mark on Lebanon and the region as a whole.

My parents and I assumed that I would go AUB for the primary purpose of preparing me for a career, whatever that might be. Our mathematics teacher in high school (whom I was fond of despite the subject he taught) tutored several of us to sit for the Oxford/Cambridge General Studies Examination ("Matriculation"), which I passed easily. That gave me the far-fetched idea of studying in Edinburgh (why Edinburgh?). I suspect that what I wanted was mainly to get out of Lebanon and live in Europe; the United States was too far away and seemed even more out of reach. The thought that I might get a better education in Europe was a secondary consideration. Moreover, I felt that the West was more secure—what belonged to you today would still belong to you tomorrow. I did not want another Iskenderun. When I told my father that the certificate I had just acquired made it possible for me to study in Europe he replied, "What's wrong with AUB?" That ended the discussion. Actually there was nothing wrong with AUB. It had a stellar reputation. My two maternal uncles, several older cousins, and virtually every professional person I knew had studied there. It cost much less than studying in Europe and was within walking distance of our house. And it was hardly a parochial institution. In 1945, when the representatives of fifty countries met in San Francisco at the Conference on International Organization to draft the United Nations Charter, there were more graduates of AUB among the delegates than from any other university in the world. Although embedded in the culture of the Middle East, AUB, like the United Nations itself, was a cosmopolitan institution. It drew students not only from Lebanon and elsewhere in the Middle East but also from as far as Africa, Europe, and the United States.

~

The American Board of Commissioners for Foreign Missions sent its first five missionaries to the Near East in 1810. Although part of the Congregationalist Church, the board also supported Presbyterian and other missions scattered over the Ottoman Empire in Turkey, Syria, and the Holy Land. The original purpose of the mission was to translate the Bible into local languages in order to preach the Gospel. However, given serious religious and political constraints on proselytizing among Muslims, most converts to Protestantism came from other Christian churches such as the Armenian Apostolic Church (as was the case with my mother's grandparents). This made the educational mission paramount for missionaries, who heeded the call to imbue native youth with Christian and Western ideals and to prepare them for service in their own communities. These considerations also applied to the Jesuits, who preceded the American missionaries by two centuries. They first arrived in Lebanon in the 1640s, but their early efforts were suppressed by the Ottomans. The Jesuits returned in 1831 and began a school in Beirut in 1875, which became L'Université St. Joseph. These two institutions—one American and one French—were largely responsible for making Lebanon the best-educated and most Westernized of the Arab countries in the Middle East. Given its significance for the region and my own personal life and career, I will digress briefly to provide a fuller account of the history of AUB.

In 1855, Daniel Bliss and his wife, Abby Maria Wood, sailed from Boston for Beirut to join the Syria mission. Bliss had studied at Amherst College, then at the Andover Theological Seminary, and was ordained a minister. In 1864, the board asked him to withdraw from evangelical work in order to plan for the establishment of a college in Beirut. Bliss went back to America to raise funds for the college, and over the next two years he traveled 17,000 miles and gave 279 addresses. Then, with sufficient funds, a charter from the State of New York, and the vital support of the board of trustees, he returned to Beirut to turn the dream into reality. In 1866, the Syrian Protestant College opened its doors to sixteen students. It had no American faculty (except for Bliss), owned no buildings, and had no academic plan other than the distant model of American institutions that had to be adapted to the region.

In the following several years, six more American faculty members were recruited, while Daniel Bliss continued to teach philosophy and ethics throughout his tenure as president. This remarkable group included four physicians who formed the core of the medical school. Their competence went beyond medicine. Dr. Cornelius Van Dyke was an ordained minister who learned Arabic and was instrumental in the first translation of the Bible into Arabic. His mastery of the language was widely admired by Arab scholars. He directed the Mission Press, which issued numerous works in Arabic, including a weekly newspaper. His colleague Dr. John Wortabet was the son of a former Armenian priest (*vartabet* in Armenian). He too was an ordained minister as well as a linguist who produced the first Arabic–English dictionary (still in print in revised form) and wrote medical texts in Arabic. Dr. George Edward Post, an accomplished surgeon, was also a prolific writer in Arabic. His works included a monumental botanical study of the flora of Syria and Palestine and a concordance for the Arabic Bible.[1] Being deeply engaged in the culture of the region saved these outstanding men from looking like fish out of water. They also ensured that students acquired an understanding and appreciation of their own cultural heritage. In 1905, AUB began to train nurses, and women began to be admitted to the college in the 1920s. (They now account for half of the student body.) AUB was founded to educate young men and women to serve their own countries; it was not meant to be a springboard for individuals to pursue their professional interests in the United States and Europe. But that is what many AUB graduates were to do, including me. Nonetheless, these alumni also provided critically important help to their alma mater in a wide variety of ways.

The task of finding qualified students was no less daunting than recruiting high-quality faculty. The educational standards in the area were low, and local schools hardly prepared their pupils for college-level work. Therefore, a preparatory school was established alongside the college. This meant taking care of youngsters, sometimes as young as twelve, who came from the neighboring countries and had to be housed and looked after. In 1936, the International College of Izmir in Turkey moved to Beirut and became affiliated with AUB as its preparatory high school. Collecting fees for tuition and room and board was often difficult. While many parents were willing and able to pay for their children's education, some others could or would not.

Those with modest means bargained over the cost (about $75 a year). Prominent families argued that their children should study free because they brought honor to the institution. One father offered to send his son as a "gift" to the college if he could attend it free of charge.

Other obstacles were ideological. The local and regional population consisted mainly of conservative Muslims and no less conservative Catholic and Orthodox minorities. Each group had its own reasons for being suspicious of handing over their children to Protestant missionaries. The Ottoman authorities in Istanbul and their local representatives were equally wary of Westerners inculcating liberal and subversive ideas in their youth. This put the AUB administration and faculty in a delicate position. On the one hand, they had to reassure parents and authorities that their religious and political views would be respected; on the other, they had to remain true to their missionary ideals that had brought them there in the first place. The college had to convince the Ottoman rulers that they were not Christianizing their subjects while reassuring its trustees and financial supporters that their Christian objectives were being pursued. The fine line that Daniel Bliss had to walk is eloquently expressed in his remarks at the groundbreaking ceremony of the first building, College Hall.[2] Living up to this ideal was not easy, and the college occasionally faltered in keeping its religious ideology from conflicting with its secular educational mission, and vice versa. Nonetheless, it generally managed to succeed in this high-wire act. Rustem Pasha, the Ottoman governor-general of the Province of Lebanon (and an Italian aristocrat by origin), told President Bliss, "I do not know how much mathematics, nor how much of history, philosophy or science you teach at your Syrian Protestant College, but I do know this: that you make *men*, and that is the important thing. I wish I had one of your graduates to put into every office in my province. I would then have a far better government than I now have."[3]

Over the past hundred and fifty years, that is what the American University of Beirut has done. The men and women it has produced have included prime ministers, cabinet ministers, ambassadors, and heads of business enterprises. Others have become doctors, nurses, engineers, lawyers, teachers, professors, writers, artists, and various other and sundry professionals and are serving their societies as educated parents and citizens.

Daniel Bliss spent the next four decades developing the Syrian Protestant College into the most highly esteemed American educational institution outside of the United States. When he finally retired in 1902, at the age of seventy-nine, he passed the leadership to his son. Howard Bliss possessed consummate diplomatic skills that proved crucial in sustaining the institution during the terrible years of World War I, when the region suffered widespread starvation and turmoil. The existence of the college became precarious, especially after the United States entered the war (but never actually declared war on Turkey, mostly at the urging of the trustees). Howard Bliss died in 1920, his health shattered by the exertions during the two decades of his presidency. By the end of his tenure, enrollment exceeded six hundred students. The name of the Syrian Protestant College was changed to the American University of Beirut in 1920 after Lebanon became part of the French Mandate. The institution was no longer Syrian, Protestant, or merely a college. As the American University of Beirut, it entered its next phase as an expanding institution of increasingly cosmopolitan character and international scope.

Daniel and Howard Bliss were exceptionally able, dedicated, and tenacious leaders. However, their efforts were unlikely to have succeeded had it not been for the unstinting support and generosity of the Dodge family. The Dodges were part of the generation that established the American tradition of private philanthropy in the nineteenth century, which was to be unmatched by any other country at any time. Their engagement with AUB began with William Earl Dodge's becoming treasurer of the first board of trustees of the Syrian Protestant College. He was instrumental in raising the first $100,000 (during the difficult years of the American Civil War) that established the college, and for the next twenty years he continued to provide generous support from his own funds as well as through the relentless process of fundraising from others.[4] His son, David Stuart Dodge, succeeded him on the board, serving as its president for fourteen years. His great-nephew Bayard Dodge became the third president in 1923. He married Mary Bliss, the daughter of Howard Bliss, thus blending the two families most closely associated with the history of AUB. When Bayard Dodge retired in 1948, he had served the university for thirty-five years in various academic and administrative capacities and made substantial financial contributions as well while drawing a presidential salary of $1 a year. I remember him from my boyhood in the 1940s as an imposing and kindly figure. He wore rimless glasses, which became associated in my mind with important Americans. His son, David Stuart Dodge, a member of the fourth generation (whom I came know on the AUB board), was acting president and then president in 1996 and served as a trustee for thirty-six years. The fifteenth and current president, Peter F. Dorman, an Egyptologist from the University of Chicago, is a great-great-grandson of Daniel Bliss. (His grandfather, Dr. John Dorman, was the obstetrician my parents consulted. And his father, Harry Dorman, taught me a course on St. Paul at the Near East School of Theology.)[5]

It is remarkable that AUB has not only survived but thrived under difficult conditions, including two world wars, several civil wars, Syrian and Israeli incursions, and endless meddling in the affairs of Lebanon by powers near and far. The ruling elements in Lebanon have treated AUB well. The Lebanese themselves came to view AUB not as a foreign entity stuck in their midst but as an institution that belonged to them and that they needed as much as it needed them. Consequently, except for a few short periods, the university never closed down and suffered relatively few serious setbacks: the kidnapping of Acting President David Dodge in 1982, the assassination of President Malcolm Kerr in 1984, the bombing of the administration building, College Hall, in 1991, and the assassination of the dean of engineering and the dean of students in 1976. At a time when some parts of the Muslim and Arab worlds view America with suspicion and ambivalence, it is important to realize that not all American institutions in the Middle East are seen as alien and suspect. When Americans wonder about how to win the hearts and minds of people, AUB serves as a good example.[6]

~

I had been on the AUB campus many times, but I hadn't fully appreciated how magnificent its physical setting was until I became a student. The campus was built on a bluff overlooking the Mediterranean with its shimmering waters stretching to the horizon. The original buildings were of hewn limestone with red tile roofs. Designed by the founding

Figure 4.1 College Hall of the American University of Beirut. (Office of Communications.)

faculty, their harmonious lines combined features of Levantine architecture with the more severe lines of the Victorian Gothic revival—a seamless blending of East and West. The graceful arches with their slender pillars complemented the massive limestone walls, conveying the historic heritage of the Old World with the forward-looking openness of the New World.

The campus was covered in a mantle of greenery, a veritable botanical garden representing the flora of both the region and faraway lands. The centerpiece was a massive banyan tree (*Ficus bengalensis*) whose vine-shaped aerial shoots reached down like outstretched arms. The Buddha found enlightenment under a banyan tree in India, and this banyan tree stood like a sentinel outside Assembly Hall with its discreet stone cross representing its Christian heritage. The gardens cascaded down terraces to the foot of the hill, leading to the green grass of the soccer field, beyond which lay the rocky shore of the blue Mediterranean.

The sea has always been an integral part of the campus. Beirut itself is hard to imagine without it. Daniel Bliss was a keen swimmer and he instituted swimming as a requirement for graduation. He wanted students to receive their diplomas by swimming to a raft on which he stood to hand them out. More pragmatic minds prevailed to drop the idea of the raft, but the requirement stood; it was still in force a hundred years later, when I was an undergraduate. This was no problem for Lebanese students, who learned to swim as children, but it was a problem for others who came from countries where they had never seen the sea. Enterprising students devised a way to get around the problem. On the appointed day, the candidate stood between two friends on the promontory at the test site. Two others took position on a rock close by. As a wave approached from the right direction, the candidate was heaved into it and fished out by those waiting on the outer rock. This took place under the watchful eyes of the formidable and always deeply tanned athletic director, Abdul Sattar Trabulsi, the pistol-shooting champion of Lebanon. Another exercise in the freshman year was having the class walk from the campus to the iconic Pigeon Rocks, with Mr. Trabulsi expounding philosophical musings along the way. To shorten the tedium of the walk and the talk, one could fall behind, jump on a tram, and then jump off as it approached the walking group. Getting on and off the moving tram was a skill one learned as a teenager to avoid paying the fare on short rides. But it was far from safe; one boy I knew lost a leg doing it.

~

The registration process was long and tedious. I stood in line clutching my tuition money for the first semester (the largest sum of money I had ever held in my hands). Finally, I received my student handbook certifying me as an AUB student and signed my name with a flourish befitting my new status. It was noon and I was hungry but I felt reluctant to go to the landmark El Faisal's restaurant on Rue Bliss opposite the Main Gate. I was intimidated by it, and it cost quite a bit. Instead I walked some distance down the tramline to a small outdoor café where I could have a plate of hummus with pickles and a loaf of pita bread. The other customers were mostly workingmen, and I felt as out of place as I would have at Faisal's.

In the afternoon, as I walked around campus getting adapted to my new surroundings, I noticed some boys lining up to put down their names for dinner at the home of Professor West, who taught chemistry. I thought it was nice of him to invite the new students, and I added my name to the list. At the dinner, I took my place in the living room and was struck by how foreign the other students looked. Professor West welcomed us and then asked us to tell where we were from. As the names of various countries began to roll out, it dawned on me that the dinner was for *foreign* students; I had no business being there. When it was my turn, I sheepishly explained that I was actually from Lebanon and that I had stood in the wrong line, but I hastened to add that since I was not *born* in Lebanon, I was *like* a foreign student. It was a lame excuse, but Professor West was most gracious and told us another standing-in-the-wrong-line story. During World War II, when so many things were rationed, a friend of his in London saw a line forming up and joined it in the hope of getting something useful, but it turned out to be a Chinese laundry. Everyone laughed, and I felt relieved.

I started college in fairly good shape. By the end of high school the worst of my psychological problems were under control. Through weight training, I had developed a strong body. I felt socially more at ease and self-confident. I was ready to throw myself into the academic,

Figure 4.2 Freshman lifeguard with no life to guard.

athletic, and social opportunities the university provided. I passed the lifeguard test, and although I was never to work as one, I was happy to wear the badge on my bathing suit. I continued with my violin lessons and joined the student orchestra. There were lingering doubts and anxieties on all fronts, but I forged ahead with optimism and trust in God.

It didn't take me long to become enmeshed in my classes. Three of these exemplified my freshman experience. At the bad end was the introductory math class. I had struggled with math throughout high school for reasons I couldn't understand. Was it because I had skipped second and seventh grades and been unable to catch up? Was it because my father had a natural gift for numbers and I didn't want to compete with him? Or, was the problem the lack of a special aptitude for math?

Whatever the reason, I belatedly came to recognize years later that there was an intellectual purity and elegance to mathematics which I failed to experience. The freshman math teacher didn't help. He had a thinly veiled disdain for those who lacked aptitude in math, which he equated with stupidity. When a student fumbled at the blackboard trying to solve an equation, the teacher's nasal voice would intone mockingly from the back of the room: "My friend, you are like a blind man, in a dark room, looking for a black cat that's not there." I realized what he was saying was pointless—if a man is blind, then the darkness of the room and the blackness of the cat would be irrelevant. That, however, didn't make his warped sense of humor any less humiliating.

The French class presented another sort of problem. French and Armenian were my choices for fulfilling the foreign language requirements on the matriculation exam, and I got honors in both. Therefore, the class was not hard, but it was mind-numbingly dull in the way it was taught. The teacher belabored one grammatical rule after another without conveying any of the sparkle of the French language. I had to take it because it was required (a legacy of the French Mandate, I suppose), and it was one of many other required courses that I had to suffer through as a premedical student. In the third and positive category were courses in psychology, which I enjoyed immensely; the fact that I got one A+ after another made them particularly interesting. The course in abnormal psychology was to prove specially significant in my choosing to become a psychiatrist.

Throughout high school and college I didn't have a single academic advisor. (Years later, when I would hear Stanford students complain about their advisors, I could only shake my head.) It never occurred to me to go to any of my teachers for advice, nor was I ever encouraged to do so; it did not seem to be part of their responsibilities. Neither of my parents had gone to college, so they hardly could advise me either. My father, however, did have definite ideas about what career I should pursue. These choices came down to three respectable (that is, lucrative) professions: business, engineering, and medicine. Had we not left Iskenderun, I would have been preordained to go into my father's business. However, since there was no family business in Beirut, my father was convinced that engineering would be the best choice. Its period of study didn't extend beyond college, and it would provide me with readily marketable technical skills with good economic prospects. The construction business was booming at the time, with the Saudis building entire cities in the desert. My father was already involved in construction, so if I became a civil engineer we could work together and rebuild the family fortunes. My cousin Jirair, who was an engineer and entrepreneur, had done quite well, and so could I. My father's reasoning was quite compelling. He was by now over seventy years old. He had so far managed to support his family fairly well, but for how much longer? I was his only child and only hope for securing their future. However, I had no interest in engineering, nor did I have any aptitude for it (the idea actually frightened me); I was even less interested in working for my father, which of course I couldn't tell him. Otherwise, it was a great idea.

Medicine seemed relatively more palatable. I could see how satisfying it would be to alleviate the suffering of the sick. However, it would take forever before I finished medical school, internship, and residency training and start earning a living. Moreover, I hated the prospect of taking the required premedical courses in biology and chemistry. They were hard, and I didn't have the slightest interest in them. Law school was not an option because Lebanese law was based on Arabic and French, in neither of which I was proficient. Had I been raised in the United States, I could well have become a lawyer and probably a good one, but it would have brought out the worst in me.

What I was interested in were the humanities, even though I hadn't even heard that term at the time. I was fascinated by philosophy and theology and somewhat less by history and literature (in later years, I would have picked art history as my first choice). I don't know where these intellectual interests came from. There were no role models in my family and precious few among the professors that I took courses from. Was my interest the outgrowth of the novels I read during my childhood illness? The residue of my contemplation of issues of life and death as a child? Or was it due to my wanting to be as different from my father as I could possibly could? I liked the idea of teaching school and I would have made a good teacher, but I had no interest in being poor. Becoming a university professor never crossed my mind. Joining the church was a possibility, but which church would I join? I couldn't see myself as an Armenian Apostolic priest, let alone a monk.

It was easier to consider becoming a Protestant minister. But I was still bound to the idea that the Khatchadourians belonged to the Apostolic Church, and it might have been embarrassing to my father's side of the family if I switched churches.

I finally reached a decision through fortuitous circumstances (like so much else in my life). In my sophomore year, our abnormal psychology class went on a visit to the mental hospital at Asfouriyeh. Before the days of psychotropic drugs, catatonic schizophrenics could be seen standing around in contorted postures. I became fascinated by them not so much out of a desire to help them but to understand their condition as a human experience. That gave me the idea of becoming a psychiatrist. It was the least medical of the various specialties and nebulous enough not to be unduly constraining. I could do any number of things and call it psychiatry (which is in fact what I ended up doing). Moreover, Dr. Antranig Manoukian, the director of the Asfouriyeh hospital, was a friend of my father, and that might make my choice more palatable to him. Unfortunately, it required going to medical school, which meant that my true intellectual interests were going to get trampled in the process of my pursuing a medical career. On the other hand, my decision to become a psychiatrist opened a career path that I could live with and that would more or less satisfy my father. Psychiatry would also be responsive to my curiosity about what made people tick. There may have been a deeper hope that psychiatry would help me understand and take care of the problems of my adolescent struggles with compulsivity whose residues were still there. For instance, I developed the habit of saying "Thank God" at every opportunity that required an expression of gratitude, as when I told someone that I had done well on an exam. One of my classmates commented on it approvingly (he must have been a Muslim, since gratitude is a cardinal virtue in Islam). If psychological problems had driven me into becoming a psychiatrist, I would not have been the only one. (Psychiatrists have been reported to have the highest rate of suicide among physicians, whose suicide rate, to begin with, is higher than that of the general population.)

At an opportune moment—if there ever was such a moment—I announced to my father my decision to go to medical school (leaving psychiatry out of it). As I feared, he was less than enthusiastic. He was glad there were doctors in the world just as he was glad there were electricians and plumbers. But they all carried the same liability in earning their living through their own labor, which put limits on how much money they could make. Even if I became the highest-paid surgeon, how many operations could I perform in a day? It made more sense to make money through the labor of others, which could then be expanded exponentially. In view of this, my father wanted to know why I would want to become a doctor. I gave him the conventional answer: I wanted to help people. My father agreed that was a good thing, but he said, "If you make a lot money you can hire ten doctors who will then go and help many more people." I couldn't rebut that argument, but I stood my ground. He finally acquiesced and was probably relieved that I didn't choose to do something even more harebrained than becoming a doctor.

~

In my sophomore year, I declared a major in biology with a minor in chemistry, as required for admission to the AUB medical school. Henceforth, most of my courses were prescribed to fulfill premedical requirements. There was room for only a few electives; I picked them playing safe by taking courses where I was sure to get an A—either because they were easy or I was good at the subject. I particularly hated chemistry. All I remember from the organic chemistry class was that the lab smelled bad. In another chemistry lab I was enchanted by the glorious red color of a cadmium compound that precipitated out of a solution. But since this wasn't an art class, the aesthetics were beside the point. Initially, I didn't even know how to study for the exams since so much of it was rote memorization rather than thinking through a problem. Before the first organic chemistry quiz, I was busy with choir practice and didn't memorize any of the structural formulas; I ended up getting a C. Insult was added to injury when the professor, who had been a classmate of my uncle Yervant, reminded me what a good student my uncle had been. (I didn't tell him that my uncle had said that he, the professor, hadn't been such a good student himself.) Biology was more interesting, but it could have been so much more so if it dealt with living and breathing creatures instead of theories of how sap rose in trees.

The saving grace of the sophomore year, if not of my entire college experience, was the year-long course in Western civilization; a course developed at Columbia University and adopted by other American colleges. We used the same two massive textbooks of readings developed at Columbia. Our course also included a substantial section on Islamic civilization, which itself was fascinating. This was to be the most intellectually exciting, challenging, and rewarding class I was ever to take. It opened new intellectual vistas and provided me with a historical and conceptual framework that I could use for the rest of my life. For a change, I actually *enjoyed* studying for it. Nonetheless, had it not been required, I am sure I wouldn't have taken the course since it entailed a huge amount of reading, and getting an A was far from a sure thing. The fact that I did very well at it added greatly to its appeal. The midterm exam was so hard and the class did so poorly that each student was given 15 additional points. That boosted my score of 92 points (the highest in the class) to 107, but I was only given the maximum of 100 points. I complained to Professor William Miller, the director of the course, that I had been short-changed. Since I didn't know how well I would do on the final exam, I wanted the extra 7 points set aside to boost my final score. Professor Miller smiled at my Levantine reasoning. He was my favorite teacher. Although dean of the Faculty of Arts and Sciences, he maintained none of the professorial distance from us that most other faculty members did. As my section leader, he taught us how to think and how to engage in intellectual discourse. He even engaged us in a debate on whether or not *ayran* (the local diluted yogurt drink) was, as we claimed, superior to American buttermilk (which we had never tasted).

All told, my college years were far more intellectually exciting and rewarding than high school. But they were also turbulent in their own way. They began with the problems of adjustment in my freshman year, followed by the pressure of fulfilling premedical science requirements. I dutifully plowed through these courses and did quite well, but it was not easy. What made the process even more stressful was that I was never sure if I was doing well enough to get into medical school, even if I clearly was. Much of this was due to my own anxieties, although the rest of the class seemed just as stressed as I was. Moreover, this bootcamp mentality was not peculiar to AUB, and I have heard similar complaints from premedical students elsewhere.

These academic problems were compounded by lingering psychological problems. I obsessively fretted over every question in every exam—before I took it, while taking it, and after taking it. These problems spilled over into my emotional life and my budding attachments to women. These conflicts figure prominently in the diaries I kept between my freshman and junior years and into which I poured out my heart. The first diary was written in a tiny calendar booklet and was grandly titled "Thoughts." It opens with a solemn request (stated in Armenian and repeated in English) for others not to read it. Yet, on the next page it adds, "It's hard to explain it, but I think this copybook will be published one day. Maybe after I die."

My freshman diary is full of admonishments to myself ("Don't lose your temper"; "Don't talk too much"; "Help those in need"), as well as reflections on life ("Life is strange, mysterious, hard, bloody, full of uncertainties"). Only rarely do I burst into exuberant expressions of joy: "I love men, women, children. I love the walls, the carpets, the sun, and music. I love books, I love games, I love fun. I love to sing. I love the stones, I love the waves, I love the dust. I love flowers. I love everything." The diaries of my sophomore and junior years were far more elaborate, laden with their own anxieties that occasionally verged on pathos ("I do not know why, but I spend many an hour feeling sad"). Many of the comments are highly self-critical ("I feel exasperated with myself for the way I talk, act, fret, and feel guilty for failing to be sufficiently grateful to God and to my parents for the many blessings I receive undeservedly"). And occasionally there is a self-congratulatory note ("You know, sometimes I think you are a nice guy").

As my language gradually shifted from Armenian to English, its tone became more reflective and analytical. The shift was not due to a newly gained mastery of English. Rather, it reflected a shift in my intellectual attitudes in becoming more "Western" in my way of thinking. Sometimes, my reflections took the form of a dialogue between two inner voices: one spoke for my "Person," the other for my "Self," which like a Greek chorus commented on the former. It was all very self-referential. And I took the trouble to point out (under the heading "Very Important") that what I write should not be taken

Figure 4.3 Graduation from college. I stand on the right, and next to me are Ali Fakhru, Michel Kahil, Ara Chalvardjian, and Ibrahim Yacoub.[7]

as a full account or representative of the important events in my life; it consisted of my random thoughts, which should not be taken that seriously. Despite this dismissive note, these diaries clearly served an important cathartic role in unburdening myself, trying to gain insight into controlling my actions and fulfilling my aspirations and fantasies.

My years in college finally came to a happy end. My performance in the freshman year had been uneven, but my record improved steadily and I received my BA "With Distinction" (AUB's version of *cum laude*). In congratulating me at commencement, Dean Miller said, "You deserve the distinction." It meant a lot to me. I had gained much in self-confidence even though I was never free of lingering doubts and anxieties about the future. I was surprised that I was ranked 23rd in our graduating class of 800 students. Had it not been for my freshman grades, I may well have been in the top ten. That would have truly amazed me.

~

I didn't have much of a social life at AUB as an undergraduate. There were no fraternities, sororities, dorm parties, or dating. Students from the Lebanese Christian elite may have had a more active social life than the rest of us from more conservative backgrounds. Moreover, unlike in high school, there were now far fewer women than men at the university. (We called them "girls" and we were "boys.") One could socialize with them in public, but more intimate relationships had to be clandestine. The university felt responsible for guarding the virtue of its young women. A stout guard with a shotgun patrolled the campus after dark, and when he came across a couple sitting on one of the outdoor benches, he took their handbooks and turned them in to the dean of students. The next day the students had to explain to the dean what they were up to. (These practices were eliminated in later years by the time I was on the faculty.)

There were a number of student organizations on campus, but I had no interest in them. I walked into one meeting of the elected Student Council out of curiosity. There was a lot of shouting and hand-waving, and I remember John Racy, the secretary of the council (who would become a lifelong friend), pounding the gavel to maintain order. Arab nationalism was in its heyday, when Gamal Abdel Nasser captured the loyalty of Arab youth. Even though Arab nationalism and political activism played no significant role in my own life as a student, they were a central part of the undergraduate experience at AUB, particularly as student groups became more radicalized in the 1970s and their relationship with the administration became more contentious. However, since these developments took place after I had left AUB, they will not be part of my account here.[8]

I wasn't that much more interested in Armenian nationalism than I was in Arab nationalism. In my senior year in high school I had joined the youth organization of the Tashnag Party referred to earlier. It was exciting to see legendary figures like General Tro, who occasionally came to our gatherings. Now in his seventies, he had been defense minister of the short-lived Democratic Republic of Armenia at the end of World War II. I was told that when joining the Tashnag Party, one took an oath swearing on the Bible and a dagger. That appealed to my heroic fantasies, but when one of the leaders took me aside and pressured me to formally join the youth organization that would lead to party membership, I quit.

My one significant social engagement in college was playing in the student orchestra as head of the second violins. It took a lot of my time, but the music director, Salvatore Arnita, arranged to have it count as student employment, for which I got paid—a small but welcome addition to my meager allowance. At the instigation of my violin teacher, I enrolled at the Académie des Beaux Arts and played in its orchestra as well. For one of our concerts a woman in her thirties showed up as the soloist for one of the Paganini violin concertos. She had maddeningly long legs and a large nose that somehow further enhanced her sex appeal. Several of us in the orchestra could have played better than she did, but I figured she must be a friend, if not the mistress, of the conductor. I had no basis for such a supposition, but this was, after all, Lebanon, and I was under the sway of the cynicism that pervaded the air I breathed.

My first serious violin teacher had been a Rumanian Armenian. He was an ailing old man who railed bitterly against the burdens of his advancing years. He liked me, and had highly unrealistic expectations of my becoming a professional violinist, when I had neither the talent nor the perseverance called for. Nonetheless, playing the violin opened up social opportunities, as I noted earlier.

My last teacher was Mr. Schwarz, a Jewish refugee from Vienna, who was an accomplished violinist and excellent teacher. He was a wonderful man and treated me with great courtesy. His wife and two young daughters (whose photographs adorned the piano) were in the United States, where his older daughter was being treated for the aftereffects of polio. He was lonely and missed his family: "Mr. Khatchadourian," he would tell me, "life is not so nice." I made the most progress with him and took part in a national competition, for which I was ill prepared. I did miserably and was embarrassed. The winner, who won a scholarship to study for a year in Paris, was an apprentice barber who was quite gifted and deserved to win. I stopped taking lessons when I started medical school, but continued playing the violin on and off, until we moved to Stanford and I switched to the viola.

I had a fuller social life outside of AUB. I was deeply engaged in the Christian Endeavor Society and became the leader of its youth group (*Badaniats*). On Sunday afternoons, I would meet with two dozen young teenage boys for a program that I mostly improvised. Its centerpiece was telling them a story in installments based on a novel like Jules Verne's *Around the World in Eighty Days*. If I had not had time to read enough of the book, I would embroider the story to make it last. It wasn't, however, all entertainment. Even though I didn't preach to these boys, I would make morally uplifting comments or expound on principles of how to live one's life. I must have become a role model, and some of these boys (now in their sixties) still tell me how important I had been for them at the time.

One afternoon when walking home, I saw one of the boys, the shy little Vicken, dragging a piece of chalk on the walls of buildings as he walked along. I stopped and told him he should not be doing that. And to teach him self-discipline, I asked that he carry the piece of chalk in his pocket and bring it to me at our next meeting. I had forgotten all about it when the following Sunday he came up and handed me the piece of chalk that had been rubbed smooth in his pocket. I commended him warmly, and he walked away smiling shyly and looking pleased with himself. Some thirty years later, I saw him again at Stanford and recognized him right away and he knew who I was. He was now a graduate student in earth sciences, with a wife and two daughters. We didn't talk about the piece of chalk, but I'm sure he remembered it. These experiences taught me how to be a mentor to young people, which became a deeply satisfying part of my career as a teacher.

The highlight of the summer was the Christian Endeavor Society conference held at the campgrounds in the mountains. We attended Bible study groups during the day and revival meetings in the evenings. My first exposure to this Evangelical tradition was unsettling. I realized that a non-Christian would have to be converted, but I couldn't understand why a lifelong Christian like me with a deep personal bond with God had to publicly declare his allegiance to Christ by being "born again." It was like having to acknowledge my own father publicly. One evening, when I sat on my bed pondering this matter, one of the counselors walked by. He was a younger man probably still in college who must have sensed that I needed to talk to someone. With some difficulty, I explained to him my dilemma. He understood readily how I felt and was warmly sympathetic. He said that being who I was, I did not need to go through a conversion experience. It was one of the most comforting encounters of my life. However, the following year, I decided it would be simpler to fall in line and go through the ritual so as not to give the impression that I was not a true believer.

We also had time for leisure activities and games such as treasure hunts and athletic events, which I was in charge of organizing. I frequently held center stage in these events, but that didn't seem to be resented by my peers. It didn't occur to me at the time that my model for taking charge was my mother. When something needed to be done, she would step up and do it better than others would have. The mantle of leadership then fell naturally on her shoulders without her having to ask or fight for it.

The conferences for the younger groups where I was a camp counselor were more rowdy. One of the favorite games of the older boys was to get up in the middle of the night and put the thumb of one of the younger sleeping boys into a cup of water with the expectation that it would make the boy wet his bed. To guard against such pranks, I had the more vulnerable boys sleep on cots closer to my bed. One particular bully was a source of endless trouble. Finally, he crossed the line when in plain view he hit one of the younger boys and took off. Another camp counselor and I ran after him in hot pursuit. As we were about to catch up with him, he went down on his knees and began to recite the Lord's Prayer. We waited for him to finish, and then the other counselor slapped him hard. However, I was also good to him. When he cut his hand while playing with a knife, I was the one who took him to the hospital in Beirut on the back of a motorcycle in the middle of the night.

One summer we had two missionaries affiliated with Bob Jones University attend the conference. One of them preached and the other played the trombone during hymns. As with all things American, they held a special fascination for us. The preacher was a captivating speaker but a dour and forbidding character. I much preferred his colleague, who was an unusually kind and joyous person. When practicing the trombone, he would spontaneously lapse into a Dixieland tune but quickly recovered himself and stopped, much to my disappointment.

The Bob Jones missionaries insisted that even born-again Christians should be "reborn" to their specifications to ensure their salvation. I was singled out by them as a particularly promising recruit. I was annoyed when my picture appeared in their newsletter as a future medical missionary and they felt bad that they had used my photograph without my prior consent. There was even some talk of my getting a scholarship to Bob Jones University. I had no idea of where and what it was about but the prospect of studying in the United States was exciting. However, for some reason it made me uneasy, and nothing came of it. (I wonder what my life would have been like had I gone there.) The Bob Jones missionaries ended up alienating the Armenian ministers by casting them as being insufficiently Evangelical, which I found hard to believe. Finally, they wore out their welcome and left the country. Two Mormon missionaries who were not associated with the Evangelical Church had even shorter tenures. One of them was a star basketball player who popularized the game in Lebanon. However, when they claimed that God had intended for the Jews to lay claim to Palestine, they were out of the country within a few days.

There were a couple of other American missionaries, but I didn't know who sponsored them. Mark was of Lebanese origin. He was particularly keen on memorizing biblical verses on little cards and pressed everyone to do likewise. I tried it for a while, without much success. It seemed like a ritual with little substance. However, Mark was quite dogmatic and used these verses to drive his message home. If you asked a question, you got a verse; if you raised an objection, you got another verse, now with a forgiving smile. The other missionary, Peter, was an unassuming, genuinely kind, warm, and wonderful person. He got ill and didn't last long but I learned much from him, including some of the baffling aspects of life in America. He said he had hoped that the Lord would lead him one day to an important public position, such as being "Secretary of State." I had never heard the term, but knowing the lowly status of secretaries, I couldn't understand why such an office should be so important to him. However, I said nothing so as not to betray my ignorance or hurt his feelings.

The most rewarding of these relationships was my friendship with John Reynolds. He was the pastor to the Protestant Student House, sponsored by the Presbyterian Church. Only a few years older than me, he became a true friend and mentor. We had wonderful talks ranging over theological, intellectual, and mundane matters. John didn't wear his beliefs on his sleeve and never preached to me, yet in his thoughtful and gentle way helped me grow up spiritually and in many other ways. He was a source of steady support and thoughtful guidance during a turbulent romance that I suffered through in my junior year. John also had a wry sense of humor. He told me the story of a taciturn American student called Brad who was at AUB on a scholarship from his church. One day, Brad bought some peanuts sold in paper cones by the Sudanese vendor outside of the Main Gate. To his astonishment, as he worked his way down the cone his own picture began to emerge on the inside of the paper. John solved the mystery. Brad had given him a magazine sent by the sponsoring church with his photograph in it. John's housekeeper had been taking and selling these magazines to the peanut vendor, who used them for his cones. Against all odds, the page with Brad's photograph had ended up as the peanut cone in his own hand.

Jim was another American friend who was a student at AUB. His parents lived in Beirut, where his father worked for an American company. We became good friends and I learned a good deal from him about the American way of life. His interest in clothes sparked my own interest in dressing well. I admired his tartan wool ties, so he ordered one for me through the mail. That was a novelty, and it made me wonder how one would bargain through the mail. When he returned from a trip to Italy he brought me a small leather box. I shook it trying to see what was inside, but it turned out the box itself was the present, which I found rather strange. (I still have the box, where I keep my collar stays.)

~

What was sorely missing in my life was female companionship and sex. Although there was no dating, let alone sleeping with girlfriends, sex for sale was readily available in Beirut. The first time I set foot in the red-light district was when I was in high school. However, it was for a wholly innocent purpose. Our household supplier of olive oil, cooking fat, bulgur, rice, and other foodstuffs was located along with other shops in the red-light district. I am surprised that my parents would send me there on a free afternoon to accompany the porter to carry the provisions home. While waiting for the goods to be loaded up, I surreptitiously looked at the balconies where scantily clad prostitutes were lounging about. They were a sorry bunch of bedraggled and careworn women, some of whom waved at me teasingly, but I pretended not to notice them.

In our senior year in high school, I began to hear accounts of visits to brothels by some of my friends. They were quite casual about it, one of them complaining about the cost (four Lebanese pounds, a little over a dollar at the time). I was amazed that such a thing would cost so little, but I couldn't imagine my taking advantage of it. I was told that one of the younger boys wanted to visit the brothel but was afraid to be seen. He asked his friends to physically carry him in as he "struggled" to get free. He could then claim that he was forced to go in against his will. The commotion they caused ended up attracting far more attention than if he had quietly slipped in.

In medical school, we were taken to the red-light district with our public health class. Some students skipped the visit, and we suspected it was because as regular customers they would have been recognized and welcomed. What struck me most were Muslim women who exposed themselves for the vaginal exam while keeping their veils on. Other women looked acutely embarrassed, blushing intensely as the government physician conducted the exams in the manner of checking goods in a warehouse. Engagements with prostitutes became more refined in college. There was a pimp who brought a young woman to a room near the campus for servicing students. Some of my friends visited the woman periodically and paid five Lebanese pounds each for having sex with her twice. Since I seemed so prudish, I was never included in these forays, nor would I have agreed to take part in them if asked, even if I might have wanted to. My sexual life was confined mainly to my fantasies. On March 1, 1951, I declared in my diary: "I think you are now mature enough to get married." This was a bold claim for a nineteen-year-old who was reticent even to talk to girls. I was invoking marriage because I could think of love and sex as being permissible only in that context. Meanwhile, I chafed under the restrictive college culture in which I was condemned to live. One diary entry reads: "I saw a movie on American college life. Boy, we are living like sheep. Even if you take out having all that freedom with girls, there are so many other nice things which we don't have. We have the work but we lack the sweetness of college life."

Most of us had very little to do with the women students outside of class. Many of them came from conservative Lebanese families. You could court them as potential marital prospects, but not otherwise. Besides, lovely as many of them were, they looked like our sisters and cousins, which made them less exciting. More exciting were the American girls. Their families typically lived in the region where their fathers worked for American firms. These girls looked different and acted much more freely. Their being under less parental supervision made us hope that they would be more readily accessible. We saw them as Hollywood mini-stars. Even when they were plain-looking, being blond made up for many deficiencies and turned them into a poor man's Marilyn Monroe. A few of the more enterprising boys managed to go out with them, but probably all they did was show them around the city no matter what else they bragged about. When I was in high school, girls from the American Community School played basketball against our school girls. The contrast was striking. The American teenagers wore delightfully skimpy shorts, whereas our girls were in miserable knee-length uniforms. During one match, when one of the American girls slipped and fell, one of their boys picked her up and carried her off the court. The sight of his *hand* on her bare *thigh* was too much to bear.

In my sophomore year, I played in the student orchestra for an operetta. Most of the girls on stage were Americans. I sullenly watched my friend Fuad waltzing with one of them, albeit at a discrete distance, while I sat there playing my stupid violin. ("There you sat in the orchestra, wishing, wishing…" laments my diary.) Finally, I had a chance to consort with these near-mythical creatures when the cast and the orchestra were taken on a picnic. On the bus, a cute American girl called Judy, who was sitting up in the front, kept turning back and smiling at me and I more or less smiled back. When we got to our destination, I sat on the grass not knowing how to approach her. I was incapable of making small talk (and still am). That left a chasm between silence and having a serious conversation. A gramophone was brought out and couples got up to dance, and some of them wandered off into the nearby woods. I waited for the initial commotion to die down and finally got up enough courage to ask Judy to dance. She jumped up eagerly. We danced in silence as I

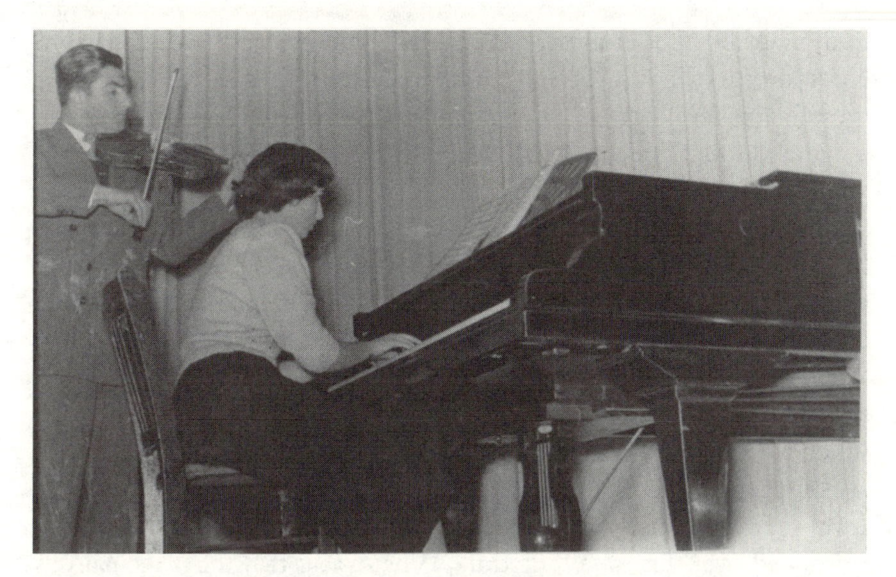

Figure 4.4 The violin as a path to the female heart.

inhaled her aroma. When I pulled her slightly closer during a turn she responded by almost hugging me. As I agonized over my next move, the music ended. I thanked her awkwardly and Judy moved away reluctantly to dance with another boy who had been lurking around. They danced and then they too walked away as I watched them go with a knot in my stomach.

I had a few ephemeral crushes in high school, but only in one case did I come to within shouting distance of having a girlfriend. Carla was the daughter of family friends. I came to know her one summer at the resort of Mrouj when she and her older sister came to stay with their cousins. I was one of two boys in our small group, and we spent much of our time sitting around and talking. One afternoon we went to a movie. Carla was sitting behind me. I surprised myself by reaching back and touching her shoe with some trepidation. She didn't pull it away so I moved my hand up and more or less (mostly less) caressed her leg. That furtive act established a secret bond between us, and I decided we were in love. When we returned to Beirut in the fall, I took my heart in my teeth and went to Carla's house. Her grandmother opened the door. Carla was not home, but the old lady very kindly invited me in. As Carla's younger sister snickered (I could have strangled her), the grandmother offered me a glass of lemonade and played a record to entertain me. That saved me from having to talk, and I felt eternally grateful for her kindness. I left as soon as the record ended, feeling like a damned fool. The visit took the wind out of my sails, and I was not about to go back to their house.

Several months later, I was astonished to find on my mother's bedroom desk an open envelope addressed to me. It was a note from Carla. She was sorry she had missed my visit and hoped to see me some other time. My mother walked in and saw me holding the envelope. She was embarrassed and apologetic. The note had been sent to my school, where the principal had intercepted it. Finding the note quite innocent, he had brought it to my mother. I fumed in silence, but mostly at the principal—it was easier than being upset with my mother. I was sad at the missed opportunity and I never contacted Carla again. I brooded over this for a long time. What was it that the principal was concerned about, and what entitled him to open a letter addressed to me? And why did my mother then sit on it instead of giving it to me? My resentment was directed not only at the individuals involved but also at the rotten society we lived in. How did I know that it could be better elsewhere? I didn't, but somehow felt that there was another world in the West where the rights of people, even those of mere boys, were respected.

My first approximation of a girlfriend was Juliette.[9] She was a few years younger than I. We attended different high schools but were both members of the Christian Endeavor Society, where we met. I don't recall when it was that I fell in love with her, but for a long time it involved mostly pining for her from a distance. I had no idea how she felt about me. We never had an opportunity to meet alone. I would get occasional glimpses of her in their family car when her father drove her with her sisters home from school. When I saw the car in the distance, I tried to be as unobtrusive as I could because at one time I thought her father saw me and scowled (which was absurd). One afternoon Juliette came to Bible study directly from the hairdresser. She looked incredibly pretty, with little curls falling over her forehead above her twinkling eyes. I could hardly take my eyes off her as she looked at me furtively; I don't think either of us heard a word of what St. Paul's letter had to say in whichever of his letters was being discussed.

We developed a closer relationship after I went to AUB and she continued to attend her private girls' school. The mutual attraction between us verged on intimacy, but our relationship ended within a year; I have no idea why. There is a cryptic reference in my diary to "the bitter disillusionment and frustration of the last affair." It had to refer to my relationship with Juliette. However, it was hardly an affair, and Juliette was such a sweet, gentle, and guileless soul that I can't imagine why I would characterize the end of my relationship with her in such harsh terms. This was the first installment of my relationship with Juliette. A second and more serious involvement would come a year or so later.

I met Talin the year after my "breakup" with Juliette. I was now twenty years old and a junior at AUB. She was an eighteen-year-old freshman at a women's college. Her family lived in another town, and our parents were old friends. I think I first saw Talin's photograph taken with one of my cousins. Then I had a glimpse of her at a church service, and we exchanged a few words in the courtyard. Sometime later, she heard me give a talk at a Christian Endeavor meeting and came up to congratulate me. That established enough of a connection for me to make further attempts to see her.

I was enthralled by Talin's deep-set brown eyes, which dominated her oval face, lit up by her whimsically crooked smile. I fell in love with her heedlessly, mindlessly, and hopelessly. She was socially more mature than I was and was being courted by older and well-established men who had cars, whereas I didn't even own a bicycle. Like Pygmalion, I created her in my imagination and then fell in love with my own creation. I fashioned her out of whole cloth to fill the aching void in my emotional life and placed her in a fantastic metaphysical web woven out of religious beliefs of my own making. In reality, she was a nice, attractive, and conventional young woman who would make a wonderful wife for a handsome, intelligent, and successful man (which is what she eventually did). The leading role in my divine comedy that I cast her in baffled her as it would have any other sane woman.

My feelings, thoughts, dreams, fantasies, frustrations, longings, anger, anguish, and self-recriminations crowded the diary of my junior year. Over the summer of 1953, I wrote a separate retrospective account of my one-sided relationship with Talin. It is a sad tale of my frustrating attempts to get close to her and the disintegration of a deeply felt yet doomed attachment. I visited her at her dorm, but we never went out together. I never touched her. That would have allowed sexual desire to sully the purity of my love for her.

The dialogue between my Person and my Self reached peak intensity in the turmoil over my relationship with Talin. My rational Self appealed to my Person: "Look," it said, "the whole thing is unreal. Maybe she likes you and wants to be your friend as she is with so many others. But that's all. The rest is your own doing. And for God's sake leave God out of it." To this, my Person had one answer: "But I love her." The central concern that I struggled with was whether Talin was enough of a Christian to dedicate her life to God totally and unequivocally and to follow me to the ends of the earth if, for instance, if I became a medical missionary. This was a purely hypothetical proposition. The idea of becoming a medical missionary had crossed my mind, but I was now on my way to become a psychiatrist, which was the last thing that the suffering masses in Africa needed. I was virtually certain that Talin would have no interest in such a prospect, but I still felt compelled to lay it out in front of her. It was a litmus test that she had to pass to qualify for spending the rest of her life with me, wherever I decided to go and whatever I decided to do.

Late in the spring of 1953, I saw Talin at the wedding of one of my cousins. Our conversation was so encouraging that I decided to make a bold attempt to see her after she returned home for the summer. My uncle lived in the same town and was a close friend of her father. It was easy enough for me to get myself invited to stay with my uncle's family. To my astonishment, a few days after I got there the two families were to go on a picnic. It looked to me to be providentially ordained. On the day of the picnic two cars came to pick us up, one driven by Talin and the other by her father. We met them on the sidewalk. Her father hadn't seen me since I was a child and asked my uncle discreetly who I was. "Efronia's son," my uncle told him, and he seemed pleased.

The picnic was deadly dull. At one point, Talin and I moved away. When walking over a fallen log across a stream, she almost stumbled. I held her by the hand and told her we could have fallen into the water. She said that would have made things more interesting. She was clearly bored, but I basked in her presence and nothing else mattered.

Thinking this would go on forever, I made no attempt to arrange a subsequent meeting with Talin. After a few days I realized my mistake, but I didn't know what to do. Simply going over to her house would have been easy enough, but it didn't even cross my mind. I became desperate. ("Not a glimpse of her. Not a ray of light. Feeling miserable. Mad and bitter at myself," I wrote in my diary.) In the middle of the week, my aunt casually mentioned that she would be attending a lady's church meeting on Saturday. I suddenly saw my chance. Through a complicated scheme, I managed to get myself invited to play the violin at the meeting. I hoped that Talin's mother would be there and she with her. That's exactly what happened.

After the meeting, Talin and I followed my aunt and her mother at a discreet distance. This was my chance. I told Talin that I had come to see her and wanted to talk to her. She was pleased and suggested that the following day we meet "accidentally" while walking down a street downtown. I was taken by surprise. I was dying to see her alone, but to my astonishment, I declined her offer: I didn't want any deception to stain our relationship. She looked baffled, then turned red for having suggested the subterfuge. I had no alternative to suggest, and that ended our conversation. I returned home to Beirut without seeing her again. To this day, I can't decide whether what I did was the pinnacle of moral integrity or the most cowardly and stupid thing I could have done.

After Talin returned to college in the fall, I tried to meet with her but she avoided me. "I have a full program," she would say, something she had not done before. I suspected that my bungled encounter with her had alienated and scared her off. However, I was determined to see her and explain myself even if it would be like jumping off a cliff. When we finally met, I laid out for her the grand vision for our life together. My convoluted account couldn't have made sense to her, but before I finished I knew I had lost her. Yet, instead of being devastated, I was curiously relieved. I felt emotionally drained and cleansed by the ordeal. I had done what I had to do, and it was over. My diary summed it up: "All the time I spent daydreaming about this was useless. All the time I spent trying to figure this out, was useless." The concluding comment was, "I am sorry I cannot write anything more. It is beyond this book, beyond everything." That ended the diary. I realized that I could no longer afford the luxury of wallowing in these fantasies. I had just started medical school and I needed every minute and every ounce of energy I could muster to see me through the year.

I have tried to understand why I put myself through this bittersweet nightmare. The fact that I fell in love with a lovely young woman requires no explanation, but the way I went about it certainly does. It may seem transparent that I was creating an idealized love that I desperately wanted but could never have. The fretting and fussing was aided and abetted by my compulsivity. The religious element is more complex. Perhaps shifting the burden to Talin for the critical decision of dedicating my life to the service of the Lord allowed me not to take that step myself. Religious beliefs, of course, are easily contaminated by psychological conflicts, and I had plenty of those.

I did not see Talin for many years. I heard that she had gotten married, but lost track of her when I left for the United States. When I returned to Beirut a decade later to teach at the medical school, I was invited to dinner by my cousin, who remembered that Talin and I knew each other. I walked into her living room, and there was Talin, looking as beautiful as ever. We spent a pleasant evening conversing like old friends. At the end of the dinner, I escorted her to the nearby apartment where she lived with her husband and their children. Talin asked me to come up for a drink. Her husband was away on a trip, and the children had been put to bed by their nanny.

We talked long into the night, telling each other how our lives had unfolded since we last saw each other. During a pregnant pause, I asked Talin if she remembered my visit to her one summer when we were in college. She said she did. This time in plain and simple words, I explained the reasons for which I had come to see her, why I had declined her offer to meet with me, and the forlorn love that I had had for her. She was silent for a while. "My father told me at the time," she said finally, "that there were only three men he could imagine as my husband. You were one of them."

5

...And Have It More Abundantly

THE SECOND HALF of the American University of Beirut's motto, which is the title of this chapter, will cover my medical school years and my first marriage. My life during these years was abundant, as well as abundantly turbulent. While I need not have worried about getting into medical school, I worried nonetheless, just as I had been when applying to college. The medical school admissions committee that interviewed me was chaired by Dr. Hrant Chaglassian, the professor of dermatology. He was Armenian and knew my father well. When he was a boy, his uncle and my father would have him carry bribes to the Ottoman officials (a story he would tell and retell me with great relish). When I entered the interview room, Dr. Chaglassian greeted me by my first name and his warm welcome set the tone for the interview. Dr. Musa Ghantus, the professor of anatomy, cheerfully asked me about my interest in medicine. I said I wanted to become a psychiatrist. What did I know about Freud? Well, I knew a lot about Freud (at least compared with him), and I launched into a spirited exposition of Freudian theory. He was duly impressed. After a few perfunctory questions by other committee members, I left the room, feeling certain that I'd gotten in. It was as easy as that.

Ten years later, when I joined the medical school faculty, I became a member of the same admissions committee with some of the very same faculty members who had interviewed me. Remembering how reassuring Dr. Chaglassian had been to me, I wanted to do the same for the candidates we were to interview. So when a young woman came in for her interview I asked her what she enjoyed doing in her leisure time. She was surprised by my question and said she liked hiking in the hills. What were some of the health concerns she would encounter climbing mountains? She said she would get tired. After she had rested, what else? She would get thirsty. She would feel lonely. These inane questions and answers went on as I tried to steer her to say that there would be lack of oxygen at higher altitudes. She didn't get it (I admit, it was far-fetched) until I asked her point blank, "What about the lack of oxygen?" Oh, yes, lack of oxygen. She agreed that that was important, and I changed the subject.

The next candidate was a Palestinian young man. Had he been to a refugee camp? Yes, he had. What were some of the health concerns he would encounter in that environment? After some false starts, he figured out that people living under crowded conditions would face health problems. Such as? He said, "Lack of oxygen." *Lack of oxygen?* He realized that that made no sense and added apologetically, "I'm sorry, sir. I said that because the girl who was here before me came out and said, 'Khatchadourian is pushing oxygen.'" Like someone under torture, he was ready to say whatever he thought I wanted to hear.

~

The first year of medical school was a living hell with anatomy at its evil heart. It was the hardest course I was ever to take. Our textbook was the famous *Grey's Anatomy*, which we had to more or less memorize. Dr. Ghantus, for whom anatomy was the center of the universe, knew and expected us to know the name of every blood vessel, nerve, and strand of tissue there was in the human body. Dr. Afif Mufarrej, who assisted Dr. Ghantus, was a practicing internist who taught anatomy part time and was more sensible. I liked him a lot, and he seemed to like me too. (He called me "*Khasha*-dourian" with great enthusiasm.) He also had a whimsical sense of humor. When a student during recitation touched the skeleton suspended from the ceiling, it would rattle. To prevent that, one had to hold it steady with one hand and touch it with the other. One day, when a student stood with one hand in his pocket, touching the skeleton with the other, Dr. Mufarrej turned to another student and said, "You! Go put your hand in his pocket, so he can use both hands." The student got so flustered having his friend's hand in his pocket that he made the skeleton rattle even more.

Ironically, the luckiest break in my first year occurred during anatomy recitation—the class I dreaded most. Early in the course, I wanted to know how long it would take to study a given section so I could recite it verbatim. I got up at three o'clock one morning and memorized the entire section on the brachial plexus—a complex tangle of nerves in the armpit. I went to class hoping to be called on to recite. To my horror, the brachial plexus was *not* the topic of recitation that morning, since we had dissected it only the day before. It was something else I had not studied for the day. I made myself as inconspicuous as possible until the danger passed as the recitations ended early. To use up the remaining time, Dr. Ghantus wanted to find out how much we remembered from the dissection of the brachial plexus before we had studied it for the class. He looked over our heads and said, "Khatchadourian." I couldn't believe it. I recovered my wits and walked slowly to the front of the class trying to look dismayed. Dr. Ghantus said reassuringly, "Don't worry. I know you haven't studied it. I just want to see how much you remember from the dissection."

How much did I remember? I remembered *everything*. The name of every nerve was at the tip of my tongue straining to burst forth. But I wasn't going to give it away so easily. I started with great effort, with knotted brows, reaching into the recesses of my memory and slowly pulled out the names of the larger branches: five *roots* merging to form three *trunks—superior, middle, and inferior*; each trunk splitting in two to form six *divisions—anterior divisions of the upper, middle, and lower trunks, and the posterior divisions of the upper, middle, and lower trunks*; the six divisions then regrouping to become three *cords*—the *posterior, lateral, and medial.* Then picking up momentum, I raised the recitation into a crescendo, unleashing the full force of the brachial plexus, like Bedrich Smetana's *Moldau*, where the river after meandering placidly through meadows comes roaring into Prague. Dr. Ghantus spurred me on with "Very good. Very good." When it was over, I stood exhausted, meekly looking at the floor as my classmates stared in disbelief, cursing me under their breath. Dr. Ghantus was delighted. "Do you see how much you can remember," he declared to the class, "if you do the dissection diligently?" I nodded respectfully, interjecting softly, "Yes sir, thank you sir."

~

The anatomy lab followed the recitation and lasted until noon. We were five students assigned to a dissecting table. My first look at a dead body was disconcerting. The cadaver belonged to a woman who appeared to be in her thirties, although it was hard to tell. She had light-colored hair. Was she a blonde, or was her hair bleached by formaldehyde? There were many Polish refugees in Lebanon at the time, and I wondered if she was Polish. Actually I didn't want to think of her as Polish or anything else other than a *body*. Could there be a person without a body or a body without a person? If I had her body, would I be her? However, this was not the time for such philosophical quandaries. Right now, I had to dehumanize her into an inert physical object, a collection of body parts to be studied. As a budding physician I had to show that I had the mettle to deal with dead bodies, blood, and gore, without losing my composure. I suppressed my revulsion and squeamishness at the sight, smell, and touch of the cadaver and went to work.

Our first task was to dissect away the skin of the back, which turned out to be quite hard. The skin was thick and leathery and the scalpels dull. By the time we were done, it was noon. I was exhausted and went to the cafeteria. I wasn't hungry, but I had to eat something, as long as there was no meat in it. I got a plate of lima beans in tomato sauce. Halfway into it, a piece of meat showed up. I stopped, pushed the plate away, and walked out. Within a week, we were having snacks while leaning on the cadaver.

Working on the same cadaver bonded the five of us together, and we remained good friends throughout medical school. Michel and Ara were from Egypt; Ali and Ibrahim were from Bahrain (see Figure 4.3). Ara became a distinguished pathologist at the University of Toronto. Ali became minister of health and then minister of education in Bahrain. Michel and Ibrahim went into private practice. During one anatomy lab, we heard a crowd of student protesters marching down Bliss Street and shouting slogans (probably protesting the French presence in North Africa). It ended with a rock-throwing melee, and the police arrested some of the protesters. Ali was an ardent Arab nationalist, like so many other students. When he heard the commotion, he began to take off his gown. He said he had to join the protesters. I told him not to be a fool—if he walked out of the lab to join the mob, they would kick him out of medical school. He said he had no choice. I told Michel to help me hold Ali down. We were bigger than he was. So he struggled a bit but gave up. Since he was physically restrained, he could stay where he was in good conscience. I wonder what would have happened if he had joined the protesters and gotten arrested.

Compared with anatomy, the physiology class was more interesting. It involved more than rote memorization and gave me a sense of the amazing workings of the human body. The lab was harder to deal with. Usually, a dog would be anesthetized for an experiment conducted by Dr. Henry Badeer. At the end of the experiment the animal was "sacrificed" by having air injected into its veins. I never had pets and didn't particularly like dogs, but I felt sorry for the poor animal. I didn't understand the point of the demonstration. If the professor had simply told me what the experiment was supposed to demonstrate, I would have believed him without having it reenacted for my benefit.

Other experiments involved frogs. To anesthetize it, we put the frog under a glass bell jar with a wad of cotton soaked in chloroform on its head. We would then pith the frog with a probe to sever the spinal cord from the brain and perform the experiment. Frogs elicited less sympathy than dogs, and I hated the touch of their slimy bodies. The first time I did this procedure, the frog was squirming so hard that I forgot to put it under the bell jar and instead kept pouring the chloroform on its head. As the fumes began to permeate the room the lab instructor yelled, "Who is trying to anesthetize the class?" I quickly shoved the frog under the bell jar and joined in the laughter, pretending it had nothing to do with me.

~

After surviving the brutal first year and actually doing quite well, my second year of medical school was less nerve-wracking. The burden of studying didn't get any lighter as we plowed through one after another telephone-directory-sized textbook in bacteriology, pharmacology,

Figure 5.1 Last to leave the library.

and pathology. What made these courses more interesting was their closer linkage with illness and the prospect of encountering patients the following year. Nonetheless, I spent a lot of time in the library, where I often seemed to be the last person to leave.

The professor of pathology, Dr. Philip Sahyoun, was a crotchety old man who never smiled except when making snide remarks about Americans ("Europe *civilized* America, and America *syphilised* Europe").[1] He was greatly annoyed when students failed to keep the lenses of their microscopes clean. He would stop by their desk, ceremoniously take out a big handkerchief, and clean the lens in silent reproach. His junior colleague, Dr. Nimr Tukan, was a flamboyant character who wore fine English shirts. He was a brilliant pathologist who turned medical grand rounds into a spectator sport by making discussants look like fools. When you asked for his help in class, he would say, "Would you also like me to give you a bath?" He was a tough grader. After a particularly difficult exam, Dr. Sahyoun chided him, saying, "Nimr, did you flunk the whole class?" Tukan suffered an untimely death when the private plane he was flying in with a wealthy contractor crashed into the Mediterranean.

Another memorable character was Dr. George Fawas, the professor of pharmacology. He had studied in Germany, had a German wife, and adored all things German. His acerbic humor was unsparing as he marveled at the stupidity of the world outside of Germany. Among his cherished targets were those who came from the northern Lebanese mountains, including some of his faculty colleagues. ("He is from the North" explained many a character flaw.) Dr. Fawas railed at the pretensions of the Lebanese. "Students," he would complain, "come to me looking for research topics. When I ask them what they are interested in, they say, 'I want to find the *cure for cancer*.'" He could not stand Alexis Boutros, the self-important conductor of the Académie des Beaux Arts orchestra. One season Boutros had the temerity to conduct Beethoven's *Missa Solemnis*. Fawas was incensed. "In Germany," he would say, "there are two, may be three, conductors who would dare to tackle this work. But when Boutros gives a concert, what does he pick? *Missa Solemnis*!" Dr. Fawas was not the only person that Boutros offended. He had insulted the men in the choir so badly that during one concert the bass section stood silently despite Boutros's frantic attempts to get them to sing.

Before the first exam Dr. Fawas would tell the class his favorite story. A professor of the Old Testament asked the same question on the final exam year after year: list the names of the kings of Israel. No one studied for the exam except for memorizing the names. His colleagues finally urged him to change the question. So he came up with, "Criticize the actions of Moses on Mt. Sinai." The students were stumped. One of them wrote down, "Who am I to criticize Moses? If you want the names of the kings, here they are. Moral of the story: Answer the question."

Years later, I got away with the same strategy of not answering the question. When I sat for the medical licensure exam in the District of Columbia, I studied one topic thoroughly for each subject area. For instance, in hematology, I picked the oxygen-hemoglobin dissociation curve. If the word *blood* appeared in a question, I wove the oxygen-hemoglobin dissociation curve into the answer. The person who corrected the exam was impressed enough to pass me.

Then there was Dr. Stanley Kerr, the beloved professor of biochemistry. He came from a family of missionaries and was an unusually kind and thoughtful person. He taught many generations of medical students at AUB and finally retired to Princeton. I last saw him there with Mrs. Kerr in 1976, when I presented a paper at a conference on psychology and Near Eastern studies. His association with the Near East Relief Foundation and Armenians in Turkey went back to the aftermath of World War I, when he lived in Marash (close to my parents' hometown of Aintab) and worked as a pharmacist.[2] Dr. Kerr did not live long enough, but Mrs. Kerr did, to see the tragic assassination of their son Malcolm, the ninth president of AUB.

Mrs. Elsa Kerr taught a course on family relations to undergraduates. It was an easy A, so I took it. She was the quintessential sweet and proper American lady of an earlier generation. The course covered every aspect of marriage except for sex. Finally, in our last class, she told us with great delicacy that there was a book on reserve in the library that dealt with that subject but that we should read it only when we felt ready. Well, we were ready right then and there, and we rushed to the library to find an innocuous book that counseled newlyweds to be kind and gentle to each other. (I hope Mrs. Kerr never saw my 747-page textbook of sexuality with its explicit illustrations of coital postures among other horrors.)

~

In the summer of 1956, following my second year of medical school, I spent a month at a work camp in Bethlehem (then still part of Jordan) sponsored by the World Council of Churches. I was twenty-two years old, and this was my farthest foray away from home. Our project was to build a playground for a local boys' school. We spent the mornings in manual labor, the afternoons in visiting local sites, and the evenings in social activities. It was a wonderful experience. The participants were mostly from Europe and one from the United States. They had never met someone of my background; the only other person from the Middle East was an Iranian woman. I was sufficiently Westernized for them to feel comfortable with me yet different enough to be exotic and interesting.

Our leader was a German Lutheran pastor, whom I came to like and admire. He wasn't much older than the rest of us and exuded a joyous sense of life. The only time he surprised and disappointed me was when he got back his gray suit from the dry cleaners and got very upset that it had been pressed out of shape. "Dammit, it's my best suit," he cried out in dismay in front of everyone. I naïvely thought that as a minister of the church he should not have let a mundane problem like that upset him. The only American in the group was a young woman from the Midwest called Sue. Unassuming and pleasant, she didn't fit my stereotype of a Hollywood mini-star. She spoke to me about her family and her life in the United States, and she was equally eager to learn about my life. It was the first uncomplicated friendship I had with a young woman. When our group went on a visit to a school for juvenile offenders, Sue had placed her wallet on a bench and it was promptly stolen by one of the boys. The stern director of the school soon found out who the boy was and was profusely apologetic. Sue felt terrible for being responsible for what happened. She knew quite well that the boy would face the wrath of the director as soon our group left the school. I tried to reassure her, but she was inconsolable. There was also a tall and reserved Norwegian young man who spoke with great deliberation, dragging out each word slowly, one syllable at a time ("My—na-me—is—Os-kar—Paul-son—I-am—f-rom—Nor-way"). He called me *Bay-rut* not to struggle with my name.

The two most colorful characters were Pekka, a Finnish Lutheran pastor, and Gregoire, a Greek Orthodox priest of Latvian origin from France. They were older than the rest of us and the only two married persons in the group. (Orthodox priests, unlike monks, can marry before ordination.) I had never met a Finn before and couldn't have imagined that one day I would be married to one. Pekka was theologically and politically liberal. He startled me by invoking Luther's admonition to sin boldly (*pecca fortiter*).[3] He loved good wine, which we were served when invited out to dinner, and he wrote to his wife every night no matter how late the hour or how drowsy he was. One of these dinners was at a Greek Orthodox monastery. A pile of delicious rice pilaf with tender chunks of lamb drenched in allspice sat in a huge platter on the refectory table. Next to it were bottles of robust red wine from the monastery cellar. The monk serving the wine got tired of filling Pekka's glass, so he put a full bottle next to him to help himself. The abbot, a portly man with a flowing beard stained by tobacco, sat at the foot of the table as a sign of humility, and as one of the youngest member of the group, I sat next to him. I noticed that he was being served from a special bottle that sat between us. When he was distracted, I filled my glass from his bottle. It was superb, so I kept refilling my glass from it as he looked on with an indulgent smile.

The liberal views of the Lutheran pastor clashed with the conservative theology of the Orthodox priest. Gregoire was no match for Pekka's incisive wit, but he was unyielding in his rebuttals. Although profusely apologetic when he argued back ("*Please* excuse me but…"), he used his convoluted Byzantine reasoning to good effect. As both of them spoke a heavily accented English, their arcane theological exchanges sounded hilarious. Given that I was the only other person who officially belonged to the Orthodox Church, Gregoire tried to recruit me as an ally, but having become more of a Protestant, I was of little help to him. Nonetheless, Gregoire and I had interesting discussions of Orthodox Christianity. I was moved by his description of elderly monks who in the evenings went around the church kissing the icons goodnight as one would old friends. He never spoke about his wife, and I couldn't imagine the sort of woman who would be married to such an odd and otherworldly man. Gregoire actually claimed to know a lot about women, and he counseled me to marry a Catholic because the devotion of Catholic women to the Virgin made them good mothers.

~

At one of the receptions we attended, I made the acquaintance of two Swedish nurses. They worked at a small hospital in Bethlehem founded by the Swedish Lutheran Church, and which was now being run by a Jordanian physician. When our program was ending, I wanted to spend more time in Bethlehem. The nurses arranged for me to work in the hospital lab, and in return I could have my dinners with them. I also got permission from the school to sleep in the vacant student dormitory under the church. It was a great arrangement. The younger nurse, Birgitta, was about my age, but far more mature. The older one, Louisa, was in her forties and had grown up on a farm in southern Sweden that had been in the family for many generations. They were lonely and felt beleaguered by the prospect of the Jordanian physician appropriating the hospital. Therefore, they were more than happy for my company. The hospital cook served our dinners in Louisa's bedroom and went to some trouble preparing them. I provided the local wine. We had delightful conversations and became good friends. After dinner, I would walk back through the night to the cavernous dormitory with rows upon rows of empty beds. It had an eerie feeling, and I never quite got used to it.

I sensed that Birgitta was becoming attached to me when we went dancing one evening. I liked her a lot (as I did Louisa), but I wasn't in love with her. She was attractive in a quiet way—a warmhearted affectionate, and able young woman (she worked as both chief surgical nurse and X-ray technician) and a good enough Christian to come all the way to Jordan to serve the sick. She was also quite athletic. One evening, when we were taking a walk, she challenged me to a race. I was reluctant to compete against a woman, but I agreed. She took off her shoes and we sprinted forward. I held back at first, but as she overtook me I ran as hard as I could to barely beat her. When we stopped, I bent over, gasping for breath, amazed that a woman could run that fast, while she looked at me tenderly, hardly out of breath, her cheeks ablaze with excitement.

I'm not sure what stopped me from responding to Birgitta's sentiments. One evening when I entered Louisa's room, I found Birgitta stretched on the bed, smiling at me. I sat down on the chair opposite and tried to engage her in a casual conversation, taking no note of her seductive posture. As we talked, a Swiss friend opened the door, saw Birgitta on the bed, apologized, and began to retreat, thinking she had interrupted an assignation. Birgitta said something like "Oh, no, it's not like that," but she said it in a regretful tone that suggested she wished it had been "like that." Was I scared of another romantic involvement so soon after Juliette and concerned over breaking Birgitta's heart as well? Or was it the awareness that I was only a medical student and hardly ready for marriage, which was the only context in which I could imagine getting physically intimate with her? Nor could I envisage how Birgitta could fit into my family given the vast differences in our backgrounds, even though she probably could have done so quite easily.

There were two other experiences in Bethlehem that stayed with me: one of them awful, the other sublime. My work in the lab consisted

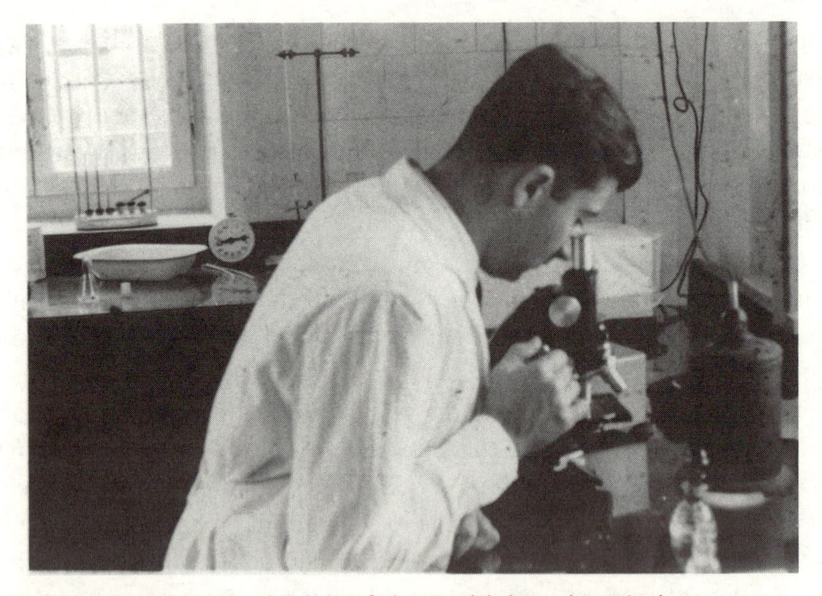

Figure 5.2 Working in the lab of the Bethlehem hospital.

of doing blood counts and urine analyses. The technician was kind and helpful, but the work became routine and monotonous. So when I met the coroner who brought samples to the lab for toxicological analysis, I asked him if I could accompany him on one of his investigations and perhaps observe an autopsy. He was happy to oblige, and the following week he called, saying he would pick me up from the lab within an hour. He was on his way to investigate a case of suspected poisoning. A woman and a handyman working for her had been found dead after sharing a meal. The husband had vanished. He was the prime suspect for poisoning his wife and inadvertently the handyman, or perhaps he meant to kill him too if he thought that the man was sleeping with his wife.

The prospect of watching an autopsy was exciting, but it turned out to be a ghoulish experience. When we got to the house, we found the woman's obese, naked body lying flat on her back on the floor of a darkened room, with flies buzzing over it. There was a lot of commotion outside as children tried to peek through the shuttered windows and adults chased them away. It was a warm day, and the coroner was perspiring profusely in his dark suit. He put down his bulging leather bag next to the woman, sat down on a chair, loosened his tie, wiped his brow, and said, "Go ahead." I didn't understand what he meant until I realized he was expecting *me* to do the autopsy. That wasn't what I had in mind when I asked to accompany him. However, my manhood was now on the line, and I wasn't about to back off. But I still had no idea what to do. He noticed my hesitation and said, "There is a pair of rubber gloves in the bag. Put them on, take out the scalpel, and make an incision from the end of the sternum to the top of her pelvis": in other words, cut open the woman's whole abdomen. I knelt down on the floor next to the woman—it was hard not to think of her as a woman, since her body still felt warm. I realized that cutting into her flesh was going to feel quite different from dissecting a cadaver. The bulging thick layer of yellow fat got in the way. I did what I was told and looked up for further instructions. The coroner told me to take out a loop of intestine, cut into it, and take a sample of its contents; same for her stomach. When I got all the samples, I pushed her organs back into the abdomen and sewed it up with crude stitches. I got up exhausted, took off the gloves, washed my hands carefully, and we headed back to the lab. "How was it?" asked the lab technician. "Fine," I said and looked away.

When I met my friends for dinner that evening, I had to give them a fuller account of the day, since they had already heard about it. Since they were nurses, I knew they could handle it, but it was hardly the best topic for dinner conversation. So I cut it short. I did, however, drink two extra glasses of wine, and they noticed it. As I headed for my quarters through the dark streets of Bethlehem, I tried not to think of the dead woman, but I couldn't get her out of my mind. I dreaded walking into the dormitory with hundreds of empty beds, with the eerie feeling that something dreadful was about to happen. Looking away from the gaping darkness, I hurriedly got into bed and tried to sleep. But I was haunted by the image of a fat, naked woman with her guts hanging out standing next to my bed. I knew it was all in my imagination, but I never turned to check if she was there. When I opened my eyes in the morning, I was relieved she was not there.

The following Sunday, I went to Jerusalem hoping to erase the memory of the awful experience. I had taken the bus a number of times before to see the city. This time I headed for the Garden of Gethsemane on the Mount of Olives. It was a glorious day. The humming of the bees in the olive trees was the only sound that intruded into the silence. I was completely alone. I sat on one of the toppled tombstones of the old Jewish cemetery and let my gaze wander over the walled city. In the brilliant summer sun the golden Dome of the Rock shone in the distance. Jerusalem seemed eternal. I sat lost in thought until the afternoon light began to wane. I walked slowly down through the olive groves to the Russian Orthodox Church of St. Mary Magdalene. It was built by Tsar Alexander III in 1888, in honor of his mother, the Empress Maria Alexandrovna, the namesake of the saint. With its seven onion-shaped brilliant cupolas bathed in the sunset, it looked as if it had been picked up from St. Petersburg and laid down at the foot of Gethsemane.

I entered the church in the pellucid light of candles and the heady aroma of incense. The iconostasis was dominated by a huge painting illustrating the legend of Mary Magdalene telling Emperor Tiberius about the mock trial and crucifixion of Jesus. It showed her offering the emperor an egg painted red, symbolizing the blood and resurrection of Christ.[4] Vespers were in progress. The mournful Russian liturgy sung by the nuns filled the air as the officiating priest in his magnificent vestments and flowing beard intoned in a deep baritone, *Hospodine pomiluj ny* ("God have mercy upon us"), to which the choir responded in exquisite harmony. I was transfixed and went out, having seen a vision of heaven. I slept that night like a child in the bosom of Abraham.

~

Third year brought me face to face with patients, and medical school redeemed itself. The clinical years were to be very different from the smelly labs and the cramming of massive textbooks. No other profession brings one closer to other human beings than does medicine. The physician sees men and women, the young and the old, in the full nakedness of their bodies, hearts, and minds. Their fears and suffering, the raw pain and anxiety, and the specter of death force one to confront one's own life and mortality. To figure out what ails the patient, the physician puts together the history of the illness with its signs and symptoms and looks for the most likely illness through the process of differential diagnosis. In modern medicine, tests and technical procedures have become crucial aids, but they have not entirely replaced the physician's own sense of sight, sound, touch, and smell in probing the patient's body. Even the physician's sense of taste used to play a role, for instance, in differentiating between the two forms of diabetes, one in which the urine tastes sweet (*diabetes mellitus*) and the other where it does not (*diabetes insipidus*). A professor of medicine once demonstrated this to a student audience while simultaneously stressing the importance of sharp observation. He held up a beaker of urine, into which he dipped his finger and licked it. The beaker then went around as students repeated the same thing with some reluctance. When the beaker came back to the professor half empty, he told the class, "You did well using your sense of taste, but you weren't observant enough to note that I dipped my index finger in the urine but licked my middle finger."

The physician's reward comes from making the right diagnosis and curing the patient, or at least alleviating the distress; the punishment is the wrenching sense of failure in making the wrong diagnosis or losing the battle to an intractable disease. Despite my initial lack of interest in medicine, it was impossible not to become immersed in it. Unlike the stench of the labs, I actually liked the pungent disinfectant odor of the hospital. It felt real and I felt alive. The long hours, the lack of sleep, the stress, and the sheer physical fatigue seemed worthwhile. What I did mattered. I was briefly tempted to pursue internal medicine or some other specialty and I would probably have made a good physician. Yet, these disciplines would not offer as good opportunities of exploring the thoughts and feelings of patients, so I never wavered seriously in my choice of going into psychiatry.

I had memorable experiences working on the wards in my third year and in the outpatient department in the fourth. There was Ara, an Armenian teenage boy from Syria, with *spina bifida* ("split spine"), a developmental birth defect where the vertebrae do not fuse, leaving the spinal cord exposed. The attending surgeon was the much beloved Dr. John Wilson (affectionately nicknamed "St. John"). The boy's parents lived far and visited him rarely. He was lonely and despondent, lying on his bed all day with nothing to do, no one to talk to, and not a glimmer of hope of getting well. My main task was to change the massive dressing over his back, a time-consuming task I did at the end of the day. The lack of any improvement made my task seem futile. It must have been even more discouraging to him. Yet, I came to mean a great deal to the boy. The change of dressing, during which we chatted and joked, was the highlight of his day. The more complicated procedures had to be done by Dr. Wilson, but Ara would ask if I couldn't do them instead.

Working on the pediatrics ward was hard. I couldn't bear to see children in pain, especially when I was causing the pain. I could put up with their screaming when I stuck them with a needle, but when a child looked at me with a tearful, how-could-you-do-this-to-me expression, I had to look away. On one occasion, I diagnosed a case of intussusception of the small intestine (where a section of the bowel is blocked by becoming telescoped into itself), a rare and serious condition. It was quite a diagnostic feat for a medical student, but the chief resident took the credit for it when she presented the case to the attending physician. I fumed silently, but said nothing. She was my immediate superior, and I didn't want to incur her enmity. I was even more upset when a resident in surgery dumped on me the task of telling a man with terminal bowel cancer that he had to have colon surgery and wear a colostomy bag for the rest of his life. The patient was a hardened Syrian army sergeant, and I lacked the stature, the experience, and the command of Arabic to convey to him the message as it should have been. The patient listened to me impassively and refused to have the operation. I blurted out that he would die. I must have said, "You die." And he mockingly imitated me, saying, "Fine, I die." The next morning he checked out, and his empty bed made me feel awful.

Another of my diagnostic triumphs was better acknowledged. In the outpatient clinic, I saw another teenage boy with a peculiar set of symptoms. He had powerful-looking calf muscles, yet his legs were so weak that he could only get up from a squatting position by pressing his hands on his thighs in a characteristic movement. I had seen the photograph of a patient with hypertrophied calves doing that very thing in a British textbook that I liked to read because it was so well written. I rushed to the library, found the book, and read the description of the condition: it was pseudohypertrophic muscular dystrophy, and it fitted my patient perfectly. When Dr. Virgil Scott, the professor of internal medicine, came to review our cases, I presented it to him in a matter-of-fact way, as if it was all in a day's work. He examined the boy and confirmed my diagnosis. He then saw the book on the table and said to me, "You did the right thing; when faced with a case like this, you go to the library." When the patient was presented at grand rounds, Dr. Scott started by saying, "At the outpatient clinic last week, Mr. Khatchadourian told me that he had a case of pseudohypertrophic muscular dystrophy." Some in the audience may not even have heard of the condition. It made my day.

One of our tasks as third-year students was to do the lab work for our ward patients. One morning the chief resident told me he needed a blood count on a patient in twenty minutes, when he would be presenting the patient to the attending physician. I had to move fast. The patient was a peasant woman with big, callused hands, and I had to prick her finger many times before I could get a drop of blood. I sucked it into the pipette, mixed it with the solution to stop it from coagulating, and was running to the lab when I dropped the pipette, shattering it into pieces. I had no choice but to go back for another sample. I approached the woman casually and told her I needed another sample of blood. "What?" she said, "Again?" I calmly told her, "The first test was for the blood from your right hand; now we have to do it for the blood from the left hand." She said, "Oh," and stretched out the other hand. She obviously had not heard of William Harvey's description of the circulatory system in the seventeenth century. I felt a little bad, but I had to do what I had to do.

It was standard practice to recommend carrot soup for children with diarrhea, especially when the parents could not afford expensive medications. When a young mother brought her infant son with diarrhea to the clinic, that is what I told her to do. However, I confused the Arabic word for carrots (*phejel*) with radishes (*sajal*) and told her to give her child radish soup. "Radish soup?" she asked incredulously. "Yes," I said raising my voice, "radish soup." "But doctor…," she began to protest, and I cut her off with, "Am *I* the doctor or are *you* the doctor?" Apparently, my voice was overheard in the corridor, and from then on, the nurses kept repeating, "Am I the doctor or are you the doctor?"

~

My interest in Juliette was rekindled during my first year of medical school. After the turmoil over Talin, I needed a less complicated and more fulfilling relationship. In our previous friendship, I had been frustrated with Juliette's failure to reciprocate my feelings for her. Now I realized that she did care for me, but she had been too young and shy to express her affection. This was no longer an issue, and she became more openly affectionate when we resumed our relationship. Moreover, she now had grown from a cute teenager into an attractive young woman.

We were part of a small group of friends within the Christian Endeavor Society. We got together occasionally, and Juliette and I began to see each other on our own as well. We usually met after school or on Saturday afternoons to have coffee and occasionally go to the beach, but we never went out in the evenings. Although our relationship was now more public, it still had to be kept within bounds. Our parents were aware of it and seemed to approve. The families knew each other and were quite compatible. It looked like we were headed for eventually getting married.

I have happy memories of our time together. Juliette was sweet, even-tempered, guileless, and cheerful. She gave me no reason to be unhappy. Moreover, unlike Talin, she was real, not a figment of my imagination. Yet, almost despite myself, I began to lose interest in her. I felt frustrated by her seeming lack of drive, ambition, and intellectual curiosity. She did not lack intelligence, but she was not an intellectual. I was hardly an intellectual myself, but as my horizons expanded, I grew more distant from her. We mostly chatted about mundane matters or gossiped about our friends, with her providing the grist for the mill. Had I tried to engage her in more serious topics, she would no doubt have been receptive, but I wasn't interested in becoming her tutor. The issue of commitment to God, which ostensibly wrecked my relationship with Talin, was still very much on my mind, and Juliette could hardly pass muster. Nonetheless, I never found her actually lacking in comparison with anyone else that I knew; it was only when pitted against some vague idealized image that she failed to make the grade.

I came up with the odd notion that Juliette should go to nursing school. That would plunge her into the real world, and we would have more in common. However, girls from her background did not become nurses, and her father wouldn't hear of it. She herself had no interest in nursing, but she may have acquiesced to please me. When our friends heard about it, they thought I had lost my mind. Be that as it may, once I started down this slippery slope, I became convinced that we were not meant for each other. I realized Juliette would make a fine wife and mother and she would try hard to make me happy, but that wouldn't be enough. I was equally concerned, and quite rightly, that I would make her life miserable. There was a hard, demanding, and driven side of me that Juliette could never satisfy. This was not my basic character, yet a significant part of it. I grew up in a tough culture where one had to drive a hard bargain to survive and thrive. I was not sure if Juliette could live up to her end of the bargain and pull her weight in fashioning our life together. I had acquired my mother's intolerance for anything smacking of weakness and lack of competence. My overly high expectations and striving for perfection compounded the problem by making small faults escalate into big ones.

I felt I had no choice but to break up with Juliette. However, since she had done no wrong, how would I justify breaking her heart? It was one of the hardest thing I was ever to do, yet the longer I waited, the harder it got. I finally took her to one of the cafés on the Corniche overlooking the Pigeon Rocks. In this romantic setting, we talked about everyday matters, but she must have sensed that there was more to come. She had realized that all was not well between us, but she couldn't understand what the problem was. I tried to explain, but it made little sense to her. When I finally told her that it would be best for us to part company, she didn't get angry or make a scene; she just cried softly and asked why we couldn't go on as we were. I was impressed by the way she handled herself in this painful situation. It was hard for me too but it must have been even harder for her. We parted amicably, albeit with heavy hearts.

~

In the spring of my third year in medical school I received a postcard from Juliette. It was sent from Egypt, where she was visiting her friend Sylvia in Alexandria. Juliette had talked to me about Sylvia off and on—her beauty, her musical and artistic talents, her voice lessons in

Zurich, her family's prominent position in the Protestant Church, and their summer villa in the Lebanese mountains. The postcard consisted mainly of casual general remarks, to which Sylvia had added two words in the margin: "Kindest regards." Those two simple words written in a bold and flowing script got a strangle hold on me and put in motion an inexorable chain of events that led to our getting married a year later.[5]

There is a great deal I could say about Sylvia, but I hope to have the wisdom not to. However, I can't tell the story of my life without referring to one of its most traumatic periods. I carry the scars of that experience, but there are no bleeding wounds. I will try not to refight old battles or settle scores. The searchlight will be on me, not on Sylvia. I will try to say nothing critical except what is necessary to explain why I left her. Moreover, the account that follows obviously represents my version of events; Sylvia's take on them is likely to be quite different.

My falling in love with Sylvia mirrored in some respects my experience with Talin writ large. There was, once again, the construction out of whole cloth of a person who lived mainly in my imagination, with my emotions tangled up in a misguided spiritual vision—a toxic mix of psychological problems and theological confusion. However, where Talin had been a passive and puzzled bystander in my Divine Comedy, Sylvia played a leading role in it. The circumstances of my life were also quite different. At the time of my relationship with Talin, I was a college student who could do limited harm; with Sylvia, I was close to graduating from medical school and the potential for damage was far greater. I proposed to Talin a wildly improbable and idealized future that led nowhere; I offered Sylvia a life together in the real world, and she agreed to marry me. My first failed relationship ended up with a bruised ego; my failed marriage shook me to my foundations.

The starting point of my fateful romance was the conviction that Sylvia was God's choice for me and her two words—"Kindest regards"—amounted to a declaration of love. There was no rational basis for either supposition. Yet, I wrote a letter to Sylvia expressing my admiration and joy over what I had learned about her Christian faith and devotion. Although I had never met her, I felt as if I had known her all of my life. While the letter was not a marriage proposal, it had strong hints of an attempt to establish a personal bond with her.

I apologized to Sylvia for my boldness in writing to her. Nonetheless, I fully expected her to answer me. She never did. I became obsessed with wanting to hear from her. I checked my mailbox every day and sometimes more than once, even on weekends, when no mail was delivered. I did this for over a month. I just wanted an answer—any answer. More than four decades later, long after we had been divorced, Sylvia and I had dinner together. When I asked her why she had married me, Sylvia told me that it was because my letter had persuaded her that I was God's choice for her. The reason she had not responded to it was that she thought Juliette was still my girlfriend or engaged to me. All this before we had even met in person. God's will? I doubt it. *Folie à deux?* More likely.

I was surprised when Juliette informed me that Sylvia was coming to Lebanon for the summer and would be attending the opera festival at Baalbek. I wondered why she was telling me this, but I took the bait, if that is what it was. I now owned a small Fiat Quattrocento, bought by my father at my mother's urging. She was worried that my life was too constricted and confined to school; a car would help to make me more outgoing and socially engaged. There was only one other person in my medical school class who owned a car, so this was a big deal for me.

I drove to Baalbek with a friend who could identify Sylvia, and during the intermission he pointed her out to me. She was standing at the buffet. I got a glimpse of her beautiful blond hair folded at the back of her head in a distinctive manner and was struck by her commanding posture; with her feet planted firmly on the ground, she looked almost regal. I walked up to her and introduced myself. She nodded. I asked if she had received my letter. She said she had. But she said it in such a chilling tone that it allowed no further inquiry. As she looked away, I bowed slightly and retreated. That was it—time to quit. If it was God's will that Sylvia and I should come together (which I now doubted), he had to make it happen. I was done with pushing the boulder up the hill only to have it roll down on my head.

A few weeks later, Juliette invited me to a party at her house and mentioned that Sylvia would be there too. I was baffled. Was Juliette trying to bring us together? Was she doing this out of the goodness of her heart, or was she out to wreak vengeance by throwing me into the lion's jaws (which would be hard to imagine coming from her). Had Sylvia told her about our brief meeting at Baalbek? Had she learned by now that Juliette was no longer my girlfriend? Was Sylvia instigating the invitation to the party? I went to Juliette's house not knowing what to expect. There was Sylvia and the ice had melted. We were introduced and conversed amicably with no reference to our earlier meeting or my letter. I didn't know what to make of it. Presumably, Sylvia had now learned that I was no longer attached to Juliette and that obstacle had been removed.

Soon after the end of the school year, I went to the Christian Endeavor summer conference. Sylvia was there too with a contingent from her church in Alexandria. They mainly kept to themselves, but Sylvia was quite friendly and drew me into their circle. They were ardent Evangelicals. Given my struggles with Talin's and Juliette's ambiguous spiritual credentials, I should have been delighted at Sylvia's unequivocal Christian commitment. Nonetheless, I felt uneasy with how she and her friends went about professing their faith. They seemed so sure of everything, with no room for doubt or dissent. Discussions with Sylvia did not proceed in a natural way nor were differences of opinion resolved in a rational manner. She made a categorical statement, and that ended the conversation. She also had a way of holding back part of what she was saying. That left me wondering about the hidden meaning behind her words. On my part, I abhorred ambiguity and tended to see things in a cut-and-dry way, which she must have found frustrating. However, our problems in communicating with each other were not simply due to differences in style. There was a deeper sense of fighting for control, or her way of asserting her dominance, or so it seemed to me.

Nonetheless, Sylvia and I were clearly drawn together. There was a mutual physical attraction between us and a certain degree of emotional warmth. We had our Christian faith in common, even if there was not much substance to this bond. Sylvia and I saw a lot of each other during that week and continued to do so following the conference. Ostensibly I was courting her, but in effect she seemed to be managing the relationship. We were soon paddling together in white water heading for the thundering falls. We never had a serious discussion beyond the giddiness of the moment. One evening, as I was driving down from Sylvia's summer house, I prayed that if God would make her mine, I would never ask for anything else. It was hubris for me to bargain with God, betting on my future and God punished me by giving me what I wanted with such reckless abandon. A few weeks later, we got engaged. On the night that Sylvia accepted my proposal, I drove down from her house in a triumphant mood, literally shouting with joy at my incredible good fortune and overflowing with gratitude. I remember being particularly pleased thinking of the great impression this would make on one of my cousins whom I thought highly of. It was bizarre. It was as if Sylvia was a prize I had won in a lottery without having bought a ticket. Surely, it could have only happened if God had ordained it in response to my urgent plea.

Yet, even during this dizzying period of jubilation, there were dark forebodings. The night that my parents and I went to meet with Sylvia's family to ask formally for her hand, a forced gaiety hovered over the evening. Sylvia's father, Mr. Yervant, who laughed at his own jokes, bantered loudly in halting Turkish with my father (Mr. Yervant too being from Aintab). I felt irritated and embarrassed. It was like being an understudy in a bad comedy in a language I didn't understand. When I had told my parents of my wanting to marry Sylvia, they seemed to be already aware of my interest, but there wasn't much discussion even though I assume they had talked to each other about it. They neither raised objections nor did they seem overjoyed at the prospect. Nothing was said about how we were going to live and who was going to support us while I was still a medical student. Sylvia's family was thought to be still wealthy, and that must have made my parents assume that they would take care of us, since they themselves could not. While I sensed no enthusiasm on my parents' part at the prospect of my marriage, my mother dutifully bought a diamond necklace for Sylvia and gave her a lovely strand of natural pearls that my father had given her many years earlier. By the time they became disenchanted with Sylvia and her family, it was too late to do anything about it.

~

When a story ends badly, its doom and gloom tend to darken the good parts as well. I will try to avoid that, since the early part of our courtship was a happy time for me and seemingly for Sylvia as well. During the summer of our engagement, I would drive up to Sylvia's house, often cutting short studying at the library (which was like having a baby pulled away from the mother's breast). Sylvia and I would sit on their veranda with its lovely view of the pine forests cascading down the mountain. We talked about various things, and I enjoyed being with her. She had a lively mind and a rich experience of life. I reveled looking at her emerald eyes and listened to her accounts of growing up in Alexandria and her travels in Switzerland, Sweden, and Greece. Although we were the same age, she was more sophisticated and mature than I was. Music was her great passion. And she was enormously self-confident. I do not recall her ever expressing any doubts or regrets over anything. Nor do I recall any serious discussion of our future together. We seemed to be acting out a script of aren't-we-happy-we-are-engaged and sweeping under the carpet whatever didn't fit the script, and there was a good deal that did not.

Sylvia had a close-knit family. I became friends with her two brothers and her uncle, whose family lived with them during the summers. We got along well, but my biggest fan was Sylvia's mother. Mrs. Elize was a strong-willed, competent, and controlling woman (someone admiringly described her as being "just like a man"). She was genuinely fond of me and at times more generous to me than to her own sons—which was awkward. I had far less to do with Sylvia's father. Given my distant relationship with my own father, I had hoped to be closer to Mr. Yervant, but that never happened. I made some clumsy overtures to establish a closer bond with him, but he seemed more baffled or irritated than interested.

This generally happy period was marred with Sylvia's unexpectedly shifting moods. Cheerful one moment, she would turn glum and sullen in the next. She seemed burdened with dark thoughts. Something was bothering her, but I couldn't tell what it was. Was there a failed romance in her past or a lost love that she was struggling to get over? Was I paying the mortgage for a debt incurred by another man? Was I being punished for someone else's breaking her heart? There were strong hints to a past relationship. Sylvia wanted me to let my hair grow long. I did not know why until she showed me a newspaper clipping of a pianist who she had referred to in a veiled fashion as a friend (or was it as a family friend?). He had long hair. Was she trying to remake me in someone else's image? When I obliquely tried to find out more, I got nowhere. I didn't dare press her harder. Maybe I didn't want to find out the truth. Instead, I searched myself for what I may have said or done to upset her, but I couldn't come up with anything.

We never argued or quarreled openly. I sometimes wished we had, since her indifference to me felt worse. She showered her pet puppy with ostentatious shows of affection instead of being affectionate to me. I wasn't jealous of the miserable little dog, but I sensed there was something seriously lacking in our relationship. Sylvia expressed no overt hostility or anger toward me. She just didn't seem to like me, let alone love me. I blamed myself for it but was at a loss, not knowing what to do except try harder. I could have readily understood if she had become disenchanted with me as she came to know me better. I could think of many reasons why she no longer liked me. I didn't like myself that much either. If she wanted to put an end to it, I would have willingly gone away without a murmur.

The prospect of marriage normally brings families together, but that did not happen in our case. There were signals that Sylvia's mother was not keen on some of my relatives, and she had made condescending remarks about them. Soon after our engagement, my mother had a reception one evening for our relatives to meet Sylvia and her family. There must have been fifty or sixty guests crowded into our living room. Sylvia and I were seated at one end when one of my older cousins spoke up and asked Sylvia to sing to the group. There was silence. My cousin repeated his request in more emphatic terms, but Sylvia would not sing. As an accomplished singer, this was not due to her being reticent or shy, but an assertion of her will. After an awkward silence, people realized there was going to be no singing and they slowly resumed their conversations. My cousin was humiliated, and I felt acutely embarrassed. This was no way to endear herself to my relatives. However, when a few days later, Mrs. Elize apologetically told me that Sylvia should have sung something and been done with it, I defended her, saying my cousin had no right to put her on the spot.

As I became immersed (mired would be closer) in Sylvia's family, I was alienated from my own. My mother was becoming increasingly concerned about my future with Sylvia. Mrs. Elize seemed to have taken over my life. Like an object she had found in an antique store, I needed some fixing but had good potential. Being young and malleable she could fashion me like a lump of clay to her own specifications. For that to happen, I had to be weaned away from my mother. At some point, Mrs. Elize had told her, "Hrant is now our boy. You all should all forget about him." That must have infuriated my mother. But since she was no wimp, why did she not push back? I couldn't have known it at the time, but after I learned about my mother's losing Ramzi, I wondered if she decided not to repeat her mother's mistake by standing in the way of her child no matter how ill-advised the path he may have chosen to follow.

~

I started my fourth year of medical school in the fall. I now worked in the outpatient clinic and could no longer spend as much time with Sylvia. Our marriage was six months away, and she wanted to go to Zurich for a last round of voice lessons. The temporary separation came as a relief to me and probably to Sylvia as well. She loved Zurich, and music was the center of her life. Getting married was going to put an end to it. I didn't realize at the time how difficult this must have been for her. Perhaps that was another reason for her unhappiness. She had rejected many suitors in order to pursue her career, and now I had gotten in her way. If I had realized that, it would have saved us much heartache and I would have most certainly not stood in her way. In her aspirations to pursue a career, Sylvia was ahead of her time. No woman in her circle or background had done that or would be expected to do, as I was soon to find out.

Sylvia and I wrote to each other sporadically, and a refreshing note of candor began to seep into our correspondence. Our lives seemed to be diverging. I had become fully aware that I wasn't ready to get married in general and to Sylvia in particular; she might well have felt likewise. Had we been free to make our own choices, we could have amicably parted company. It would have allowed her to continue her musical career, and I would have gone back to my studies, from which I had become seriously distracted. However, we never openly broached the prospect of ending our relationship but merely skirted around it.

One evening, shortly after I had heard from Sylvia about these matters in a particularly candid manner, I went for a walk with her mother. I told Mrs. Elize in carefully couched terms where matters stood. I said I didn't want to stand in Sylvia's way in pursuing her career, but I didn't say that I no longer wanted to marry her as I should have. This put the burden on Sylvia. Mrs. Elize became very upset and told me that Sylvia's singing was never meant to become a career. "You put a baby in her arms," she told me vehemently, "and that will be her career." I looked down and said I couldn't do that. The following morning, I was called back to see Mrs. Elize. I found her in bed. She must have had a rough night and looked wan and worn out. Yet she was determined as ever. She wanted me to go to Zurich and bring Sylvia back. I meekly agreed. Yet I was baffled by why Mrs. Elize had become so distraught by even a hint that our engagement might be in jeopardy. Was I such a desirable catch for her daughter that she could not countenance losing me? Or was it because I could repair some damage that her daughter had suffered that could not be fixed otherwise? At any rate, Mrs. Elize bought my plane ticket and I got permission to absent myself from the outpatient clinic for a few days. My mother couldn't understand why it was necessary for me to make this long journey simply to accompany Sylvia home. She reluctantly acquiesced, and as a sign of goodwill she gave me a package of lahmajouns, which Sylvia liked. Food would solve all problems.

It was the first time that I had been on an airplane, and I found the experience exhilarating. I marveled at the snow-covered Alps and the view of Zurich from the air. Sylvia met me at the airport. She seemed delighted to see me, and we embraced affectionately. We took a taxi to the house where Sylvia rented a room. That evening, we ate the lahmajouns and talked about the time we had been apart. Our faltering relationship seemed to have miraculously mended itself. We slept under the same billowing quilt like a pair of innocent children, which is what we were underneath our veneer of maturity.

I said nothing to Sylvia about my conversation with her mother, and nor did she say anything about the conversation she must have had with her. There was no reference to the doubts expressed in our correspondence. It seems incredible that we didn't even touch on these subjects, let alone discuss them openly, when we had the perfect opportunity to do so. Despite my mounting reservations about our impending marriage, I still loved Sylvia. When she embraced me with open arms, my doubts vanished. Why would I spoil the moment when everything looked so full of promise? I still believed that our bond was divinely ordained, and having it reaffirmed prevented a crisis of faith in my ability to discern God's will. I would also be spared the loss of face that I would have suffered if our engagement had been broken off. I can only guess at Sylvia's own apparent change of heart. Did absence make her heart grow fonder? Did Mrs. Elize make her aware of the imminent danger of her losing me? Did she chastise her into falling into line? In any case, none of these considerations on both of our parts justified our failure to confront the problems we faced in our relationship, to either repair it—not just paper it over—or break it off. We missed that chance, and once we returned to Beirut we had passed the point of no return.

Over the next few days in Zurich, Sylvia busied herself with preparations for returning to Beirut. We spent a whole afternoon in a music store, where she went through countless sheets of music with an accompanist deciding which ones to take back to Beirut. I was bored stiff, hardly realizing how important it was for Sylvia to do this; it was like storing provisions for the endless winter that was about to descend on her life.

The following day, I went to visit the renowned mental hospital at Burghölzli where Eugen Bleuler had introduced the term *schizophrenia* in 1911. Bleuler with his assistant Carl Jung had made psychiatric history in the early twentieth century, turning Burghölzli (like the Salpêtrière in Paris) into one of the great psychiatric centers in the world. Bleuler's son, Manfred, was now the director. An elegantly dressed and distinguished-looking man with white hair and steady gaze, he looked like a picture in a book on the history of psychiatry. Although I was just a medical student, he treated me with great courtesy. I was fascinated to attend their staff meeting, where Dr. Bleuler interviewed a schizophrenic woman. I told him that by looking at her facial expressions, I could detect the split between her thoughts and feelings, the key symptom that had been the basis of his father's calling the illness schizophrenia ("split mind"). Not knowing German, I don't know how I could have done that, but Dr. Bleuler was pleased, and there were appreciative nods from the other respectful psychiatrists. The visit made an indelible impression on me. (In later years, I could boast, "When I was visiting Manfred Bleuler at the Burghölzli...")

What is amazing is how I could have become so distracted to forget the reason I had come to Zurich and the looming disaster of my impending marriage. It was another example of my ability to compartmentalize—to focus on one issue and shut out everything else. While I sat in that hallowed psychiatric shrine, there was no Sylvia or anything else to worry about. This ability to turn parts of my mind on and off helped me to stay focused on what I was doing even if the world around me was going to hell in a handbasket. But it also made it easier for me to deny what I should I have confronted and dealt with.

On our return to Beirut, we were caught up in preparations for our marriage. Two months later, we got married. Our wedding was an elaborate affair with dinner at the Bristol Hotel, all of it arranged by Mrs. Elize. When my mother saw the menu, she became concerned that the guests were not going to have enough to eat. I was incredulous. Was food all that she could think of when I was about to make the worst mistake of my life, as surely she must have known? Perhaps she focused on the food because she could not focus on anything else. Of course, there was plenty of food and the guests seemed to enjoy themselves. I did not. I felt trapped. In the only wedding photograph that I have, Sylvia looks radiant in her magnificent flowing white gown that fills up the photograph. Standing next to her in my tuxedo, I look like a lost boy about to fall out of the frame.[6] Yet, I had not lost all hope that Sylvia and I could be happy together. Maybe being married would change our hearts and minds. Maybe the finality of our commitment to each other would bring us to our senses in how we related to each other. And no matter how unhappy I might be with Sylvia, I was sure that I would be spending the rest of my life with her if that's what God wanted me to do, which I still thought to be the case.

~

It might appear from this account that Sylvia's mother was responsible for our getting married, with its unfortunate consequences. Mrs. Elize certainly played a key role in averting the breakup of our engagement when she sent me to Zurich. However, placing the entire blame on her would not be fair. Sylvia and I should have stood our ground and done what we wanted to do. And I should have been the one to take the initiative to break our engagement. It would have been easier for me to walk away, since I would no longer have had to deal with Mrs. Elize whereas Sylvia had to live with her mother. I don't doubt that whatever Mrs. Elize did was well intended. Yet, when parents take it upon themselves to manage their children's lives even with the best possible motives, they still can do a lot of damage. That having been said, Sylvia and I bear full responsibility for our getting married and what we did in our marriage. We were neither passive spectators nor hapless children. However, even if I might consider Sylvia to have been primarily responsible for the deterioration of our marriage, I will focus here mainly on my own part in all this. To do otherwise would be difficult without putting her on trial. To a secular-minded person, the very idea that there is a God whose will must be ascertained as a guide to one's actions would sound implausible if not absurd. However, what matters here is that I believed that to be true, and it is in that context that my actions must be assessed.

There are two sharply contrasting ways in which I can try to explain my motivation for marrying Sylvia. The best construction I can place on my actions is that I was genuinely enthralled by what I had heard about the depth and sincerity of Sylvia's faith. Besides, her beauty, intelligence, talent, and privileged background were extra gifts from God. My desire to get married when quite young was predicated on my having adopted the moral code that restricts sexual and romantic relationships to one's spouse. Therefore, I wanted to put this aspect of my life in place and move on. Thus Sylvia fit into a script that was already in place. Viewed in that light, my actions do not appear to have been unreasonable. The problem was that my assumptions about Sylvia were based on skimpy and secondhand evidence. I should have used these assumptions as starting points in getting to know her personally and determining firsthand who she was, how I actually felt about her, and how she felt about me. Instead, I had already made up my mind by the time I met her. As for Sylvia's own conviction that I was God's choice for her, all I can say is that the way she treated me would seem inconsistent with such a belief.

The worst construction I can place on my motives for my marrying Sylvia is that I was interested in gaining access to Sylvia's personal and social assets and was playing the Christian card to win her over. Would I still have thought that it was God's will for me to marry Sylvia if she had been ugly, stupid, lacking in talent, and from a poor family? I would like to say yes, but I will have to say no. Given the presumed wealth and status of Sylvia's family, marrying her would have provided a shortcut to recovering Iskenderun. If that is what I did, then I allowed my personal desires and objectives to masquerade as the will of God, and God rightly punished me for it.

A secular and psychologically based alternative for explaining my motives for marrying Sylvia would point to my deep attachment to my mother. If I had been my mother's substitute for Ramzi, then Sylvia and other women I idealized might have been substitutes for my mother. Does that mean that my irrational and self-defeating actions were determined by the repressed desires of my childhood buried deep in my unconscious? Was I, like Oedipus, subject to a complex I didn't know existed and about which I could do nothing after the gods had sealed my fate? Transferring to Sylvia the unequivocal commitment and unquestioning obedience I had toward my mother turned me into a hapless child with no freedom of choice and no capability to break loose and pursue the wishes and desires of my own heart. There is also the cultural context in which these events took place fifty years ago. We lived under the heavy hand of convention. Family honor and personal reputations were at stake. Sylvia and I were pawns in a scripted game. Perhaps, none of this would have happened in today's world, even in Lebanon.

Whatever the reasons for my initial conviction that I should marry Sylvia, I also need to explain why I continued to cling to the conviction that it was the right thing to do despite mounting evidence to the contrary. When people began to comment on how my character had changed—how it had become so much more docile, subdued, and passive—why did I not listen to them? Would God be so cruel as to toy with my life? Or punish me so harshly for the pride of presuming to know what I wanted most in life? These questions troubled me, but I still let the fear of offending God cloud my judgment and paralyze my ability to act. As the reality of what I gotten into began to sink in, I wanted the whole thing to just go away. But it didn't go away. Moreover, the course of my relationship with Sylvia did not go downhill inexorably. It continued to have its ups and downs, its good times and bad times. Yet my feelings of regret persisted to the day of our wedding, by which time I knew I was making a grave mistake. Yet I felt I could do nothing about it. Having so ardently wished to marry Sylvia and having had my wish granted, I could not walk out on her without also walking out on God.

~

Sylvia and I went to Egypt for our honeymoon. We stayed at their old apartment in Alexandria, as I remember. Sylvia wanted to show me the remnants of her life there. We had dinner with one of her Egyptian school friends and her husband at their palatial mansion, which had seen better days. They were from the Egyptian elite whose circumstances had been greatly reduced by Nasser's nationalization policies (which also seriously affected Sylvia's family). We went on to Cairo, where I had an exciting time climbing to the top of the Great Pyramid of Khufu (which one was allowed to do at the time). What looked like steps from a distance turned out to be massive blocks one had to clamber over. That gave made me a greater appreciation of the enormity of the effort the ancient Egyptians exerted to lift those blocks into place. The view from the top was overwhelming. From Cairo, I went to Upper Egypt. Sylvia had been there and didn't want to come along. I went around gaping at the great monuments, with no idea of what they were about. Like others who live close to ancient sites, I had little appreciation of their history and significance. This wasn't the first time that I would be looking at the wonders of the ancient world with little comprehension. The remnants of the temple of Jupiter at Baalbek in Lebanon are the largest standing Roman ruins in the world. Yet, apart from filling me with awe by their size, they too held no meaning for me when I was growing up, nor was I taught anything about them in school. A big American car with shining chrome bumpers was more exciting than all that rubble. (Little did I know that fifty years later I would be lecturing on such ancient sites.)

I spent the whole day with an old Egyptian guide. He got a mummy case opened for me by bribing the guard at the site. Otherwise, I learned nothing from him. The following morning, when I went to take the train for the return trip to Cairo, the guide was waiting for me on the platform. Since I had already paid him well, I didn't know why he was there. He came over and launched into a panegyric about what an outstanding young man I was and how I surpassed in nobility all the English gentlemen that he, his father, and his father's father had guided in Upper Egypt for a century. I didn't believe a word of it, but it still sounded nice, and I thanked him profusely. As I was about to board the train, he took out a dirty handkerchief with a green scarab in it. He thrust the scarab into my hand, and holding onto it tightly, he told me in a hoarse whisper that it had been discovered by his grandfather in the tomb of the Pharaoh so-and-so and that he wanted me to have it. Taken aback, I tried to extricate myself, saying I couldn't possibly accept it. He looked at me—now on the verge of tears—and said my refusal would wound him to the depths of his heart. As the conductor began to call out for the passengers to board, I feared I was going to miss the train. On an impulse, I took out the largest bills from my wallet and pushed them into his hand, took the scarab, and jumped on the train. Quite moved and not a little relieved, I sat looking out at the desert as the train began to roll on. Then I took a closer look at the scarab—it was an obvious fake.

~

My final year in medical school and the first year of our marriage are a blur in my mind. As I rotated through various services on the wards, the pace became even faster than in the preceding year. When I got to surgery, it became frantic. We were supposed to be three interns on the service, but one got sick, so only two of us were left. That meant being on call every other night. The day started at seven in the morning, when we scrubbed for surgery, and it didn't end until twelve hours later. And when I was on call, it could go into the small hours of the morning. By the time I got home, all I wanted to do was to sleep. Even the brief respite of hurried meals at the hospital cafeteria felt like a blessing from heaven, and the nurses in their white uniforms looked like angels to my sleep-deprived eyes.

In all this frenzy, I barely saw Sylvia. Her parents had rented a furnished apartment for us and given us some money to live on. My father told me we could live with them but he couldn't help us beyond that. I could hardly blame him for it. Yet he said it in such a harsh tone

Figure 5.3 Graduation from medical school.

that it hurt my feelings. As I labored at the hospital, Sylvia spent most of her time at her parents' house. I have no recollection of our ever going shopping, cooking, or having dinner at home, let alone having friends over, which we didn't have in any case. We ate either at her or my parents' house (where the food was much better). In either case, they were cheerless occasions despite our attempts to be civil and keep up appearances. Between my marriage and the hospital, I felt like living under siege with my life put on hold.

~

I had a brief respite in the spring when I arranged to spend several weeks at the Asfouriyeh mental hospital. That was not part of our regular rotations, but I was able to arrange it since I was going into psychiatry and the director, Dr. Manoukian, was a family friend. It was neither an interesting nor an instructive experience but an opportunity to enjoy a relaxed pace of work and life for a short while. I didn't have much to do and spent much of my time idly roaming the grounds while lost in fantasies. I took to smoking Marlboros because it made me feel more grown up. One of the patients was a voluptuous young woman from a prominent family. She was a drug addict and began to pursue me, but I stayed away, having enough sense not to play with fire.

Our lives became further complicated with the outbreak of the first Lebanese civil war of 1958. We lived through anxious times, although we were never in real danger. Nonetheless, it was a harbinger of what eventually led to the second civil war, which exploded two decades later, in 1975, and would have a profound impact on our lives. The first civil war of 1958 started shortly after the unification of Egypt and Syria as the United Arab Republic under Gamal Abdel Nasser.

As a wave of pan-Arab nationalism swept through the region in the 1950s, some Lebanese factions wanted the country to join the United Arab Republic. The Maronites and other groups resisted, and the pro-Western president of Lebanon, Camille Chamoun, called on the United States, Britain, and France for help to save the independence of Lebanon. President Eisenhower responded by authorizing Operation Blue Bat on July 15, 1958—the first application of the Eisenhower Doctrine of intervening in countries threatened by Communism. Close to 14,000 American troops, including 6,000 Marines, were dispatched to Lebanon. We were told that the Marines never walk ashore—they jump into the surf and charge in full combat gear. When they did that near Beirut, they ran into a crowd of cheering, bikini-clad bathers and a horde of boys yelling, "Mister, mister, you want Chiclets?" (Over fifty years later, my wife Stina and I were having dinner at a Paris bistro with an American couple sitting next to us. It turned out that the man was one of the Marines who had charged the beach and then camped there until he was shipped back home.)

The presence of American troops quelled the conflict, and the Marines withdrew in October. Those were simpler days, and it was a nearly bloodless mission, with only four casualties: three American soldiers died in accidents and one was shot by a sniper. President Eisenhower's envoy, Ambassador Richard Murphy (a remarkable diplomat, whom I came to know on the AUB board), persuaded President Chamoun to resign in favor of General Fuad Chehab, the commander of the Lebanese army, which had maintained neutrality during the conflict. The country settled down with a sigh of relief, having narrowly avoided an all-out sectarian war.[7]

~

I graduated from medical school on June 23, 1958. Despite the enormous demands and distractions of my turbulent marriage, I had managed to do quite well when I would have been lucky just to make it through. Once again, it was largely due to my ability to compartmentalize my life and keep running on one leg while the other limped. I had also built enough momentum to keep me afloat. For reasons of security, commencement ceremonies were canceled. Instead, the medical school graduating class gathered, dressed in gowns with green hoods, in the garden of Marquand House, the residence of the president. There were no guests. As Dean McDonald gave his address, gunfire broke out in the neighborhood. He dismissed us, saying the medical school couldn't afford to lose an entire graduating class all at once.

I hurriedly made my way back to my parents' house to a reception where I was hugged and kissed by my relatives. There was the usual commotion as my mother and Lucine kept a stream of refreshments and hors d'oeuvres coming. It should have been a great day for me. I had worked hard and graduated with distinction, ranked second in my class, and elected to the Alpha Omega Alpha medical honor society. Yet, despite my putting on a good front, I felt sad and resigned to my fate. As I sipped a drink, my heart felt heavy under the burden of my marriage. This is not how things were supposed to have turned out. The future looked exciting professionally, but deeply troubling personally. My relationship with my parents had become strained. It was particularly hard on my mother to see my barely concealed unhappiness. My parents and Lucine were getting old. I worried about them, and they worried about me. I seemed to have lost control of my life, and there was nothing to be done about it. These circumstances made it even harder to contemplate our impending separation, when Sylvia and I would be gone for at least the next three years and live on the other side of the world during my residency training at Rochester, New York.

It was at a turning point in my life. An important decade had ended. It started by my going to college and ended with my becoming a physician. I began as a confused and troubled adolescent and grew into a confused and troubled married man. The next decade would see me complete my training as a psychiatrist and return to Beirut to teach at AUB and then emigrate to the United States to start my forty-year teaching career at Stanford, my divorce from Sylvia, my marriage to Stina, the birth of our two children, and the death of my father.

6

To the Land of Milk and Honey

MY MOTHER HAD A CRUSH on America. After her tragic experiences during World War I, she would have dearly loved to leave Turkey behind and emigrate to the United States. Finally, she got her wish years later. My own feelings for America were shaped by my mother's love for it. Long before I even knew where exactly America was, I thought of it as the Promised Land—the land of milk and honey.

A friend of my mother's who lived in the United States had given her a gift subscription to the *Ladies Home Journal*. I was about twelve years old and could not read English, but I looked in wonderment at the pictures. My favorite was the ad for Life Savers with its colored circles rolling out of the package in a dreamlike cascade. When someone gave me my first glass marble, I was amazed at the swirl of color at its center. I couldn't imagine how it was put inside of the tiny sphere. When I was told that it was made in America that explained it: they could do anything in America!

The idea of going to the United States became more concrete after I decided to become a psychiatrist. The psychiatrists at the Asfouriyeh mental hospital had gone to Britain to be trained. Normally, I would have done the same and then gone back to work at Asfouriyeh or taken up private practice. However, AUB wanted to establish its own psychiatric service and its own faculty trained in the United States. To that end, John Racy was handpicked to be the first faculty member. John was an excellent student who had attended the International College and then AUB and graduated from medical school two years ahead of me. For his residency training, John went to the Department of Psychiatry at the University of Rochester medical school in upstate New York. He did so at the suggestion of Dr. Virgil Scott, who was a graduate of Rochester medical school and knew Dr. John Romano, chairman of the Department of Psychiatry. Two years later, when I was trying to decide where to go for my residency training, I wrote to John Racy. He wrote back telling me about the department and suggested that I come to Rochester for my residency training. It was an excellent choice and a crucial step in the development of my career. Rochester proved to be such a good fit that I cannot imagine myself having gone anywhere else. John was to play another crucial role when he helped me four years later to join the AUB faculty and we became lifelong friends.

I needed a letter of recommendation from an American professor and went to see Dr. Calvin Plimpton, chairman of internal medicine (and president of AUB twenty-six years later). Cal Plimpton was a Boston Brahmin with a fixed, skeptical expression. He asked why I wanted to become a psychiatrist. I told him because I liked to be intimately involved with my patients. He arched his eyebrows and said that the psychiatrists he knew were the physicians least intimately involved with their patients. This rejoinder took the wind out of my sails, but he must anyway have written a good letter on my behalf (and John Racy must have put in a good word), since shortly thereafter I was accepted into the Rochester residency program.

The prospect of going to the United States sustained me during the difficult last year of medical school. I was reconciled to being married to Sylvia for the rest of my life. But I hoped that being on our own and away from our parents would bring us closer. Another important consideration was that the Eastman School of Music was at Rochester and Sylvia could continue her music education there. She was pleased about that. And I was glad to help her advance her career and hoped that it would make up for her loss of Zurich. I think this boosted my standing in her eyes more than anything else.

I decided to leave for Rochester ahead of Sylvia. Ostensibly, I wanted to have a chance to find a place for us to live and make things ready for her. Actually, I didn't want to have Sylvia to worry about as I moved into my new life. Residency training was a huge step for me professionally, and I wanted to start it with my right foot forward. It didn't occur to me, nor did anyone suggest, that Sylvia could have actually helped me settle down. At any rate, she agreed to join me a few months later, probably welcoming the prospect of being free of me for a while herself.

Leaving my family was very hard. My parents and Lucine were getting old. When would I see them again? Who was going to help them if they needed anything or got sick? They were in turn worried about me. How was I going to manage on my own? Was Sylvia going to stand by me or become a thorn in my flesh? I asked my mother not to make things harder by having an elaborate farewell party. Nonetheless, a dozen or so of our closest relatives showed up at the airport. As they lined up to say goodbye, I first hugged my parents and then Lucine, Elvira, Nora, and their husbands, and finally came to the end of the line. Until then, we had managed to keep our composure. The last person to shake my hand was Mr. Nuri, the husband of one of my cousins. He was the most distant of my relatives present, yet there he stood, tears streaming down his face. I broke down and began to cry, too, as did everyone else. I don't remember Sylvia or her family seeing me off, although I can't imagine their not being at the airport.

~

I left for London to spend the night before flying on to the United States. My flight from London to New York was due to leave in the afternoon, so I had some time to walk around. I was dazzled and overwhelmed by the great city. There was a ponderous elegance to its buildings and a striking formality in the way men in the central district were dressed and carried themselves (the women didn't make as much of an impression). There was no time to visit any of the famous historical sites. Besides, I was on a special quest. I had seen a pair of English-made shoes in Beirut with a special copper sheen that I became enamored with. I had been unable to buy them at the time, and if I were ever to find them again it would be in London. I went from one shoe store to another, lugging my suitcase along, and saw many fabulous shoes, but not the particular pair I was looking for. One exasperated salesman told me he could not even understand what I was looking for. I suppose it was the Perfect Shoe that didn't exist.

During this fruitless quest, I ambled by the Dunhill store, with its amazing display of pipes and cigars in the window. Although I didn't smoke, I was intrigued enough to go in. A formally dressed man with graying hair, distinguished enough to be the Speaker of Parliament, approached me and asked with measured politeness, "May I help you, sir?" I felt compelled to respond, so I said, "Do you have any pipe tobacco?" He discreetly glanced at the mounds of pipe tobacco piled on the tables, bowed his head slightly, and replied in his clipped accent, "Yes, sir. We have fifty-seven kinds. Which one would you like?"

This was the second of my pipe humiliations. During our honeymoon in Cairo, while waiting for Sylvia in the lobby of the Hilton, I noticed a tobacco shop. I went in to look at the pipes and asked the man hovering over me if smoking a pipe was complicated. No, not at all. There was pre-packed tobacco in cellophane wraps; all I had to do was plug one into the bowl and light it up. I thought a pipe would make me look more sophisticated, so I bought one with pre-packaged tobacco and a disposable lighter. I went back to my leather armchair and did what the man told me. But when I inhaled the smoke, it felt awful. I thought I could instead gently blow into the pipe and let the smoke curl out. It worked. I began to blow harder until the cellophane caught fire and the glowing tobacco flakes scattered all over. People turned around wondering what I was up to. I put down the smoldering pipe on the ashtray with the pre-packaged tobacco and the lighter on its side and walked away.

By mid-afternoon, I had given up my quest for the mythical shoes and looked for a taxi to go to the airport. To my horror, it was rush hour and there were no taxis to be had. I panicked: what if I missed my flight? I couldn't begin to imagine the consequences if that were to happen. In desperation, I started toward a policeman for help when a black London taxicab pulled up to the curb. I threw in my suitcase, jumped in after it, and leaned back with a sigh of deep relief.

The propeller-driven Pan Am plane took forever to cross the Atlantic. I remember reading in the in-flight magazine what travel was going to be like when commercial jets would come into service in the near future: the glorious sunsets at 30,000 feet, the thrill of flying on a magic carpet, the gourmet food, and so on. It all sounded so wonderful. Maybe I too would be on one of those flights one day.[1] We finally landed at La Guardia. I went into the terminal bleary-eyed looking for my connecting flight to Rochester. As I looked at the milling crowd, I was puzzled. These people didn't look like Americans; no one wore rimless glasses, and they spoke English with a funny accent I had not heard before. Had a large plane just landed and disgorged its foreign human cargo?

On the flight to Rochester, I wondered why so many of the passengers were dressed in short-sleeved shirts. Wasn't Rochester supposed to be very cold? John Racy who had also first gone there with that same expectation had walked out of the plane into the muggy June afternoon with his heavy coat, scarf, hat, and gloves at the ready. A short taxi ride took me from the Rochester airport to the Strong Memorial Hospital. It was a Sunday, and there were few people around. I found my way to the residents' quarters, where a room had been reserved for me. I unpacked and spent the rest of the day walking on the hospital grounds and in nearby Genesee Park. Everything was so lush and green. I felt happy to be spending the next three years there.

I returned to my room from a new direction and came to a swinging door with a sign that read "This door must be closed at all times." To me that meant it should never be opened, so I went out and around the building to get to my room. I did this for several days, but since everyone else was going through the swinging doors, I began to follow suit. I realized that in America words did not always mean what they said.

~

Bright and early on Monday morning, I presented myself at Dr. John Romano's office, in the Department of Psychiatry. His long-time secretary, neatly dressed and coiffed, greeted me with guarded cordiality. (I was to learn that she was the only person in the department who dared to talk back to him.) She asked me to take a seat, and a few minutes later a tall and imposing man came out of the adjoining office. He introduced himself, we shook hands, and he ushered me into his office. I sat down and glanced at the wall facing his desk. It was covered with framed diplomas, certificates, licenses, and commendations; it was all Romano-Romano-Romano. I tried not to gape at them, so I turned my attention back to Dr. Romano. I noted his deliberate

Figure 6.1 John Romano.

movements, his slightly drooping right eyelid, and the mild tremor of his left hand that he seemed to display rather than suppress. Romano had an imposing Roman nose. He was dressed in a moss green tweed jacket with complementary gray slacks under a white doctor's coat. His lighthearted conversation was finely balanced with a touch of gravitas. The man was clearly the master of his universe (no wonder they called him Lorenzo the Magnificent). I had heard that he was authoritarian and people either loved or hated him. As the son of an authoritarian father, I would know how to live with him and be fully prepared to love him as I hoped he would love me. John Romano was to become my first mentor, career model, and lifelong friend—the most important influence in shaping the contours of my new life. I hesitate to say this, but he became like a father to me.

In Lebanon we wore suits, and I had brought several of them with me. I soon discovered that one did not wear suits at Rochester except on formal occasions, but sports coats and slacks. Romano's sports coats were all in shades of green. On the next weekend, I went downtown to buy a sports jacket and a pair of slacks. I picked a tweed jacket—in moss green. It became like a uniform that I wore for many years, eventually having leather patches stitched to its threadbare elbows. I finally gave it away after fifty years and came to regret parting with it.

~

The three years I spend at Rochester were to be one of the most significant and transformative periods of my life professionally and personally. Although within a decade I would no longer be actively practicing psychiatry, the modes of thinking and relating to people that were inculcated in me by my training remained an integral part of my professional identity. This was an equally critical period of personal maturation during which I became emotionally more independent of my parents and free of my bondage to Sylvia. It also entailed my integration into American society, which eventually led to my becoming a part of its culture.

The first two months before Sylvia joined me at Rochester were an exhilarating time. I lived at the residents' quarters, ate at the hospital cafeteria, and had no one to worry about but myself. I could totally immerse myself in my work as well as plunge headlong into life in America. I corresponded sporadically with Sylvia and my parents, but tried not to think too much about them or the world I had left behind. This was the happiest I had ever been since I was a child.

John Racy was one of the three chief residents, and I was happy to have a friend who knew the ropes and would help me find my bearings. My chief resident on the ward was Elmer Gardner, who took me under his wing. Elmer had served in the army and gone to medical school on veterans' benefits. He was quite a bit older, was married, and had several children. He proved to be a patient and effective mentor. A few weeks after I arrived, I was pleased and flattered to be invited to dinner at his house, in one of the small towns bordering Rochester. Elmer gave me clear directions, but I still felt apprehensive, knowing my dismal sense of direction. Soon after leaving the boundaries of the city, I knew I was lost. I worried about being late. That would be rude, but also I had learned already that there was no such thing as making a mistake; all mistakes were "over-determined" and had deeper symbolic significance. I began to wonder how my being late would be interpreted. As the lights of houses on the sides of the road grew less frequent, I drove faster and faster, getting more and more lost. To light up the gloomy darkness looming ahead, I switched on my headlights, and a sign came into view. It read "Egypt." I knew I was lost, but *Egypt*? As I passed the sign, it said "Population 124" and I breathed a sigh of relief. I was reminded again that in this country words don't always mean what they say. Actually I was not that badly lost and soon found Elmer's house. As I apologized profusely, Elmer put a cold beer in my hand, and he and his pretty wife made me feel at home. I told them about my encounter with "Egypt." They thought it was hilarious, but I was too shaken to join in the laughter.

I was warmly welcomed by my fellow residents. I attended my first clambake and loved the taste of the fresh corn on the cob (and ate at least four of them). Then there was a lamb roast, when my friends placed a cowboy hat on my head and put me to work turning the spit. Warren Seides took me to the Tanglewood music festival in Massachusetts over a weekend. I didn't realize how far it was (almost like going from one end of Lebanon to the other) and was surprised that one would drive all that distance just to attend a concert. However, it was worth it, and during the long ride I learned much from Warren about growing up in New York City. The music played by the Boston Pops was interesting, but I was more fascinated by the crowd sitting on the grass listening to it.

I spent a lot of time with my single friends, especially Ned Freeman, whose ebullient mood always cheered me up. A special treat was having a ham and cheese sandwich at *Don and Bob's* after work. When Ned began to date Arden, who worked at the hospital, she would often come along and I had a chance to observe an American courtship from up close. One weekend, several residents from internal medicine took me to Niagara Falls. I thought waterfalls would be in the wilderness, not next to Buffalo. After we viewed the awesome thundering spectacle from the American side, my companions crossed the bridge to view the falls from the Canadian side. Since I was not a U.S. citizen, I stayed behind. As I waited in the immigration hall, I noticed the American flag next to the Canadian flag in the middle of the bridge to Canada. Assuming that to be the border, I walked up to the American flag and came back. The U.S. immigration officer at the checkpoint asked me where I was born. I said "Iskenderun." He asked me for my passport. Why? He said I had crossed into Canada. But I had not crossed into Canada. I had gone to the *middle* of the bridge *up to* the *American* flag and turned back. My explanation was too outlandish to be a lie, but the officer called the Strong Memorial Hospital anyway and verified that I worked there. He let me go with a stern admonition to stay off the bridge next time and forget the flags.

When another fellow resident, Cliff Reifler, and his wife, Barbara, drove to Chicago to visit her parents, they took me along. I sat in the back of the car and played with their two little girls by pouring milk into our palms and drinking from it. One evening, Cliff took me to my first burlesque show. We sat at the bar and one of the scantily dressed women came to sit next to me. As we exchanged a few words, she picked up my accent and asked where I was from. When I told her I was from Lebanon, she was ecstatic—her parents were from Syria! I was like a long lost cousin. She wanted to take me to a room backstage. I regretfully declined, and she was crestfallen. On another trip, George Nesbitt took me to the oyster bar at the Grand Central Station. I had never had oysters, and I found their slippery feel in my mouth disgusting. (I like them now.) George introduced me to his parents, who lived in New York, where his father was a prominent clergyman. I was to see more of George and his wife, Maureen, in Washington DC two years later.

My next trip to New York was to attend one of Jacob Moreno's psychodrama workshops that was in vogue in the heady days of the '60s. Warren Seides arranged for me to stay at the Presbyterian Hospital dormitory on West 168th Street. I drove to New York in my pockmarked green 1954 Ford with only a vague idea of where I was going. No sooner had I entered New York when I was hopelessly lost. I stopped at a bridge to ask a burly policeman for directions. He glared at me and said, "You have thirty seconds to get out of here." He blew his whistle to stop the traffic as I swiftly turned around and drove away. By the time I finally managed to find the dormitory, it was well after midnight. There was nowhere to park. I ended up finding a spot next to the Hudson within sight of the George Washington Bridge. I got to bed at three o'clock in the morning. The next day, I moved to a cheap hotel downtown and went on an orgy of sightseeing. In a single week, I saw five museums, three shows, and half a dozen sights such as the Empire State Building. When I went to bed, my legs would be aching from the endless walking. One very cold night I got stranded in Greenwich Village without a coat. I could not find a cab, and had to go into a bar to get warm. I ended up spending the night in a small hotel across the street. I had no luggage and had to pay cash before being let in. It did not occur to me to ask the barman to call a cab.

The trip to Moreno's workshop was sponsored by the department, and on my return to Rochester I gave a presentation on psychodrama. I described the stage and how Moreno conducted his workshop. Among other trivia, I pointed out that Moreno sat on the second of three steps leading to the stage. Dr. Romano, true to form, wanted to know why the *second* step. I had no idea and replied, "So he could rest his arm on the third step." As everyone burst out laughing, Dr. Romano said, "You ask a stupid question and you get a stupid answer." After the meeting, George Mizner gleefully told me that this was the first time that someone had made John Romano play second banana. *Second banana?* What could that possibly mean? I laughed anyway.

~

My life was exciting and felt wonderful after the stress and drudgery of medical school and despite the clouds over my marriage. I had spread my wings in a free new world in ways that would have been unimaginable to me a few months earlier. However, Sylvia was on her way to join me. She was coming by sea and would be arriving in a few weeks. Her ship was to dock at Hoboken, New Jersey. I found the place fairly easily despite its strange name. I waited for Sylvia to disembark and was surprisingly happy to see her. We spent the night in a hotel in New York and drove to Rochester the next day. The trip gave us a chance to catch up, and for me to tell her about the new life ahead of us. I had rented a modestly furnished one-bedroom house on Genesee Street, which was within walking distance of the hospital. I was relieved that Sylvia found the house adequate and she seemed happy. To her credit, she was never to complain about our strained circumstances. The fact that we were now alone and far from our parents seemed to make us more at ease with each other. I became hopeful that our problems were perhaps now behind us.

A few days later, I took Sylvia to meet Dr. Romano. He was most gracious to her. I had told him about Sylvia's interest in attending the Eastman School of Music. When I brought it up again, he called his friend Howard Hansen, who was the head of Eastman, and Sylvia was admitted on the spot. I knew she had the voice and the talent to study at one of the top music schools in the country, but I was not sure if she had the paper credentials for admission. That, however, did not seem to matter. This was not the first, nor the last, time that John Romano would go to bat for me. Sylvia was pleased in getting into Eastman and wholeheartedly threw herself into her studies. I was proud of her. She was not only taking voice lessons but studying theory and other complex subjects as well. (I didn't realize at the time how hard it was for her to take the bus to school and get there early enough so she could practice on the piano.) My friends and their wives welcomed Sylvia as cordially as they had welcomed me. She was quite different from the other wives, who worked or took care of small children instead of going to school. Nonetheless, Sylvia fitted quite easily into my circle of friends. The fact that she was my wife was important, but her self-confidence and social graces also helped.

The rest of that first year was uneventful, and I remember little of it. I suppose we did more or less what we thought young couples did in America. We were both busy and happy with our work. We kept in touch with our families but never wrote to each other's parents. Otherwise, we didn't have much to do with each other except over the weekends. We had a small black-and-white TV set, and I recall watching Jack Paar in the evenings. I don't recall our going to movies or concerts. We went to a neighborhood church one Sunday and were startled to find that the entire congregation was black. We must have been a curious sight, for after the service the wife of the minister, who was white, caught up with us in the street to make sure that we felt welcomed. We never went back, nor did we become a part of any other community outside of the Department of Psychiatry. I don't remember ever meeting any of Sylvia's classmates or friends from Eastman.

I am still mystified by how we lived from day to day. I don't remember either of us shopping, cooking, cleaning, or doing the laundry. We obviously must have done so since we no longer had our parents to take care of us. I often ate my meals at the hospital cafeteria; I assume Sylvia did the same at her school and probably studied there as well. On most evenings, when we got home, we already had had supper. I recall occasionally our eating out in a modest restaurant. On one occasion, we were having a hamburger or sandwich with another couple. I ordered a coke and Sylvia asked for warm milk. The waitress looked baffled and I thought Sylvia was trying to embarrass or irritate me but which might have been a reflection of my own frame of mind regarding her.

One reason we didn't go out much was because we didn't have the money. My starting salary was $1,100 for the entire first year (in 1958 dollars). I occasionally worked for pocket money at the hospital pharmacy filling out prescriptions in the evenings. Since I had never learned how to type, making the labels one letter at a time was the most difficult part. The other residents, who were American citizens, had fellowships from the National Institutes of Health, which paid $7,000 a year, and some of them lived in university housing (which I knew nothing about). I wished Dr. Romano had supplemented my salary, but he may have thought that we came from affluent families, since we might have looked as if we did. The fact that I had to get his permission for working at the pharmacy should have suggested to him that our situation was otherwise, but I assume he thought I should work for my keep as he had done when he was in my position. Sylvia's parents must have paid her tuition, since I don't remember doing so myself and could have hardly afforded it. On one occasion, her father sent us a few hundred dollars, which I refused. He was apparently well impressed. This was the only occasion I know when he had something positive to say about me. Had Sylvia's parents settled a substantial amount of money on us when we got married, I would have been happy to accept it, but I didn't want handouts. Moreover, their declining fortunes in Egypt probably made it no longer possible even if they had been inclined to be generous.

Our financial prospects were another issue that was never discussed openly. When I set out to marry Sylvia, I had no means of supporting myself, let alone supporting a wife. Neither of our parents sat down with us to discuss how we were going to live. No one wanted to take the initiative to discuss it so as not to become stuck with the responsibility. The one exception was the lavish trousseau. A seamstress worked on it all summer. Unfortunately, Sylvia was marrying a medical student and not the Shah of Iran. There was no opportunity for her to wear even a single one of her sumptuously embroidered gowns. The money would have been better spent for us to live on. I couldn't bring myself to ask for money from my parents. I knew that they had limited means, and my father was working into his seventies. At this stage in their lives, I should have been taking care of my parents instead of having them take care of me. In a deeply satisfying gesture, I sent my father my first paycheck as a token of appreciation for all he had done for me. I got a letter from Lucine—a rare event—saying she had found my father at home crying. He had shown her my letter and told her his tears were tears of joy.

~

Residency training was mainly an apprenticeship. The first year was spent on the inpatient wards. For the daily management of patients the chief resident was the primary supervisor, model, and teacher. He reported to the faculty clinical director, with whom we had weekly meetings when a resident or a medical student presented a case from the ward. The faculty member would then interview the patient and discuss it. The star of these presentations was Dr. Sidney Rubin, a senior psychoanalyst who was a master interviewer with a gift for making sense of the underlying psychodynamics of the case.

Then there were the grand rounds chaired by Dr. Romano. These were open to all members of the department and featured cases of particular interest. On one occasion, the discussant was Dr. Rubin. He offered a stock psychoanalytic interpretation of the patient's illness.

Romano took him to task saying one could have read what he said in Fenichel's standard psychoanalytic text. He wanted to know Rubin's own views based on his personal clinical experiences. Sid Rubin turned red and said, "My personal experiences?" It was the standard tactic of answering a question with a question, but it didn't work in this case. It was an awkward moment, and I felt torn between my loyalty and affection for both men—a bit like watching one's parents quarrel.

The three psychiatric wards (Y2, R3, and R4) had a different mix of patients, and rotating through them gave us a broad range of clinical experience. I started on R3. Before the era of effective antidepressant drugs, severely depressed patients were routinely treated with electroconvulsive therapy (ECT). Typically, these patients stayed about three weeks in the hospital, which was the period insurance policies covered. After the first week of shock therapy, the patient's sleep pattern began to improve and the depression to lift. It seemed to work like clockwork.

For outpatient supervision we had a faculty tutor with whom we met weekly. These patients were less severely ill, and we treated them with some form of psychotherapy. My tutor was Dr. Bill Hart, who also supervised George Mizner. George had been a first-year pediatrics resident before switching to psychiatry and had more clinical experience than I had. It took me some time to find my bearings, but Bill Hart was a kind and patient tutor who helped me along. One of the first lessons we had learned in psychotherapy was to let the patient talk without interruption. The therapist intervened only when necessary, to clarify an issue or to offer an interpretation. No matter how awkward the silence, you had to let the patient sweat it out. On my first intake interview, I carried this principle to an extreme. I was seeing a new patient while Bill Hart and George observed from the other side of the one-way mirror. The patient was a glum and taciturn man in his early forties. After we sat down, I nodded to have him start talking (I may have also said, "Yes?" or something similar). He didn't respond. I looked at him expectantly, but he still said nothing. I waited, but there was only silence. We seemed locked in a contest of wills. I thought to myself, if he won't talk, neither will I. This went on for half an hour, after which the patient got up and left. Clearly, that was not the way to go, but I caught on fast.

I seemed to have a flair for putting patients at ease. They trusted and opened up to me. Listening with a "third ear," I became adept at understanding what was behind their thoughts and feelings. I had been a good observer of people, and now I was able to attach labels to what I was observing. If being a good therapist depended as much on what kind of person the therapist is as on knowing about therapeutic technique, then I was ahead in the game.

There was relatively little by way of didactic work. We had a psychiatric literature seminar with assigned readings. Bill Hart urged us to read as many of Freud's papers as we could, but I never got around doing it. Psychoanalytic theory was the dominant theoretical orientation, and the ideology we absorbed from our mentors. I soaked it up avidly and became fascinated with Freud. His overarching theory encompassed the whole range of the normal and the abnormal, the psychological and the cultural. Its all-embracing reach appealed to my passion for comprehensiveness and order. It explained everything (hence, perhaps nothing). In a sense, it was more like a religion that required a lot of faith, but that was not a problem for me. Orthodox psychoanalysis (especially at the New York Institute, with which Rochester was affiliated) had a reputation for being dogmatic: if Freud said it, then it had to be right. This did not especially trouble me, since I preferred certitude over ambiguity. I especially liked the way Freud wrote, and it must have been wonderful hearing him lecture. I could almost hear him talk while reading him. Freud's view of the world, with his mixture of compassion and clear-eyed realism (if not cynicism) about human nature, appealed to me because it confirmed what I already sensed. I was vaguely aware of Freud's critical views on religion, but since these were not an essential part of his clinical work, I was not particularly troubled by them.

Psychoanalysis was in its heyday in the late 1950s, when I came to Rochester. Although it had started in Europe, the psychoanalytic movement spread and reached its apogee in America. Sigmund Freud made only one visit to the United States. He was invited in 1907 to Clark University in Worcester, Massachusetts, by G. Stanley Hall (president and eminent psychologist) to receive an honorary degree and to give a series of seminal lectures. It was an important occasion for

Freud and the first major public recognition of his work. Nonetheless, he didn't like America and never came back.

The American Psychoanalytic Association and the New York Psychoanalytic Society were both founded in 1911. From then on, the psychoanalytic movement grew steadily but it was not until World War II, when mostly Jewish psychoanalysts fled from central Europe, that psychoanalysis came to dominate American psychiatry. Consequently, virtually all major academic departments came to be headed by practicing psychoanalysts. Rochester was an exception. John Romano had completed his didactic analysis in Boston and attended the seminars at the Boston Psychoanalytic Institute. He thought he had learned a great deal and profited particularly from his association with a number of distinguished psychoanalysts. However, he could not complete his work with patients before he left Boston to chair the Department of Psychiatry at the University of Cincinnati. On coming to Rochester, Romano wrote to the Boston Psychoanalytic Institute and requested permission to complete his training with Dr. Sandor Feldman, the training analyst who was already at Rochester. His request was rejected in a high-handed manner. It strained Romano's relationship with psychoanalytic institutes, and he became dismissive about how much he had learned from analysis.[2] He remained critical of some aspects of psychoanalytic theory and practice, and his critical outlook contributed further to his being viewed as a renegade, even though much of what he took issue with is now more widely accepted. And he remained unfailingly supportive of his faculty's pursuing their own psychoanalytic training.[3]

Given the enormous influence that John Romano had on me professionally and personally, I will say a few more words about him. Romano was born in 1908 in Milwaukee, Wisconsin, to an Italian immigrant family. His father was a violin teacher, and Romano never forgot his modest roots ("My cup runneth over," he would say). He went to Marquette University and entered medical school at the age of twenty. To support himself and help his family he simultaneously worked at a variety of jobs, including caring for an ailing Jesuit priest as a private night nurse, besides managing the medical school store.[4]

During his internship in Milwaukee, Romano decided to go into psychiatry despite its benighted reputation. But he couldn't get into a residency training program. Instead, he got accepted into a second year of internship at Yale. While in medical school, John had been courting Miriam, who was still in college. After she graduated, they got married in the depth of the Depression and moved to New Haven. Miriam was the daughter of the head of the YMCA, so her family was a notch higher than Romano's. I remember Mrs. Romano as a tall, gracious, and formal lady with distinctive blue-framed glasses. John was respectful and fond of her as he was of their son David, who became a classicist.

The newlywed couple couldn't live on John's meager salary and Miriam was unable find a job, so she went back to Milwaukee to work for the rest of the year. John finally obtained a position at Colorado as a resident with a Commonwealth Fund Fellowship. The following year, they moved to Boston, where John met Soma Weiss, the Hersey Professor at the Peter Bent Brigham Hospital—the most prestigious medical chair at Harvard. Weiss invited Romano to join him as a full-time member of the faculty. Romano often referred to Soma Weiss in almost reverential terms. As a mentor and role model, Weiss was to Romano what Romano was to become to me.

In June 1942, John Romano, at age thirty-three, became chairman of the Department of Psychiatry at the University of Cincinnati School of Medicine. Alan Gregg, of the Rockefeller Foundation, who had been supportive of John's career, cautioned him that it would be too soon for him to take on the administrative responsibilities of running a department. It would stunt his professional growth as a researcher and a clinician. John accepted the position anyway.

Three years later, Romano was recruited by George Whipple, the eminent Nobel Laureate at the University of Rochester Medical School, to head the newly established Department of Psychiatry. The position offered broad opportunities, including the building of a new wing funded by the Rivas family (hence called the "R" wing for Mrs. Rivas, not named after Romano, although it was easy to think so). John devoted the rest of his professional life to building the department to a high level of excellence. He played a key role in educating generations of psychiatrists and medical students and in promoting a bio-psycho-social model of medicine with his colleague George Engel.

Figure 6.2 Speaking at the dinner of the department's twenty-fifth anniversary.

Romano was insistent that psychiatric theory and research be firmly based on clinical experience with patients within a wider medical framework, not abstract speculation, into which psychoanalysts sometimes lapsed. He always wore a white doctor's coat whenever he was going to interact with a patient. Romano's dedication to administrative and teaching commitments got in the way of his own research and scholarship, just as Alan Gregg had predicted. At the celebration of the twenty-fifth anniversary of the founding of the department, I went to Rochester from Stanford to be one of the speakers at the banquet. I turned to John and told him to look around—we were his books. To hear this must have been deeply satisfying to him, but it may also have been difficult. My former fellow resident John Grinols told me I would never know how good my talk was.[5]

John Romano's influence on me went far beyond my training as a psychiatrist. At the time, I was aware of his impact on me, but it was not until years later that I realized the full extent to which he had shaped my professional identity and intellectual outlook. My dedication to teaching and mentoring the young, the importance I placed on maintaining a multidisciplinary perspective, on being a generalist rather than a narrow specialist, and on aspiring to become a broadly educated and cultured person could all be traced back to him. Like Romano, I saw myself as someone who took care of people and managed them. That originally meant treating patients, then teaching students, being a university administrator, and finally addressing the social problems of the world through philanthropy. He instilled in me the importance of staying close to those one is trying to help, to listen to them and to gauge their needs in their own terms and not one's own. Ultimately, one had to be a broadly informed and intellectually engaged person—"broadly civilized" was the term he loved. By contrast, as John Racy recalled, Romano's disapproval of people came at three levels: "limited" meant not very good; "pretentious" was bad; "mountebank," the worst.[6]

I learned from Romano to listen to others but not run things by consensus. Someone had to bite the bullet and take charge. This didn't endear him, or me, to everyone, but that never seemed to bother either of us. It allowed me in later years to step into positions of responsibility where the tasks were unclear and the lines of authority hazy, and make it work. John was not easily awed by people (nor am I). This didn't mean being arrogant, but being aware that people are flawed. One may be great in one respect and not so great in another. A Nobel Laureate may be a genius, but a lousy parent. Picasso was arguably the greatest artist of the twentieth century, but he treated women badly. One needs to have humility, not hubris. One of John Romano's favorite phrases was "Never underestimate people's ability to act irrationally." Adults are not children, but when they act childishly they have to be treated with a mixture of kindness and firmness. One can be paternal without being paternalistic.

Mentors come in various forms. Some turn their followers into disciples as extensions of themselves. Others limit their influence to the professional or personal lives of the young, but not both. I learned from Romano how to integrate my professional responsibilities to the young with my personal bond with them. Like Romano, some of my students became lifelong friends, and helping them to become the best they could be a central aim of my own professional and personal life. And rather than stamping my influence on them, I tried to present to them alternatives derived from my own experiences that they could accept or reject according to their own needs. (But I may have ended up leaving my stamp on some of them anyway.)

The reactions of people to Romano were not always positive. He could come across as authoritarian and did not take kindly to being challenged. While generous, kind, and good humored, he could also be harsh and intimidating. Once he had formed an opinion about a person or an issue, it was hard to change his mind. He could be unforgiving when someone crossed a line (which he drew). He occupied center stage as a matter of course. No wonder his nicknames included "The Great White Father" and "The Greatest Roman of Them All."[7] At the start of a hospital meeting, a student had pointed to a chair and told him to please sit at the head of the table. Romano had replied, "Son, where I sit is the head of the table." Notwithstanding his many accomplishments, Romano spread himself too thin, thereby making his contributions more ephemeral for those who were beyond his personal orbit. Some of Romano's more problematic characteristics may have also rubbed off on me.

Like all good mentors, John Romano not only became my lifelong advisor but also exerted an enormous influence in furthering my career. I never made an important move without consulting him first. He gave a huge boost to my self-confidence and provided crucial help at critical stages of my career. I was one of Romano's boys, and that made a huge difference. John was instrumental in my going to the National Institute of Mental Health, where I met David Hamburg. He too was to make an enormous impact on my career and reinforce some of the same intellectual perspectives, besides providing critical help. He essentially took over where Romano left off. I will have much more to say about him further on.

~

There were others who also made a significant impact on my career at Rochester. George Engel could have been one of them but I did not have a chance to work closely with him. He was the bright light in psychosomatic research and, next to Romano, the leading figure in the department. Having been trained as both an internist and a psychoanalyst (without going through a psychiatric residency), Engel had an unusual background. His research team was engaged in groundbreaking investigations in psychosomatic medicine focusing on the interaction of psychological factors and physical illnesses. However, this area was peripheral to our training except for those residents who had a particular interest in it: I was not one of them. Most of us were there to become clinicians, not investigators or scholars.

Engel was assigned to be my tutor in the second year, but another resident who was unhappy with his tutorial assignment had us switched. I ended up with a tutor who was clinically depressed (and committed suicide in his office the following year). Nevertheless, Engel was such a central presence at Rochester that it was impossible to escape his influence. He had a great sense of humor and told many stories about the tricks he and his identical twin played on people. For instance, one of them would go to have a haircut and complain that his hair grew so fast it had to be cut every week. The barber would laugh and say should that happen he could have the second haircut free. The following week the twin would go and get the free haircut. When presenting a paper at a conference, Engel would place blank sheets of paper in his text. As he turned over the blank pages, the audience would be pleased to note that he was cutting his talk short. A year later, I ran into George's twin on coming out of an elevator and there he was standing at the lobby. His resemblance to George was uncanny, down to his mannerisms. The only way I knew it wasn't George was that his brother looked at me and didn't know who I was. After working closely for decades there was a falling out between Romano and Engel, which was both baffling and distressing. No one seemed to know exactly what went wrong. During the last decade of his chairmanship, Romano seems to have become more difficult to deal with. But if I were forced to choose sides, I would have definitely been on Romano's. So far I was concerned, he could do no wrong.

~

During my second year of residency training, my professional and personal lives began to diverge more sharply. As my work became increasingly more rewarding, the problems in my marriage got progressively worse. Even if Sylvia and I were not entirely happy during our first year, we were at least not overtly unhappy. Perhaps we suppressed our feelings, since expressing them would have brought out our unhappiness to the surface. Absorbed in our work, we led parallel lives—with me at the hospital and Sylvia at Eastman. Neither of us knew much about what the other was doing, nor did we seem to particularly care. We shared a house and a bed but were hardly intimate at a deeper level. Perhaps that was the best we could do to keep the peace. However, we could not maintain this precarious balance indefinitely, and things got progressively worse as our unhappiness surfaced into the open during the second year.

Although I have written a fair amount about Sylvia, she may still seem opaque and does not come through as a real person. Instead, there is a sense of emptiness about who she actually was. Some readers have felt frustrated by the lack of concrete facts beyond my feelings about Sylvia.[8] What did she actually *do* that brought about my disenchantment with her? The reason I have not provided more concrete details about Sylvia's character and behavior is that it will be difficult to do that without putting her on trial and in a bad light. Moreover, the reason I cannot give a fuller picture of Sylvia as a person is because I never understood her well enough to do that—just as, I suppose, she didn't understand me. It may not seem possible to live with someone for years and not to know who that person really is, but that is how I felt. It was as if she stood on the ramparts of a castle looking down on me while I looked up from the bottom of the wall with no way to reach her. Our inability to see eye to eye became one more dispiriting and discouraging aspect of our relationship. I blamed her for it, but possibly the fault lay with me. Perhaps she was not perched high up but was standing right next to me, but I would not look her in the eye. Maybe our feelings for each other were out of sync. I started by being in love with Sylvia (whatever that meant) when she was not in love with me (or loved me in ways I could not understand). Later in our marriage, Sylvia became more affectionate, and on one occasion she actually said that she loved me but by then I no longer cared. It was too little, too late.

The most damning aspect of our relationship was that God never seemed to be part of it. I had ostensibly married her in fulfillment of

God's will, and presumably so had she. Why then was God not at the center of our relationship? Why would we not turn to him together for help with our shortcomings and to heal our wounds? While we continued to maintain our intimate bond with God individually, were we so angry with each other that we could not stand together in God's presence? Did we want God to be an ally in our battle with each other rather than a friend we had in common? Or, was it because our marriage was not founded on the rock of our faith but on the quicksand of our personal problems?

By the end of the year, our relationship had gone from bad to worse. By this time, Sylvia's family had emigrated to the United States or was in the process of doing so. They must have expected us to stay and be reunited with them. This may not have been an unreasonable expectation. However, I had had no part in that plan. I had promised my parents that I would return to Lebanon, and that is what I intended to do. Sylvia must have communicated this to her parents, and it must have hit them hard. I had always complied with their wishes, so my newly found defiance was quite out of character.

It was out of character because my character had changed. In learning to help my patients, I had learned to help myself. I had become more self-aware and recognized that there were reasons behind my actions, thoughts, and feelings that I was not aware of. I became more assertive as I realized that I was not doomed to be unhappy for the rest of my life no matter how wrong the choices I had made. My exposure to psychiatry had not undermined my faith in God, but it made me question my understanding of what constituted God's will. Free of that paralyzing constraint, I could now think and act more freely and take responsibility for my own life. These developments put us on a collision course.

Sylvia became increasingly distraught. Every letter from her family (I assumed her mother) plunged her into tears. She got little sympathy from me, and I became increasingly intolerant. I finally wrote to Mrs. Elize and told her that if her letters were going to continue to have this effect on Sylvia, she should stop writing to her. That was not a reasonable request and must have infuriated her. Meanwhile, my parents were losing their patience with me. In an effort to keep them free of accusations of meddling (which were leveled at my mother anyway),

I kept them in the dark. My letters to my parents became reduced to single-page aerograms with no substance. Finally, my exasperated father wrote me an angry letter saying that if that was the best I could do, I might just as well not write to them at all.

Once the deceptive calm and tenuous equilibrium of our relationship was shattered, I knew that we could not go on as before. But I had no idea in what ways our lives could change. I was getting no sustenance from Sylvia, just as she was getting none from me. It no longer made sense for us to stay together, yet the idea of divorce was so foreign that I could not contemplate it. Meanwhile, the issue of returning to Lebanon was a good way for me to draw the line in good conscience. It became an anchor to which I could attach myself.

~

While my marriage crumbled, my psychiatric training forged ahead. During my first year on the wards, I had dealt mainly with patients who were depressed. These patients were interesting initially, but one severely depressed patient ended up looking like another. Occasionally, there were more interesting cases. One man in his early forties suffered from an incapacitating obsessive-compulsive disorder, which was of particular interest to me both professionally and personally. One of his rituals involved his shaving. To see exactly what he was doing, I sat in his room to watch him shave from start to finish. He began by taking the can of shaving cream and reading the instructions, which he must have read a hundred times before. He got stuck on "Lather and shave." He read it over and over with different intonations to get it right. ("*Lather* and shave." "Lather and *shave*.") I couldn't stand it any longer and I told him to move on. Quite reluctantly, he began to lather his face and kept lathering, and lathering, and lathering. When he finally began to use the razor, it became almost too painful to watch. He concentrated on a small area of his face and went over it up and down, down and up, left to right, right to left, until his face began to look like a lobster. Remembering the compulsive rituals of my younger days, I could not but look at him with deep compassion. (There but for the grace of God went I.)

In the second year, we switched from the wards to the outpatient department. The patients here were less disturbed and more amenable to psychotherapy. My rotations through various services included working in the emergency room on the pediatrics unit and at the Rochester State Hospital. There was also a stint in clinical psychology where I learned to administer Rorschach tests, which became useful later on. I had many memorable experiences in the emergency room. Patients who came in the middle of the night, especially on holidays, were particularly interesting. One New Year's Eve when on call, I was roused out of bed to see an elegantly dressed, attractive and slightly inebriated couple. They had had an argument and the woman had thrown a temper tantrum that was still simmering. She was a tall and voluptuous blonde in an exquisite ankle-length mink coat. As I drowsily asked what brought her to the emergency room she replied dismissively, "I don't deal with little boys. Call John Romano." She finally consented to talk to me and we went into one of the examining rooms. As I closed the door and turned around to face her, she had taken off her fur coat and stood stark naked in her full glory. It was like facing a battleship with all its big guns trained on me. Now fully awake, I managed to patch up their quarrel and sent the couple home with the admonition to see a therapist, which I was sure they would never do.

My encounter with a potentially homicidal patient was almost as unnerving. He was a slightly built, unshaven, and disheveled man with a peculiar smile who walked in at midnight and asked to see a psychiatrist. Before I could say anything, he took out a bullet from his pocket and said, "What do you think?" I tried to deflect the question, but he stopped me short with, "Cut the bullshit. What do you think?" Fortunately, he didn't have a loaded gun and I managed to talk him into signing himself into the hospital without having to call security. Then there was the drug addict who had been seen by every resident who had rotated through the emergency service. Each of them had written a note in the chart more or less saying the same thing. John Racy's note was the most succinct: "The patient asked for drugs—I refused. I offered hospitalization—he refused."

My rotation through child psychiatry was particularly frustrating. Some of the children on the ward were autistic, while others had serious behavior problems. I didn't know what to do with any of them, nor could I understand what the child psychiatrists were doing to help them. Albert at age twelve had the mind of a master criminal. He regularly escaped from the locked ward. One day, he disappeared into the heating ducts in the ceiling. Another time, he was discovered in the open vault of a downtown bank. No one could figure out how he did it. The boy had a diabolical imagination. He managed to take off the master key from Dr. Romano's keychain and put it on his therapist's key chain. He urinated into the electric outlets, shorting them. He had me so exasperated one time that I flicked my finger at his head (as a poor substitute for wringing his neck). He looked up indignantly and said, "You goddamn Katchadourian. Didn't they tell you not to hit the kids?"

Working at the Rochester State Hospital filled me with an even greater sense of futility and sadness. Three thousand patients were packed in cavernous dormitories that were like warehouses for lost and unclaimed human beings. Most of them were chronic schizophrenics who had been there for years and would spend the rest of their lives there. Dedicated nurses and nursing aides cared for them with patience. The staff psychiatrists relied on newly emerging antipsychotic drugs, like Thorazine, to treat them. I had been taught to scoff at using drugs rather than psychotherapy, since drugs treated the symptoms of the illness rather its cause (even if they were unknown). I tried to introduce some of the same treatment principles we used at the Strong Memorial Hospital. For instance, even though I could not spend time individually with the hundreds of patients on one of the wards, I had some of them lined up around the room so that I could ask their names and shake their hands as a way of establishing human contact. The patients looked baffled and the nurses stood around shaking their heads.

My bleakest memory was being awakened at night when on call by an apologetic nurse. One of the patients had struck another in the face and there was blood coming out of one eye. The nurse was terribly sorry to wake me up; perhaps it could wait till morning. It was clear that it could not. I walked in the chilly night through the gloomy buildings looming out of the darkness and found the ward. The patient was sitting on a chair with his head down. I lifted up his chin to find his left eyeball hanging out of its socket. I asked the nurse to bring me the first-aid kit, and she returned with a flashlight and some gauze. I told her to call an ambulance, but she said only the director could do that.

I told her to wake up the director, but she would not do it. I called the director myself and an ambulance soon came to take the man to the emergency room. I never saw him again.

During the 1960s, one of the well-intentioned but utopian domestic programs enacted under President Lyndon Johnson involved the closing down of state mental hospitals. The hope was that antipsychotic drugs would make it possible for patients to come out of asylums and live in the community and be cared for in day-care centers. It was a good idea but it led to considerable numbers of the mentally ill ending up in the streets or in jail. As George Engel put it acerbically, while psychoanalysis offered something to a few, community psychiatry offered nothing to everyone.

The most rewarding part of my second year was getting to know Dr. Sandor Feldman, the grand old man of psychoanalysis at Rochester. He was a kindly and avuncular man in his sixties with gray hair and bushy eyebrows. A Hungarian Jew who had left Hungary when the Nazis came to power, he retained his Old World charm and thick accent even after living for several decades in the United States. As he was the first training analyst at Rochester, most of the younger members of the faculty had had their didactic analysis with him. There were many stories about Dr. Feldman, including his loud alarm clock that went off ten minutes to the hour, startling the daylights out of the patient lying on the couch.

Dr. Feldman ran a seminar for the residents that was based on his cases. We sat through lengthy readings from his yellowing notes, consisting mainly of the-patient-said-this-and-I-said-that. It was hard to make much sense out of it. But then he would come out with an insight that would take your breath away.[9] What was most impressive about Dr. Feldman was his total lack of self-consciousness. We asked him how he could be so at ease with himself. He said people felt anxious because they pretended to be someone other than who they were; it was the fear of being found out that caused the anxiety. He had overcome that by being who he appeared to be—the ultimate truth in packaging. This idea had come to him from an actress in Budapest who had no stage fright even with Emperor Franz Joseph in the audience. One of Feldman's colleagues knew her, and they took her out to dinner to find out the secret of her self-composure. The actress said that before

Figure 6.3 Sandor Feldman.

coming on stage, she imagined herself to be totally naked—she then didn't have to worry if her slip was showing.

Given his transparency, there was no need to "interpret" what Feldman said or did. During a departmental meeting, he would take out a banana from his briefcase, peel it slowly, and then eat it. The symbolic ramifications of the act in that setting were mind-boggling. Yet, if Feldman was eating a banana, then he was just eating a banana; there could be nothing more to it. (Or as Freud put it, "Gentlemen, a cigar is sometimes just a cigar.") If someone thought ill of him, then, *Honi soit qui mal y pense* ("Shamed be he who thinks evil of it").[10] This is not to say that Dr. Feldman's life had been free of sorrow. His only daughter committed suicide.

People noted that there was something rabbinical about Feldman (his father was actually a rabbi). He exuded a tremendous sense of integrity and compassion. There was a story about his going to a bank to cash a $10 check. The teller refused to give him the money without proper identification. Feldman called for the manager and told him, "Look at me. Do I look like an honest man?" The manager responded, "Yes, you do," and he authorized the check. The cashier handed him a $10 bill, and when putting it in his wallet Feldman noticed another $10 bill stuck to it. He handed the second one back to the chastened cashier with a smile.

My regard and affection for Sandor Feldman was quite different than what I felt for John Romano. I wanted to be like Romano, while I wanted Feldman to be my grandfather. One of my great pleasures and claim to fame as a resident was impersonating Feldman in our skits at the departmental Christmas parties. (Racy played John Romano.) My accent made it easy. As I slowly shuffled on stage fitted with fake bushy eyebrows, an ill-fitting jacket, and a long scarf trailing to the floor, everyone knew who I was supposed to be. As I took out a cigarette holder, broke a cigarette in half, and lit it, people began to giggle. Then when I stretched out my hand and said "Hov do yuu doo," the house came down. George Engel, who taped one of these performances, claimed that people could not distinguish my voice from Feldman's. My good friend Ned Freeman played the guitar and I accompanied him on my violin as he sang cowboy songs that parodied faculty members. It was considered an honor to be made fun of even if sometimes we cut it too close for comfort.

Many of the faculty members and my fellow residents were Jewish. As I noted earlier, I had been quite oblivious of Jews living in the periphery of my life when I was growing up in Beirut. In medical school, we had Dr. Nachman, who was a professor of pediatrics, and Dr. Buchbinder, who headed the student health service. But these names sounded European to me and there was nothing Jewish about them, as there would be with names like Isaiah and Jeremiah. My Jewish American fellow residents and friends were all secular and even further removed from Old Testament Jews than were the Lebanese Jews. Actually I would have hardly realized that they were Jewish if they had not told me so. Then I came to realize that they were quite conscious of their ethnic identity, just as I was of being Armenian, even though they varied in the importance they attached to it. Given the historical similarities between Jewish and Armenian diasporas, there were many commonalities between us, so much so that some of my friends had initially thought I was Jewish too, since I came from "that part" of the world. The fact that they could not tell that I was Armenian surprised me. How could I have possibly been anything else? I suppose they were just as ignorant about Armenians as I was about Jews. In due time, I became particularly aware of the intimate association between Jews and psychiatry, particularly psychoanalysis, to the extent that I thought one had to be at least a little Jewish in order to be a real psychiatrist. Consequently, I began to sprinkle my language with Yiddish words like *schmuck* and *schlemiel* while being careful about their fine distinction. (The waiter who spills soup on a customer is a *schmuck*; if he spills the soup on himself, he is a *schlemiel*.)

~

My third and final year at Rochester was especially significant both professionally and personally. It culminated in the completion of my residency training and my separation from Sylvia—a graduation of sorts on both counts. I came fully into my own when I became chief resident on R4. The position entailed a good deal of clinical, administrative, and teaching responsibilities. It was a pleasure to follow Cy Worby in that role, and he too was happy to have me as his successor. Cy was a year ahead of me and had a distinctive flair. He and his wife, Marsha, treated me warmly, had a good sense of humor and it was fun to be with them. When I was getting ready to take over the responsibility for the ward from him, he invited me to dinner to talk over the transition. As we approached his car, I inadvertently headed toward the driver's side. Cy stopped me with a smile and said, "Not yet." It was vintage Worby.

The attending physician on R4 was Dr. Robert Atkins, another Boston Brahmin. Many others smoked pipes, but Bob turned it into a special ritual with a special type of tobacco custom-mixed for him. It induced me to make another attempt smoking a pipe myself, although my heart was never in it (nor did I ever forget my misadventures with

Figure 6.4 With fellow third-year residents in 1961. I am at the center in the back row. To my right is George Mizner, at the extreme left is Don Grinols, and seated in front of me is Cliff Reifler.

it). Bob and I had a cordial relationship, and I also made friends with his second wife, Louise Moran, one of the senior social workers and the first American I knew who came from the South. It was a special pleasure to work with Sue Graham, who was the specially pleasant and competent Canadian head-nurse. Working with all them added much to my experience on the ward.

Being chief resident became the central concern of my life. This was the most rewarding period of my residency, and it provided me with a respite from my personal problems. It gave a huge boost to my self-esteem: even if I failed as a husband, I was not failing as a psychiatric resident. I was at the peak of my powers as a clinician, administrator, and teacher. The confidence these roles instilled in me was as important as anything I had learned during my three years of training. The first-year residents and students under my supervision took to emulating my accent. (I would tell them, "In this business, anyone who has something to say, says it with an accent.") I was a good teacher and mentor who cared about his students and made it fun to work together.

I was also a hard taskmaster. When medical students were to present a case to Dr. Romano, I had them rehearse their presentation with me. I would ask them the sort of unexpected questions that Dr. Romano was fond of springing on them, such as asking for the name of a patient's dog. One morning, when we were administering electroshock therapy, I had one of the medical students inject the drug that would temporarily paralyze the patient's muscles to attenuate the force of the convulsions. As the student held up the syringe, I noticed that it contained a lot more of the drug than it should have. It may have killed the patient by paralyzing her breathing muscles. I raised my hand silently, took the syringe, squirted out the excess slowly, and gave the syringe back to him without a word. He managed to give the injection despite his trembling hand. It was a harsh lesson, but I hoped it would help him to never make the same mistake again.

My tutor for the third year was Dr. Paul Dewald. He was also Cy Worby's tutor, which created an additional bond between us. Dewald was a dashing character (at least for a psychoanalyst) and a superb teacher. Most of the psychotherapy I learned was from him. It is a skill that cannot be learned from books alone. It requires a combination of personality characteristics, self-knowledge, and the skills acquired through apprenticeship to a master. Dewald fulfilled my high expectations. Unfortunately, my respect and fond memories of him were to become compromised when two decades later I saw Dewald at a psychiatric conference in Helsinki. After attending his talk, I went up to greet him warmly. He didn't recognize me. I was deeply disappointed and rather hurt. I thought I had been one of his star residents; how could he not even remember me? He sensed my disappointment and offered a lame excuse saying it had been many years and he had known many residents. He had left Rochester under a cloud and in retrospect it confirmed my suspicion that there was something not quite right about the man. Or perhaps I was merely trying to soothe my wounded pride.

~

In the fall, Sylvia and I had to give up our rented house because the landlady needed it for her daughter. Considering that I was to leave Sylvia within a few months, it is surprising that I didn't move out at that time. Perhaps it was because Sylvia would have needed a place to live in, whereas I could have moved back into the residents' quarters at the hospital. Ironically, we were fortunate to find a much nicer house on the edge of a city park. It was better furnished and had a spacious living room with a picture window that opened on a lovely forest of birches and maples. Their brilliant colors with the turning of the leaves in the fall were spectacular. I thought it was a pity that we moved into this delightful setting at a time when our marriage was finally on the rocks. As I walked through the park, especially on a misty day, I couldn't decide if all that beauty surrounding me made me feel better or worse.

By that time, Sylvia and I barely talked to each other. We retreated into our shells and sulked in silent reproach. I mostly felt sad and resigned rather than angry or resentful. Sylvia looked glum, but I don't know how she actually felt, for she said nothing and I never asked. I don't know if it ever crossed her mind that I might leave her. I never told her I was going to until I couldn't see any other way out. I had no conscious desire to hurt Sylvia although I realized that my leaving her was going to be painful and difficult for her. Nonetheless, I wasn't going to mortgage my life simply to avoid causing her pain.

We did not seek help individually or as a couple, maybe because there was no one we could turn to. My being a psychiatrist should have facilitated our getting professional help, but on the contrary it became an obstacle. Seeking help would have hurt my professional pride and reflected poorly on me. Sylvia may have been too proud to seek help, or perhaps she didn't trust psychiatrists because I was in their camp. There comes a time in a broken relationship when it is no longer possible to undo the damage—just as a spring that has lost its memory can no longer have its coils pushed back together. A chasm had opened up between us that could not be breached. Rather than succumbing to a fatal ailment, it looked as if our marriage eventually starved to death.

In bringing this sad tale to an end, it is tempting to ask what went wrong. If we compare marriage to a card game, there are two basic components that decide its outcome: the hand we are dealt and the way we play that hand. It would seem like Sylvia and I had been dealt a good hand. We both were highly eligible and brought many assets to the table. However, if we look deeper into our temperaments and aspirations in life, then we were badly mismatched from the outset. It was like having a fine pair of shoes that pinches your foot. We also played our hand badly. I was not ready for marriage and I doubt if Sylvia was. And neither of us was right for the other even under the best of circumstances. We were at a point in our lives where we couldn't make a deep and lasting commitment (I certainly could not, even if I thought otherwise). Without such a commitment, no marriage will work. We also lacked the willingness and ability to identify and deal with problems as they came up in our relationship and find someone who could help us. All told, I don't know which factor would be more charitable to blame—the mismatch or the mismanagement. There were times when I thought if given half a chance I could have made our marriage work, but perhaps I was deluding myself. Finally, there was nothing else that I could have done without compromising my integrity, my obligations to myself, and the prospect of my future happiness. The impact on Sylvia did matter to me, but not enough to sacrifice the rest of my life for her sake.

~

As Christmas approached, I began to make plans to move out. There was no particular crisis to precipitate this; I had not been and wasn't involved with anyone else, nor was Sylvia. My level of unhappiness didn't go higher; it was my level of tolerance that got lower. I had no thoughts or plans for getting a divorce. I just needed to get away from Sylvia and take it from there. When I finally told Sylvia that I was going to move out, she didn't have much of a reaction, which surprised me and came as a relief. She only asked me when I was leaving.

Sylvia's father was visiting us at the time. Perhaps he had come to be with Sylvia in anticipation of my leaving. I hadn't seen him or communicated with him since I had left Beirut three years earlier. I don't recall our talking to each other during his visit, let alone discussing our situation. I got the impression that it would be beneath him to talk to me about these matters. Nonetheless, his presence made it easier for me to make the final move instead of leaving Sylvia alone. On the last day, as I was walking out with my suitcase, I saw Mr. Yervant sitting at the window reading the newspaper. I looked at him but he didn't look at me. I closed the door behind me and walked out. I was never to see him again.

~

I went to see Dr. Romano and told him about my having left Sylvia. I don't know if he had suspected that I was having problems in my marriage, but he did not seem surprised. He listened carefully and I was much relieved that he didn't seem upset, nor did he tell me what to do. I then saw Dr. Rubin and was rather cavalier in my account of our breakup, but he must have seen through my bravado and realized that I needed help. He referred me to Dr. Feldman.

I met with Dr. Feldman five or six times. He took me on rather reluctantly. He had no time available and could see me only when one of his patients canceled a session, which made my "therapy" rather disjointed. I was surprised when at the second session he asked me to lie on the couch. I wasn't expecting to go into psychoanalysis. Nor was I clear about what I was hoping to get from Dr. Feldman. It was a difficult time for me. I needed help to bolster my sagging self-esteem and practical guidance on how to manage my life during this painful transition. Perhaps I was looking for vindication for my leaving Sylvia. Had Dr. Feldman told me I had done the right thing, it would have greatly eased my mind. But had he told me I should go back to Sylvia, I don't think I would have. However, he did neither. Nor did he provide any guidance for what I should do. He was a psychoanalyst, and that's not what psychoanalysts did.

The main benefit in seeing Dr. Feldman was my getting a glimpse of what it felt like to be in analysis. This was more of a professional rather than personal gain. I recall reporting a dream during which I touched my hair and it came off like a toupee, leaving me bald. Dr. Feldman asked what I thought the dream meant. I came up with a convoluted explanation. He countered with his own simpler interpretation: My "hair" symbolized my marriage to Sylvia, which I was about to "take off." Being "bald" reflected my concern over how I would appear to others as a divorced person. I knew instantly that he was right. As Goethe said, the right answer is like an affectionate kiss.

When I left Sylvia, I still had six more months left before finishing my residency training. Three of my friends lived in a large house with a spare room that I could rent, and I moved in with them. I immersed myself in work but I didn't have much of a social life. I was occasionally invited to dinner by friends. I went out a few times with a woman who worked at the hospital. During our second date, she told me she would sleep with me if she were not afraid of getting pregnant. She also told me that after I got a divorce, we could get married. It felt as if I had opened a window slightly ajar to be hit by a blast of wind. I was not ready to get involved with anyone even casually, let alone think of marriage anymore than I would have thought of walking into the eye of a tornado.

My further contacts with Sylvia were sporadic. She had become associated with a local Evangelical church and had made friends there. The minister of the church called me to see if there was any prospect of reconciliation. We met with him once. Sylvia said little, so it was mostly a conversation between me and the minister. He barely knew me, and I had little to say to him beyond the fact that I was not interested in going back to Sylvia. Nor could I see the point of having a serious discussion about our future. It would be riddled with recriminations, and nothing useful would come out of it. I just needed to be away from her. The rest would have to wait.

My key concern at that point was figuring out what I would be doing the following year. I was determined to go back to Lebanon, but I was not yet ready to do it right away. I wanted to spend an additional year in the United States, but I didn't want to do it at Rochester. My psychiatric training was completed, and little would be gained by my spending an extra year there. Besides, Sylvia would still be at Eastman, and living in the same city would be awkward.

More than anything else, I wanted to be *free*. To be on my own as I had never been before, whether living with my parents or with Sylvia. I had spent my whole life trying to please my mother and then to please Sylvia—to comply with the wishes of my parents and then with those of her parents. I wanted to live my own life without any of these obligations and constraints—to do what I wanted to do when I wanted to do it and how I wanted to do it. At last, I wanted to be an adult, not a superannuated adolescent. I hoped for a new beginning that would usher the dawn of a new day in my life—a prospect that was exhilarating and unsettling at the same time.

7

Passing Through the World of Research

I WAS NOW TWENTY-EIGHT years old. I had been a student all my life. My psychiatric training was coming to an end at Rochester, but the idea of having a regular job, let alone a career, was still new to me. It evoked all of the uncertainties about who I was and who I wanted to become. My immediate concern following the end my residency was the coming year. But it was hard to think about that without looking beyond to long-term career prospects. My fellow residents had a clearer sense of where they wanted to work and live. I remember Cy Worby talking about a place called Stanford in glowing terms. An excellent university with a fine medical school, it was located in Palo Alto, close to San Francisco and the Pacific Ocean. It had delightful weather in a bucolic setting of fruit orchards. However, it was out of his reach. This was the first time I had heard of Stanford, and it sounded as if it certainly would be beyond my reach as well. Besides, the West Coast was so far away that it seemed like the end of the earth.

More pertinent to my life were John Racy's plans. John had stayed at Rochester for an extra year as an instructor. He was going to join the faculty of the medical school at AUB, where he had an appointment in the department of internal medicine. There was yet no department of psychiatry; John was expected to start one. He and I had discussed the prospect of my joining him after he had returned to Beirut. We agreed that after John had settled at AUB he would try to help me get an appointment as well. That was very reassuring, yet I still felt reluctant to return to Beirut right away. My relationship with Sylvia was in limbo, and I wanted some resolution to our stranded marriage before leaving the United States. I was not interested in getting back together with Sylvia, nor was I ready to get a divorce. I needed more time and hoped that things would somehow sort themselves out. Meanwhile, I dreaded the prospect of facing my parents and relatives. No one in my extended family had been divorced, and I hated to be the first.

The idea of staying on at Rochester on a more permanent basis crossed my mind. I would have enjoyed living there. It would have allowed me to enter into psychoanalytic training. I might eventually get a divorce and marry a nurse or someone else working at the hospital, since I knew no one outside it. I had been happier at Rochester—despite my marital woes—than I had ever been in Beirut. I would feel safer and more at ease living in America than in Lebanon. I spoke English far better than Arabic. Why wouldn't I want to spend the rest of my life in the United States? It was after all the land of milk and honey. The only compelling reason for me to return to Beirut was to be with my parents and Lucine. I could not abandon them in their old age, especially after the grief I had caused them with my ill-conceived marriage. However, I could still spend one more year in the United States before going back.

I met with John Romano to discuss these prospects and get his advice. He listened to what I had to say, as he always did, and offered me a one-year position in the department as an instructor before I returned to Beirut. I was touched and thanked him profusely. It felt awkward to turn down his offer, but he understood why I needed to get away from Rochester. He would help me find a position wherever I wanted to go. I thought we could start with New York. John arranged for me for meet with Dr. Lawrence Kubie, the chairman of psychiatry at the Columbia Presbyterian Hospital. Kubie was civil to me but less than enthusiastic. They had no need at the time for someone like me, but he was willing to hire me to do John Romano a favor. He offered me a one-year fellowship and suggested a few projects I could work on; they all sounded like make-work. Besides, the salary of $7,000 was modest by New York standards, and New York was no place to be poor.

Shortly thereafter, David Shakow, who was head of the psychology unit at the National Institute of Mental Health, gave a talk at Rochester.

I had not heard of the NIMH but was very impressed by his talk and liked him personally. I introduced myself and asked him about the prospects of getting a position with him. He said since I was a psychiatrist and not a psychologist I should get in touch with Dr. David Hamburg, who headed the division of Adult Psychiatry. I went back to Romano. He thought it was a good idea. I would have a different sort of experience at a research institute that would complement my training at Rochester. He contacted David Hamburg, who invited me for a visit. I drove to Bethesda and found my way through the maze of buildings to Hamburg's office. He was much younger than Romano and quite different in his low-key manner. The furniture in his office was government issued, and the wall displayed no diplomas.

Dr. Hamburg was very kind and cordial and told me of his high regard for Dr. Romano and Dr. Romano's high regard for me. After a relaxed conversation about my interests and about some of the research projects in his unit, he invited me to spend a year with him as a visiting scientist with a stipend of $13,000. I accepted his offer on the spot, and my plans for the following year were settled then and there. On my way out, Hamburg introduced me to Dr. Lyman Wynne, one of the senior research psychiatrists on his staff. The NIMH and Washington looked like the prefect place for me to spend a year. The Romano magic had worked again. And even though I didn't know it then, I was to acquire another critically important mentor and friend in Dave Hamburg.

~

There is not a single photograph from the year I spent at the National Institute of Mental Health in Bethesda, MD, close to Washington DC, starting in the fall of 1962 (hence no photographs in this chapter). I wonder if that was simply because I didn't own a camera or because I didn't want a record of that year. That year was to be a hiatus, a parenthetical period in my life set apart from my past and my future, even if in reality the past would still be with me and the year would have momentous consequences for my future. Nonetheless, I wanted to feel free of the constraints of past and future, at least in my personal life, and live in the present for a year.

A few weeks after my meeting with David Hamburg, I drove back to Bethesda to formalize my appointment and to look for a place to live. The sensible thing to do would have been to rent a furnished room in a house. It would have been cheaper and more convenient, but that prospect didn't cross my mind. One of the social workers told me that a number of people he knew lived in the Pooks Hill apartment complex off Wisconsin Avenue, within walking distance of the NIMH. I met with the manager, who told me that the choices ranged from fourth-floor corner units with great views at $110 a month to smaller units with no views at $60. I should have rented a cheaper unit, but instead I took one of the top-floor corner apartments.

When I left Sylvia, I had walked out with nothing but my suitcase. We actually owned very little, but later on I went back to take a small carved table with a brass tray from Damascus and an Italian silver *repoussé* representation of the Last Supper. It was a wedding present, which I sold to an unscrupulous antique dealer for a pittance. I don't remember the details of my final move to Bethesda, but I do remember how disconcerting it was to find myself in an empty apartment and realize that I needed furniture, sheets, towels, pots, pans, cutlery, and all else that is required in a house. First of all, I needed a bed. I bought a cheap inflatable plastic camping mattress with four long sausage-like segments. In a few days, one sausage sprang a leak; then the second one gave out. That put half my body on the floor. I was relieved to find an ad for a bed, couch, and lounge chair for an amazingly low price. The owner was a middle-aged woman living in a large house. She no longer had any use for the furniture and was essentially giving it away. It was a gift from heaven. I bought and transported everything strapped to the roof of my beat-up car. Then, after many evening visits to a department store, I got a folding table with two chairs for my dining corner. I couldn't afford to buy a TV. Otherwise, I was as settled as I would ever be for the rest of the year.

The only friends I had in the area were George and Maureen Nesbitt, who lived nearby in Chevy Chase. George had left Rochester after the first two years to complete his training in Washington DC. They were kind to me and it was nice to have them close by. George was an expert carpenter, and he made me a small bookcase for the few books I had. However, the Nesbitts had their hands full with their young children and I couldn't impose on them. The tenants in my apartment complex were mostly married couples and some singles.

There was a swimming pool where people congregated mostly over the weekend. I managed to get into a few casual conversations and asked a young woman, who turned out to be a nursing student, if she would like to come up to my apartment for a cup of coffee and what-not. She said, "Tell me more about the what-not," and I changed the subject.

I realized that I would now have to take care of myself in ways I had never done before. My salary was a vast improvement over Rochester, but it didn't stretch far enough especially as I started taking out dates to expensive restaurants. I usually had breakfast and lunch at the NIMH cafeteria, but the dinners and weekends were a problem. My Sunday treat was to eat at the International House of Pancakes, where I went down the menu country by country. My big splurge was broiling a porterhouse steak topped with mushroom and parsley drenched in butter. I longed for the food that my mother and Lucine used to cook and that I had taken for granted for so long.

~

My work at the NIMH took some getting used to. An institution devoted solely to research was a very different place than a university-based teaching hospital. The National Institutes of Health (NIH), of which the NIMH was a part, was a huge complex of buildings with thousands of employees—an impersonal federal institution where layers of bureaucrats did everything by the book. There was no Romano in charge. The Adult Psychiatry branch itself was mercifully smaller. In addition to its regular staff of half a dozen researchers, there were a number of younger public health officers who were discharging their military obligations. They were essentially my peers, and I shared an office with one of them.

When David Hamburg invited me to join his division, I didn't know that he was slated to leave to become the first professor of psychiatry and head of the Department of Psychiatry at Stanford. He was still at the NIMH when I got there, and he assigned me to three research programs: a study of hormonal factors in depression, family studies of schizophrenia, and a study of adolescent development. At the time, I wasn't that interested in psychiatric research although I recognized its importance. The research that I was more attracted to was carried out by humanists such as historians, who mined the past, philosophers, who thought through ideas, theologians, who probed the mysteries of faith, and literary critics, who analyzed texts. Statistical significance was not my way of distinguishing between truth and falsehood, the trivial and the profound.

The project that dealt with endocrine studies of depression was relatively more appealing since it was the most clinically oriented. It consisted of monitoring the blood levels of select hormones and correlating them with the clinical symptoms of depressed patients. Some of my responsibilities on the ward were not that different from my responsibilities at Rochester. However, these patients were to be studied, not treated. They couldn't be given drugs, nor was there any attempt at psychotherapy. Instead, this basic research would hopefully lead to better forms of treatment in the future. Patients came to the hospital because it was free, and they must have hoped that they would somehow be helped by being there. A few of them were quite intriguing. One woman had a manic-depressive cycle that shifted on a twenty-four-hour basis. It was like clockwork. I would leave her in the depth of depression at the end of the day and find her flying high in a manic state the next morning (I could hear her laughing and yelling before I entered the ward), only to get depressed again by the end of the day.

Gradually, my main engagement at the NIMH shifted more to the family studies of schizophrenia. The primary investigator was Lyman Wynne, who succeeded David Hamburg as head of Adult Psychiatry. Wynne was a highly trained psychiatrist, psychoanalyst, and psychologist, whose exclusive interest was in research. He was a pioneer in family studies of schizophrenia and became an eminent figure in the field of family-based therapies.

The family studies were baffling. They included group sessions with families who had a schizophrenic offspring. This was quite different from the individual therapy I was trained to do at Rochester. Nonetheless, I adapted to it readily and became quite good at it. Family members were given Rorschach tests, which were analyzed by the psychologist Margaret Singer, Lyman's colleague at Berkeley who had the uncanny ability to predict whether or not a family had a schizophrenic offspring on the basis of a blind analysis of their Rorschach protocols. She did this by identifying the peculiarities of their cognitive style rather than the more standard method of interpreting these tests.

One of the families that I worked with closely were the Rosenbergs.[1] Their only daughter, Patty, was in her early twenties and had been diagnosed with schizophrenia a few years earlier. She was a heavyset young woman who was quite talkative, but it was often hard to make much sense of what she said. Her father was very fond of her, and we thought he was aiding and abetting his daughter's illness by denying it. Mr. Rosenberg was a well-dressed, mild-mannered, soft-spoken, and unprepossessing man. His thin mustache and whimsical smile gave the impression that he was subtly laughing at you. His wife was of tougher fiber and unsentimentally realistic about her daughter's mental dysfunction. She seemed as exasperated by her husband as she was by her daughter. Her imposing, ample figure towered over her puny husband, making the couple look mismatched. Mrs. Rosenberg would be especially annoyed when her husband repeatedly claimed that there was "nothing wrong with Patty." It was not clear if he really meant it or said it to comfort his daughter or provoke his wife.

Although I was hardly helping Patty, the Rosenbergs were most appreciative of my efforts and unfailingly cordial to me. In late spring, when I informed them that I would be leaving the country shortly, they invited me to dinner at their home. On the appointed evening as I waited at the curb, a black Cadillac sedan stopped to pick me up. Mr. Rosenberg was at the wheel, his enigmatic smile dimly visible in the luxurious interior of his car. Their house was equally grand. Whatever the cognitive peculiarities of Mr. Rosenberg's Rorschach test, he was an obviously successful businessman who made more money selling men's clothing than the combined staff of the Adult Psychiatry unit. The Rosenbergs were gracious hosts. The food was delicious, the wines carefully chosen, and the china exquisite. Even Patty acted more sane at the table. Mr. Rosenberg proposed a toast and expressed heartfelt thanks for all I had done for their daughter. He then ceremoniously presented me with an elegant, wafer-thin gold Ulysse-Nardin wristwatch. I was speechless and felt like apologizing to them for the indignities I had visited on them in our family sessions. (The NIMH higher-ups allowed me to keep the watch, and I still wear it on special occasions.) My experience with the Rosenbergs was another demonstration of how strongly the social context of our interactions influences our psychological perceptions. In the group sessions the Rosenbergs looked like a dysfunctional family; in their home, they did not. Perhaps the social setting helped conceal the underlying dysfunction or we as psychiatrists saw dysfunction because that is what we were looking for.

In working together closely, Lyman and I became lifelong friends (especially after he and his family spent a year in Beirut the following year). He never became a role model for me the way John Romano had been and David Hamburg was to become. Nonetheless, he was unfailingly supportive of me and directly responsible for getting me involved in my own psychiatric research projects in Lebanon, which proved crucial for my career. Coincidentally, Lyman went on to chair the Department of Psychiatry at Rochester after he retired from the NIMH. His style of administration was very different from John Romano's. Whereas running the department was Romano's primary concern, Lyman was a far more reluctant administrator. I remember his telling me half-seriously that the reason he took on administrative responsibilities was to stop others from interfering in his research. Just as John kept his research interests from getting in the way of his running the department, Lyman never let his running the department get in the way of his research. Lyman's wife, Adele, who was a gifted painter, also became a family therapist and Lyman's collaborator. Together, they established the Wynne Center for Family Research at Rochester, and Lyman continued working there after he left the chairmanship. I kept in touch with him to the time of his death in a retirement home.

~

Sylvia knew that I was going to Washington for the year, but I didn't call her to say goodbye. Nor did I call her from Washington. I assumed she could reach me through the Department of Psychiatry if she needed to talk to me. Late in the fall, my friends at Rochester urged me to attend the departmental Christmas party and play Sandor Feldman again. I told them I would come but wouldn't take part in the skit. I no longer had the heart for it. Shortly before I left for my visit to Rochester, I heard from Sylvia. She was in the United States on my visa as my wife and wanted to make sure that she could stay in the country after I left for Lebanon. Would I be willing to sign some papers at her lawyer's office at Rochester to facilitate her getting a new visa? Of course I would, particularly since I would be going to Rochester anyway. The day after I got there, I met with Sylvia at her lawyer's office in downtown Rochester. The lawyer was a conservatively dressed man in his early forties with a clean, scrubbed look. Being in a lawyer's office made me feel slightly apprehensive but I was reassured to see on his desk a stack of small cards with biblical verses. The man was a Christian and would do me no harm.

The lawyer ("Jim") started by asking me questions about where our relationship stood and the prospects for a reconciliation. I couldn't understand what this had to do with Sylvia's visa, but I told the lawyer anyway that I saw no likelihood of our getting back together at that time. He called in his secretary, and a jittery young woman holding an official-looking paper came in. She mumbled something and handed it over to me. I turned to the lawyer with a mixture of alarm and incomprehension. He said it was a document that would prohibit my leaving the United States until certain legal and financial matters were resolved between me and Sylvia. I was flabbergasted. (Was I going to jail?) When I refused to take the document, the lawyer told me that once the papers had been served it would make no difference whether I took them or not. I said I had agreed to come to help Sylvia but it looked like I had walked into a trap. Sylvia said nothing. The lawyer must have felt bad (maybe he saw me looking at the biblical verses) and said he was sorry to do this but he would have been legally liable if he did not take action. (Liable in what way? What did that mean?) I took the paper—I didn't know what else to do—and got up. So did Sylvia. We left the building together in silence. I was visibly shaken and she looked sullen. As we stood on the sidewalk, I asked Sylvia in an almost plaintive tone if we could talk this out. She looked away and said there was nothing to talk about. At that moment, I had an almost physical sensation that something that attached me to her snapped. We walked away in opposite directions. When we reestablished contact several decades later, Sylvia referred to our meeting with her lawyer and disavowed all responsibility for it. She wanted to make clear to me that she was not a conniving person and that she did not condone such actions. I had never thought of Sylvia to be conniving or being personally responsible for what transpired in the lawyer's office. However, since she had evinced no surprise at the time and expressed no regrets right afterwards it was hard to think that she was not complicit in what had happened. Moreover, I was genuinely frightened of what might happen to me, and I had a sense of betrayal I had not experienced before. The symbolism of the snapping of the cord that attached me to Sylvia was my way of condensing a long process of alienation into a single point. The cord was already badly frayed. Had it not snapped then, it would have some other time. This incident also provided me with a final justification to get out of my relationship with Sylvia. It was easier for me to feel that I had been pushed out of the door than walking out of my own volition.

I headed straight to John Romano's office. I explained to him what had just happened. I was trying hard to stay calm, but he could see I was very upset. He waited for a few moments and said, "You need to talk to a lawyer." He picked up the phone and called his friend Anthony Gucci, the top divorce lawyer in town. He told him, "I have my good friend Dr. Katchadourian here and he needs some legal help. Could you see him?" "Yes, of course," said Gucci. "Send him over." In my overwrought state, John's referring to me as his friend overwhelmed me. I fought back my tears, thanked him, took the address, and left.

Anthony Gucci was in his early sixties, about Romano's age, and a fellow Italian American. He was a shortish man with graying, grisly hair. His expression made him look both reassuringly calm and faintly menacing. As a divorce lawyer, he had learned to deal with the emotional turmoil of his clients as much as their legal tangles. He shook my hand firmly and said John Romano was a dear friend (I heard it as "Since you are Romano's friend, you are my friend too.") He took a cursory look at the legal papers and said, "This is nonsense. Don't worry about it, doctor. Let's go see Jim." ("You are now under my wing. No one can touch you.") His calling me "doctor" boosted my self-esteem and referring casually to Sylvia's lawyer as "Jim" defanged him.

We crossed the street and walked up to Jim's office, passing the point where I had stood with Sylvia an hour or so earlier. With a perfunctory nod to the jittery secretary, Gucci went in ahead of me. Jim was not expecting us. He looked at Gucci, and the color drained from his face. In a cool yet friendly tone, Gucci said, "Jim, you know this means nothing," and he handed him the legal document. Jim took it silently and we left. That was the end of it.

We walked back to Gucci's office. Relieved but still shaken, I thanked him profusely. He smiled and said, "Doctor. Remember, this also shall pass." *This also shall pass*. At that moment, this was the most profound and prophetic statement I could have possibly heard. Socrates and Confucius could not have said it better. I asked what I owed him. He looked at me as one would at a child who has scraped his knee and said, "Nothing at all. Give John Romano my regards." (I thought of sending to Mr. Gucci the watch given to me by the Rosenbergs but I decided it meant more to me than it would to him.)

The Christmas party that evening was bittersweet. I was happy to see my old friends, but the ordeal of the day had been draining. They urged me to dance with their partners, but I declined. I was in no mood for dancing. The skit was disappointing. Ned wasn't participating in the program, and I was relieved I wasn't either. Without Barnum and Bailey, the circus would never to be the same. By the end of the evening, I realized that I was no longer a part of Rochester.

~

My last encounter with Sylvia, painful as it was, had a profoundly liberating effect. After coming to Bethesda, I had met a few young women but felt constrained in becoming intimate with them. Now I felt free. The Ecclesiastical Court in Beirut, no doubt, still considered me to be married, but I no longer did so in my own heart and mind. The only court whose dictates I would obey from now on was Kant's Tribunal of Conscience, which resided within me—and it declared me to be free. I realized I still had to get out of the legal tangle of my marriage, but I was in no hurry. I would wait until I returned to Beirut.

My previous romantic relationships had been predicated on the eventual prospect of marriage. The only sexual relationship I had ever had was with my wife. I had played the game by the rules, but the rules had let me down. I had gotten nothing but grief for being virtuous. From here on, there would be no precondition of being in love or the prospect of marriage to get sexually involved with anyone. I would not be dragged into a relationship any deeper than I wanted to, and I wanted no one to take over my life. All transactions were going to be on a cash-and-carry basis: no line of credit, no mortgage, and no long-term debt. However, I was not sure how these principles would be translated into how I conducted my personal life in my new conception of morality. It's one thing to issue a revolutionary manifesto and quite another to produce a coherent constitution with which to run a country—or one's own life.

At a more practical level, the first challenge I faced was not knowing what to do with the time on my hands. I was busy at work during the week, but the weekends left me at loose ends. I knew very few people outside of my work. I had little money left after paying my lavish rent. I lived in a suburb with no nightlife. There were a few bars, but I never set foot in any of them. I would not have known what to say to anyone standing at a bar. I didn't drink except for an occasional beer or a glass of wine if someone offered it to me at a dinner. I don't know if there was a movie theater in town and I didn't have a television set. I didn't read the daily paper. Freedom began to smack of boredom and loneliness. Had there been a woman in my life, all of these problems would be solved. But there wasn't one, and I had no idea where or how to find one. I occasionally thought of contacting Sylvia but banished the thought. I was not that desperate. I needed an aspirin for my loneliness not narcotics.

Newly divorced people complain that it's hard for them to get back into the dating game. It was harder for me since I had never been in the dating game to begin with. The sexual revolution of the 1960s had not yet come in full force. The dating patterns were those at the tail end of the 1950s. The man who was supposed to be my guide in this thicket of the world of women in Washington was Kevin. He was the brother of a friend's friend and proved to be a sorry guide. He was about my age, in his late twenties. He looked quite ordinary but was full of enthusiasm, which made him more noticeable, and he was intelligent enough to sound interesting. Kevin worked for a publishing house, but the central purpose of his life was seducing women. I thought women were attracted to him because he gave them a standing ovation no matter how they performed, especially if they were starved for attention. Kevin had rented a room in a well-known hotel in downtown Washington.

That made it easier to invite women for a drink in the lobby and then get them up to his room. His bed was placed on a platform and fitted with an ornate cover. Looking over it was a life-size statue of a smiling Buddha. It was like a temple in a cheap movie, but its novelty would have made even the most ordinary woman feel something like a goddess for an evening.

When I had dinner with Kevin the first time, he was very pleased to have me pay the bill. He told me that he had the names of over a hundred conquests (literally) inscribed in his little black book. He dangled the prospect of introducing me to women, including the divorced daughter of a former Chinese general who had served with Chiang Kai-shek, but I didn't see hide nor hair of her. There was something disingenuous underneath Kevin's slick surface. His statue of the Buddha said it all. He took to calling me over the weekend to talk at great length about his quandary over marrying his Guatemalan girlfriend, whom he kept in reserve for a rainy day as he grazed in sunnier pastures. One Sunday, the conversation lasted six hours. I felt sorry for him for the first hour, then bored stiff for the next, and then it became a challenge to see how long I could listen to him. In my last meeting with him, I gave him an ornate silk smoking jacket, an odd gift from someone that I had never used. I told him it was in appreciation of all the women he had promised to introduce to me but never did. He didn't get the irony and thanked me extravagantly. Years later, I heard that he had become an alcoholic.

I knew that the Kevin model would not work for me. The problem was not the number of women he bedded, but his feigning affection, and acting under false pretenses to get them into bed. At the same time, I did not want to return to my previous, sexually restrictive script, so I had to find some acceptable alternative to either example. Meanwhile, women kept making overtures to me but they were not the sort of women I was looking for. A fellow psychiatrist in her forties invited me to a play at a theater in the round. I think it was Berthold Brecht's *Mongolian Chalk Circle.* I thoroughly enjoyed it. However, when she suggested that we go for a drink afterwards, I declined. I was not attracted to her, and I did not want to get involved with a colleague. One of our secretaries was an openly predatory married woman, but being married did not stop her from pursuing other men.

I heard that at a party she had taken one of the younger men to a room upstairs and slept with him while her husband was having a drink in the living room. She began dropping increasingly overt hints of her interest in me and ended up standing at the door of my office, refusing to leave until I talked to her. She finally gave up. If what I wanted was sex, why would I not respond to her when she was offering it to me so freely? For one thing, her being married troubled me. And being in her crosshairs made me uncomfortable. I knew that the day after her conquest the entire office would know about it. I also sensed that her interest in me might not have been purely sexual. Perhaps she wanted more and that scared me off even more.

I was in a quandary. What was I looking for? My romantic engagements with women had been complicated and weighty affairs, topped by my oppressive marriage. I had missed the lighthearted fun of being young. I felt as if I was owed something for all my years of self-imposed celibacy and marital strictures. Owed by whom? God? The Evangelical Church? My mother? Sylvia? Myself? Who was going to pay me the debt? It looked like the source of restitution had to be me. However, the payments had to start coming slowly. I could not plunge into Kevin's sinkhole or the grasping arms of lonely or voracious women.

As I began to be more adept at meeting women I found more appealing and less threatening, I had a number of fleeting relationships. One was with a German au-pair girl who lived with a family in Georgetown. I met her at the Washington Ski Club. There was no skiing anywhere near Washington; it was essentially a singles club. I spotted her at a dance from afar. She looked like a young Sophia Loren. I persuaded her to give me her telephone number, and we began to date. She told me about her life in Germany. Her father had been an SS officer, and people got upset on hearing that. But he was her father and she loved him; what was she to do? Then there was the dance instructor at the local Arthur Murray Studios who had been a teenage country singer in Tennessee. She agreed to go out with me on condition that we did not go dancing—she didn't want a busman's holiday. But dancing was my primary interest in her so it didn't work out. At a skiing trip in Vermont I met two interesting women who were intellectually more compatible. One worked for *Time* magazine in New York. She gave me a book on American art, which sparked my lifelong interest in art. The other was from Northern California and worked for the State Department. She was smart and had a sharp edge to her. We went out for a while, but I stopped seeing her when she named her puppy "Siggy" after Sigmund Freud and I took it personally. With some relief, I gradually realized that not every interaction with a woman had to have a romantic or sexual agenda; I could just be friends with them, and friendship had its own rewards. Knowing them opened for me new vistas into lives I wouldn't have known about otherwise. I should have learned that by then, but I suppose I hadn't because of my delayed maturation.

For most of my life, almost all my close relationships were to be with either older men or younger women; rarely did I have a close friend of my own age of either sex until I was in my fifties. The older men were typically my mentors, and I myself became a mentor to the younger women. My relationship to older men may have been to compensate for my distant relationship with my father. Having younger women as friends perhaps boosted my ego. I generally got along well with men but rarely sought their friendship for its own sake. Our associations were typically related to work. There was a competitive side to such relationships that I both relished and wanted to avoid. I also was not that interested in the sort of things that many men are passionate about, such as spectator sports.

There were a few women with whom I could have developed a more serious relationship had the circumstances of my life been different. Catherine was a Canadian occupational therapist I met at the World Congress of Psychiatry in Montreal in 1962. I first saw her literally standing head and shoulders above the crowd at a reception. She was tall and stunning like a supermodel. The next day I found her waiting in line for lunch with an older woman, and I stood behind them. The older woman looked strangely familiar, but I couldn't place her even though she too kept looking at me. I finally got up the courage to ask her if we had ever met. She laughed and said she was wondering about the same thing. Did she live in Washington? No, she lived in Ottawa. Had she ever been to Rochester? No, she had not. Surely, she had never been to Beirut. Yes, she had! She actually worked for a few years as an occupational therapist at the mental hospital at Asfouriyeh. That is where we had seen each other. We introduced ourselves. Her name was Mary, and she introduced me to Catherine (whom I had studiously pretended to ignore). We had lunch together, and I invited them out to dinner. Toward the end of the evening, Mary said she would take a cab and leave the two us to enjoy ourselves. We protested (mildly) and then went straight to Catherine's hotel room and to bed, where we talked through the night. I told her about Sylvia, and she told me about her uncertainties over what she should to do with her life. I left for Quebec in the morning, and we didn't see each other again. Two weeks later, I got a letter from Catherine. She was interested in me and wanted to hear more about the state of my relationship with Sylvia. I was in a bind. I didn't want to lose her, but something about her didn't fit. She seemed larger than life. It was if I was being offered a Maserati when I had just gotten my driver's license. I sent back an evasive answer and never heard from her again.

Claudia was a psychiatric nurse who worked on one of wards in our unit. She had a wonderful character that enhanced the lovely dark hair and eyes of her Italian heritage. I took her to a formal ball at the Corcoran Gallery. She looked ravishing in her long dress and was delightful to be with in that magnificent setting. There were three orchestras playing waltzes, Big Band, and Dixieland music. At midnight, we were served scrambled eggs with champagne in a gallery surrounded with Aubusson tapestries. It was an enchanting evening except for the fact that Claudia didn't know how to dance. Not only did she not know the steps, but when I tried to steer her in one direction she perversely went in the other. The real problem, however, was that Claudia was ripe and ready for marriage and I was not. We both knew it and parted as friends. She wrote to me the following year saying she had married a Harvard lawyer. I congratulated her and hoped that he deserved her.

Julia and I were introduced at a dinner two weeks before I was to leave for a journey around the world on my way to Lebanon. She had graduated a year earlier from Amherst and worked for a federal agency. Her keen intellect and sterling character endeared her to me. She had an elemental goodness and sense of integrity. However, there was too little time and too many pressing details for me to attend to in order to allow for the forging of a real bond. I had to wrap up my responsibilities at work, prepare for my long trip, and close down my apartment. A week before my departure, when I last saw her, I had moved out and was sleeping on the couch in my office.

Despite my firm intention not to get emotionally involved with these women, they became attached to me. This was particularly true for Julia. I knew she liked me, but I didn't realize that she had fallen in love with me until her letters made that amply clear. I hadn't told her about Sylvia, because I thought there wasn't enough time to get into that. However, I must have dropped enough hints for her to sense that there was something in my past that was holding me back. Since I liked and respected Julia a great deal, I was loath to break her heart, if that is what I did. These experiences eroded my confidence in managing my newly found freedom without causing damage. The fact that I did not lie, feign affection, or make false promises to these women did not prevent them from feeling disappointed or hurt. What they wanted was to be loved at a time when I could not love them. The limitations I had placed on how far to go in my relationships were frustrating. It was like being forced to listen only to the first movement of one symphony after another. It made the experience truncated, incomplete, and unsatisfactory. While I had no abiding regrets over my experiences in Washington, I realized that this transitional phase of my life had to gradually wind down and I had to find the person with whom I would want to spend the rest of my life.

~

Parallel to the changes in my personal life were changes in my attitudes toward my profession and career. It was not until I left Rochester that I realized how narrowly focused I had been on the insular views of orthodox psychoanalysis, even without becoming a psychoanalyst. The psychoanalytic orientation at Washington was far broader, and its leading light had been Harry Stack Sullivan. Its focus was on interpersonal relationships rather than intrapsychic, unconscious conflicts at the individual level. Sullivan was born in the United States, and his brand of psychoanalysis was more distinctively American than the central European Freudian import. The guardians of orthodoxy considered it suspect, if not downright heretical. I didn't abandon Freud, but he became more fallible in my eyes.

Some of the leading psychoanalysts in Washington worked at the renowned psychiatric hospital at Chestnut Lodge at Rockville, Maryland, near Bethesda. It specialized in the treatment of schizophrenics through modified forms of psychoanalytic therapy. Treating these cases year after year with no substantial improvement, let alone a cure, took heroic patience. I was astounded to learn that one of the senior therapists had taken on his last new patient six years earlier. The reason that the hospital could stay in business was because many of the patients came from affluent families who could afford to keep them there, hoping for some amelioration if not cure. Besides, there was no reasonable alternative. It was certainly better than letting the patient languish in the bottomless pit of a state hospital.

I would occasionally be on night call at Chestnut Lodge to make some extra money. It usually meant just being there with nothing to do. One evening, when I was making my rounds, a young woman with a startlingly masklike expressionless face approached me. She asked me who I was. Instead of telling her my name, I answered her question with a question. She looked me in the eye and said, "Who the hell do you think you are—Aristotle?" Revealing nothing about oneself to patients felt like being in a straightjacket, even if it was necessary.

One of the most eminent psychotherapists at Chestnut Lodge was Dr. Otto Will. Soon after coming to Washington, I went to see him as a patient to get help in dealing with the aftermath of my separation from Sylvia and the future of our marriage. What I got instead was the Washington version of my experience with Dr. Feldman. Instead of making me lie on a couch, Dr. Will asked me to sit on a chair in one corner of his office while he sat at the opposite corner. We could not have sat farther apart, but I did not dare ask him the purpose of this strange arrangement. After I told him briefly about the problems in my marriage, Dr. Will asked me the sort of questions one usually asked schizoid characters. I was astonished. Whatever else I might be, I was not a schizoid character. However, since his specialty was dealing with schizophrenics, it seemed as if he had to put me in a similar mold to be able to deal with me. The only comfort I got from him was his remark that single men were sought after by Washington hostesses as dinner partners for single women. It was nice to hear that, but of course no such invitations came my way. Nonetheless, it was worth my hard-earned money to have the experience of being Dr. Will's patient and getting a firsthand look of his style of treatment. Another memorable character was the medical director of Chestnut Lodge. He was the heaviest chain smoker I had ever met. To save the time it took to take a cigarette out of its packet, he had a special receptacle built for a pile of cigarettes that sat on the dashboard of his Cadillac. All he had to do was reach out and take one cigarette after another while he drove. I wondered how a psychiatrist could be so blind to his own addiction.

I required special dispensation from the NIMH to be on night duty at Chestnut Lodge. Seventy years ago, when the federal government proposed to locate the National Institutes of Health in Bethesda, the physicians in the area were worried that the influx of hundreds of renowned specialists would drain away their patients. Therefore, it was mandated that staff physicians could not see patients outside of the various institutes unless they provided medical services in areas that were underserved. This provision made it possible for me to also spend one evening a week seeing patients at a low-cost Arlington County clinic in Virginia. It was a good opportunity to do what I missed most—seeing patients in therapy—and to earn some sorely needed cash. It took over an hour to drive there and back, but it was worth it. Besides, I had nothing else to do most evenings.

One of my patients was a charming middle-aged Italian immigrant who had problems with impotence. This didn't happen when he was making love to his wife, but it did when he was sleeping with his mistress. He denied feeling guilty, which would have been the likely explanation for his problem. I periodically reported my cases to the clinic director. When I told him about the Italian patient, he chewed me out. Why was I treating this man? Was this a good use of the clinic's scarce resources? Should the man have been cheating on his wife in the first place? My defense was that as a physician my duty was to alleviate suffering and not to pass moral judgment. I had adopted as my motto a phrase from the Roman playwright Terence, whom John Romano never tired of quoting: *Homo sum: humani nihil a me alienum puto*—I am human; nothing human is alien to me. One of Romano's detractors accused him of being the type of person who uses Latin phrases without knowing Latin. I didn't like to hear that, but had to admit that the point was well taken. That, however, did not stop me from using the phrase myself.

Another of my patients was a young man who worked at a women's shoe store. He hated his mother and took it out on his middle-aged female customers. He was very slick and could talk women into buying shoes that were smaller than their size by telling them that it made their legs look sexy. He then took sadistic pleasure in watching them hobble out of the store in their tight, ill-fitting shoes.

~

My experiences at the NIMH led me to wonder about the value of the psychiatric research being conducted even at the most prestigious research institution in the country. I couldn't see how these investigations were going to lead, at least in the foreseeable future, to more effective methods of therapy or a better understanding of mental illness. Years later, a distinguished British civil servant told me that when he was on the governing board of the National Health Council, there was a survey of the impact of social science research projects funded by the government. The only significant benefit that could be identified was the impact of the grants on the advancement of the investigator's career. I think that might have been true in my case as well. My association with Lyman Wynne and the year I spent in research at the NIMH were to prove critically important for the advancement of my academic career, irrespective of any contributions to the advancement of psychiatry.

I tempered my skepticism by ascribing it to my lack of knowledge and experience in psychiatric research. The reason I failed to see the big picture that would eventually emerge from it was because I was merely passing through the world of research and could not be a good judge of its true merits. However, these doubts spilled over into my engagement in the field of psychiatry as a whole. Was that the right field for me? Should I have gone into some other specialty? Or should I have even gone to medical school in the first place? Psychiatry no longer seemed to be where my heart and talents were. Was it too late to switch to some other field? What would that be? I had no idea. My youthful fascination with theology and philosophy was still there. However, the humanities looked like a remote wilderness about which I knew very little. What was it that made them relevant? Doctors saved lives and relieved suffering. Even a psychiatrist could do some good. What good did humanists do? Did I know enough about these fields to make an informed judgment? Was I deluding myself with some idealized image of becoming a modern Socrates? Besides, the cost of such a drastic move would be daunting. It looked like a path to poverty, and I had no stomach for being poor. By contrast, psychiatry was like a paved road that led to a more predictable and secure future, even if not a very exciting one. Why not bide my time and see how it all turned out?

That was not the end of the matter. I could not decide if the problem was with the field of psychiatry or with myself. I had a nagging feeling that the problems troubling me were due to my own inadequacy as a psychiatrist. What could be more exciting than exploring the thoughts and feelings of people and to relieve their anguish? But was I a good enough therapist to do it? Did I have the right temperament and training for it? I was impatient with the progress my patients made. By the same token, psychotherapy took so long and was riddled with so many uncertainties. Even at best, how many patients could I hope to help? Dr. Rubin had detected this impatience when I was a resident and told me not to be "beguiled by numbers." I might help only a few people, but one of them might end up saving the world. I knew what he meant but still found it difficult to accept. I myself wanted to be the one saving the world and not help someone else do it. I had felt confident of my calling when surrounded by my tutors and fellow residents at Rochester, where I was part of a congregation of believers. Now that I was on my own, I was assailed by doubts over whether I was doing any good for my poor patients who faithfully kept coming to see me. Twenty-five years later, I had some vindication in a letter I received from a former patient I had seen at the Arlington County clinic during this time of discontent.[2]

~

On the more personal side, the year I spent in Washington was enormously important in expanding my cultural horizons. There was no art museum in Beirut when I was growing up there. Beirut had an important national archeological museum and AUB had a smaller collection of antiquities, but I never visited either of them. Rochester had a good art museum, and I enrolled in an evening painting class for one year. I was convinced that I couldn't draw, but painting in oils seemed more within my reach. We had a good instructor, and despite my limited talents I derived much enjoyment from it.

My exposure to great art had been confined to occasional trips to New York, but these visits were rushed and fragmentary. Now living close to the fabulous collections in the National Gallery of Art and other Washington museums was a whole new experience. I soon realized that knowing more about the art I was looking at would make the experience much more rewarding. I read and reread the one book I owned on American art and assiduously studied each painting. At first my pedantic side focused on matching paintings with painters. However, I finally got into the habit of standing in front of a single painting at the National Gallery, usually from the Renaissance, absorbing all I could and then leaving the museum. The Kennedy Center did not yet exist (it was to open a decade later, in 1971), so the musical performances that I attended were mainly the free concerts at the National Gallery on Sunday afternoons. I occasionally went to hear the National Symphony Orchestra, including a memorable performance by Mstislav Rostropovitch.

I avidly absorbed these experiences, and they had a profound impact on me. However, at one point, I may have gotten a surfeit of high culture. After a symphony concert in Baltimore, when I was heading for my car, I passed by an old-fashioned striptease theater. I went in with some hesitation. The place was packed with self-conscious, sweaty men. The performer was a heavyset middle-aged woman who undressed only to her silk underwear. However, she had such a compelling presence that she held the rough crowd in the palm of her hand. After the refined and aseptic atmosphere of the concert hall, I found the vulgar setting with its stuffy smell of stale tobacco alarmingly exhilarating.

I regret that I didn't make better use of my leisure time during that year. My evenings and weekends were usually free. It was the perfect opportunity to take evening classes at a university or read on my own at home. During my residency training I had had some opportunity to delve into the psychoanalytic literature, but there was so much more to read besides that. I could have also read novels and books on art.

It's a poor excuse, but true nonetheless, that my mind was focused elsewhere at the time. I wanted to broaden my horizons socially more than intellectually. I remember reading Emily Post's original book on etiquette and learned how to behave in polite society, to which I had no access. Nonetheless, I matured a great deal during that year and gained in social competence. I learned how to dress well, to dance, to ski, and to carry on a casual conversation on an ordinary topic. My Old World persona got a New World polish.

A deeper sore point is that I lost touch with God. I went once or twice to the Sunday service at the National Cathedral, which was fairly close, but I did so more to look at the architecture and the stained-glass windows than to worship. I struggled with the problem of how I had gotten married—ostensibly by the will of God—and how that marriage had gone awry nonetheless. I still felt responsible and remorseful over what had happened, but when I examined myself, as honestly as I could, I was unable to see where I could fault myself. Nor could I blame God for my plight. Consequently my alienation may have been my way of gaining distance, a moratorium of sorts, until these problems sorted themselves out. I knew I could always come back to God like the prodigal son. Being part of a Christian congregation may have helped, but I felt reluctant to be around Christians. I had lost faith in them, even if I had not lost faith in God.

~

By the middle of 1962, I had obtained a faculty position at AUB starting in the fall. I would join John Racy as an assistant professor of internal medicine, albeit acting as a psychiatrist. My parents and Lucine were overjoyed. Consequent to my separation from Sylvia, my relationship with my parents was back on track. I no longer needed to keep them in the dark. They had never urged me to leave Sylvia, but I'm sure they were relieved to see me break free. They were delighted that I was returning to Lebanon and proud that I would be an AUB professor.

Shortly thereafter, Lyman Wynne spoke to me about the prospect of our collaborating on several research projects in Lebanon. He wanted me to conduct cross-cultural studies with Lebanese Arabs and Armenians replicating some of the work he had been carrying out at the NIMH. We would hire an anthropologist with experience in the

Middle East to work with me. He would help me to get an NIMH grant to carry on this work. I was more than willing to accept his proposal, even if I was unclear about what it all meant. Bringing research funds to AUB would bolster my standing; what I would do with the funds was less important. Lyman and Stanley Diamond, an anthropologist on his staff, wrote an elaborate grant proposal for $230,000 (a large sum in 1962) to the NIH with me as the principal investigator. It was a convoluted project that included research instruments I was only vaguely familiar with. I met with the visiting advisory review committee in a hotel room. It was an impressive group that included eminent psychologists like Starke R. Hathaway, the developer of the MMPI.[3] The committee members had finished a long day reviewing proposals and were relaxing with glasses of scotch. They offered me a drink, but I politely declined. I felt confused enough without befuddling my head any further. They asked me a lot of questions about the proposal. They were particularly interested in and rather amused by its bewildering array of psychological tests. Fortunately, the MMPI was not one of them, and it spared me from justifying its use to its originator. I was clearly in over my head, but I improvised the answers as best as I could. I knew how to swim in stormy waters and not drown.

Two weeks later, I went to Lyman's office to hear the verdict. The proposal had been turned down. However, the review committee had been well impressed with me personally. They proposed giving me a grant of $50,000 over three years to use more or less as I saw fit in carrying out my work. There was so little psychiatric and behavioral science research from the Arab countries in the region that whatever I did might be of some value. This was not what Lyman was hoping for, but he took it well and readjusted his expectations of what I would do in Lebanon. I was very pleased, even if I didn't realize fully at the time how significant this was going to be for my career. In those days, few AUB faculty members received grants of this size, and I was just joining the faculty in a junior position. I received a letter from Calvin Plimpton (my old professor of medicine and now chairman of the board) congratulating me and saying how proud he was of me. I was surprised that he even knew about my joining the AUB faculty, and this added to its significance.

The grant gave a huge boost to my standing even before I had arrived at AUB. It would pay for 70 percent of my salary and would free me from most of my obligations for teaching and seeing patients. Since the grant was ten times my starting salary of $5,000, I would be basically paying my way. The fact that the money was given to me personally also meant that I would no longer be beholden to Lyman. However, I knew full well that I couldn't possibly have gotten the grant on my own. I also wanted to work with him—I wouldn't know what else to do with the money. The NIMH grant was also to change my benighted perception of psychiatric research. Now that I was a researcher myself, I could appreciate more fully the challenges that investigators in psychiatry faced in doing what could be done. It put me on a path that was to prove critical to further developing my academic career and eventually obtaining tenure at Stanford.

~

I am not sure when is was that I first thought of going back to Lebanon the long way, across the Pacific. It must have been after I saw a brochure for the *SS Arcadia* of the Peninsular and Orient Lines. It sailed from Vancouver to Barcelona via Port Said, where I could get off and go on to Beirut. The voyage would take about six weeks and cost $1,000. It was a journey not to be undertaken lightly, and I didn't have the money for it. However, I had not had a real vacation for years. My life had taken a turn for the better, and I owed it to myself to go on this adventure. Even though I had never before asked my parents for money, this was a good time to do it. I wrote to my father and he agreed to send me the fare for the journey. His generosity helped assuage the smarting I still felt over the puny allowance I got when growing up.

I first needed to get myself to Vancouver to board the ship. I asked Lyman to have the NIMH pay for my journey from Washington to San Francisco. I needed to meet with Margaret Singer in Berkeley to learn from her how to administer the Rorschach tests in Lebanon for her analysis. Lyman agreed and I got the funds for the trip to the West Coast. I flew to Salt Lake City and then took the California Zephyr across the Sierras to San Francisco. After meeting with Margaret, I would drive to Seattle and on to Vancouver to board the ship.

I spent two days in Salt Lake City. The Mormon Tabernacle was impressive, but I was not allowed to go in to see the interior. I went to the Great Salt Lake and swam in it (unless I am confusing it with the Dead Sea, were I did swim some years later). I was collecting experiences with a voracious appetite for seeing and doing everything I could and wherever I went. The train journey was spectacular. I had never imagined the size and splendor of the country I was going through. Hour after hour, I looked out of the window at the sweeping landscape and let my mind run free. San Francisco was fascinating, and I walked all over the city. Margaret Singer and I spent quite a bit of time together. I began to realize that giving the Rorschach tests in Lebanon was not going to be easy. It actually looked like a harebrained idea, but I kept my doubts to myself. I would figure something out.

Margaret remembered that David Hamburg, now at Stanford, had brought me to the NIMH and she invited us both to dinner at her home in Berkeley. I couldn't fathom why David would come all the way from Stanford to have dinner with me. I thought he would hardly remember me by now. Yet, there he was and after the dinner, as we were parting company, he said to me, "If you get tired of Lebanon, I hope you'll think of Stanford." I had heard of Stanford, but that was the extent of it. I am not sure I even realized that the Stanford David Hamburg was referring to was what Cy Worby had so admiringly told me about the year before. I thanked David and went on to my hotel in San Francisco without giving the matter further thought.

In San Francisco, I met with Herbert and Judith Williams. Herb taught anthropology at San Francisco State and Judith was a psychologist. The Williamses had carried out a study in a Lebanese village some years back, and Lyman had hired Herb to work with me in Lebanon. We had dinner at Omar Khayyam's restaurant in deference to its famous Armenian owner, George Mardikian. It was the beginning of a half-century-long friendship.

I rented a car and drove to Seattle for the World's Fair with its famous Needle. At an Argentinean restaurant at the fairgrounds I was served an inedible chunk of meat consisting mostly of bone and fat. I should have refused to pay for it, but I said nothing and didn't even touch it. I then drove through a national park to a remote post on the Canadian border. I had been told in Washington that I needed to get a sailing permit showing that I had paid my taxes during my years in the United States. I went to the central IRS office to get the permit. The nondescript agent turned out to be a shark feeding on small fish like me. He figured out that I owed $270 in back taxes and seemed to take perverse pleasure in telling me that. Given my meager income, that sounded incredible, but I had no choice except shell out the money even though I needed every penny for my trip. When I stopped my car in front of the small booth at the Canadian border crossing, I had my sailing permit at the ready. The gruff and grumpy guard hardly looked at me and said, "Passport." I gave it to him. He stamped it and handed it back. Would he like to see my sailing permit? He didn't answer. I said I had it right there in my hand. He shook his head. I told him it had cost me $270, sounding a bit hurt. He glared at me, pointed to the road and growled, "Go!"

The *SS Arcadia* was docked at the Vancouver harbor. It was by far the largest ship I had even seen. Hundreds of passengers and crew swarmed on the decks. The quay was packed with people who had come to bid farewell. They threw streamers and frantically waved and yelled to catch the attention of travelers on board. I knew no one. As the ship began to pull away, a band struck a mournful tune. People began to cry. I was so moved that my eyes began to well up as I waved slowly at nobody in particular.

The voyage on the *SS Arcadia* was a once-in-a-lifetime experience. I was one of seven hundred and thirty-five tourist-class passengers and shared a cabin with three others. The food was bland "English" cooking with overcooked roast beef alternating with nondescript Yorkshire pudding, but the freshly baked buns sustained me. The orchestra could hardly keep the beat, but it was nice to have music to dance to. The majestic Pacific made up for all shortcomings.

Many of the passengers were Australian and quite a few about my age. (And they smelled very different from the Australian soldiers I remembered from World War II.) Quite a few of the younger passengers were returning home after working and traveling in the UK and North America for a year or two. I became part of a small group consisting of a married couple, two sisters, a Communist high school teacher, and Jennifer—a nurse and part-time model. Every evening after dinner, we gathered in the lounge to sip brandy, talk, dance, and while the time away. The brandy was not cheap so I bought a bottle during a port-call and after buying the first glass, I would go back and forth to my room and replenish it. During the days, we sat around the pool and idled away our time.

I also spent a lot of time by myself on deck staring at the endlessly fascinating ocean that constantly changed its color and character. Calm days with lazy swells alternated with rougher seas and occasionally howling storms. The Pacific looked more like a collection of seas than a single immense body of water. As I looked at the flying fish and dolphins following the ship, I reflected on my life since leaving Beirut four years earlier and looked ahead to the life awaiting me. The ocean looked as mysterious and disquieting as the future, yet strangely comforting in its vastness.

My relationship with Jennifer could have easily turned into a shipboard romance, but it did not. We became good friends without setting ourselves apart as a couple. Jennifer was engaged and was going home to get married. She had doubts about her fiancé, who was overly fond of beer and soccer. I sensed that she might have given him up for me, but the issue never came out into the open. When we docked at Sydney, Jennifer took me home to have lunch with her widowed mother. Her father was an anesthesiologist who had died some years earlier. We said goodbye rather wistfully and I re-boarded the ship. Had events taken a different turn, I could have easily spent the rest of my life in Australia.

It would be hard to give a detailed account of this amazing journey. When one is constantly on the move, there are too many distractions to allow epiphanies. However, even as a passing tourist, the exposure to cultures so different from my own had a profound effect on me. In my previous, more limited travels, I had mainly focused on the sites I was visiting. In this case, it was the changing sea of people that held my interest. Just being with them without knowing their language or culture gave me a profound sense of the enormous diversity of God's many children. A sea voyage has the additional advantage of providing a respite between port calls and the time to catch one's breath in familiar surroundings to reflect on and absorb what one has seen. The six weeks I spent on the *SS Arcadia* were to remain one the most memorable periods of my life.

The journey started with our sailing from Vancouver to San Francisco and Los Angeles. We then headed for Honolulu, where I walked from one end of Waikiki beach to the other. I attended a luau where the hula dancers outdid the roast pig, or whatever it is that we ate. (I could not have imagined that I would have a son who would spend many years on Maui). Japan opened a whole new world I didn't know existed. We docked at Yokohama and visited Tokyo, which I found bewildering. A few us then took the train to Kyoto, with its fabulous temples, and where we spent two nights at a country inn before boarding the ship again at Kobe. I took my first Japanese bath at the inn. I knew these baths were hot, but I couldn't have imagined how hot until I dipped into one. I would have instantly jumped out had others had not been in the pool. In the evening, I wore my newly acquired blue silk kimono with red lining, and it made a great impression on the Japanese at the inn.

We then went to Hong Kong and to yet another new world. The streets were teeming with people. I had been told that the Chinese tailors could make a suit overnight. I didn't really need a suit, but the idea of having it made overnight appealed to me. On the morning we docked, I found my way to one of the tailors and I ordered a jacket of "bleeding" Madras with a pair of slacks. When I went to pick it up, the jacket fitted surprisingly well but one leg of the slacks was an inch shorter than the other. The tailor pulled the short leg down to make it look even and kept saying, "Very good, very good." I asked him if he was going to travel with me to hold down my trousers, but he didn't get the joke. I refused to pay him for the slacks, but he assured me in his breathless, halting English that he would fix it and bring it to the ship that evening. He did, but the slacks still didn't fit. He was so distraught over how he was going to feed his "ten children" that I didn't have the heart not to pay him and he left with profuse thanks. I gave the slacks to the attendant of our cabin, but kept the jacket. As its colors bled over time, I came to like its bleached look and wore it for many years.

I knew even less about the Philippines than I did about Japan or Hong Kong. I was surprised to hear people speak Spanish in Asia. Jennifer and I went to a nightclub where people checked in their side arms before going in. There was a fabulous orchestra playing the *paso doble* and we danced into the small hours of the morning. The ship then headed south on our enormous swing around Australia and stopped at all the major ports—Sydney, Melbourne, Adelaide, Perth. We were merely skirting the southern edge of the continent, whose enormity was hard to comprehend. After the last Australians disembarked, the trip no longer felt the same.

Our next port of call was Bombay (now Mumbai), where I roamed the streets feeling perfectly safe among the bustling crowds of strangers who were remarkably kind and gentle. Despite the masses of people thrown together, there was no shoving and yelling, which made it easier to disregard the garbage and squalor. A shoeshine man with a small, worn-out brush offered to shine my shoes. My shoes didn't need shining, but I couldn't turn him down. As he labored at it, he offered me the glass of tea he was sipping from. I felt a real connection with him. Had the accident of birth not set us apart, we could easily have been in each other's shoes. When I paid him with a generous tip, he knew it was meant as a gift and acknowledged it with a smile. That evening, as we resumed our journey, I noticed that I had been moved to another dining table. When I asked the purser about it, he told me that another doctor by the same name had come on board. Since I don't have a common name, I was incredulous until I actually met the man. He was a dentist from Bombay in his sixties and had in fact the same name.

We then went on to Aden and Port Said, both small, dusty towns of no charm. But it was fascinating to go through the Suez Canal. I disembarked at Port Said and took the train to Cairo. I didn't realize how dusty the trip had been until I tapped my shoulder of my jacket and a cloud of dust rose from it. Once off the ship, I also realized how oblivious I might have been of the risks I had taken during the trip. When on port calls, I would leave the ship in the morning and wander off into a foreign city until an hour or so before the ship was going to leave. What if I had gotten lost, robbed, and failed to get back in time? How would I have caught up with the ship again? I had little money on me and there were no credit cards at the time. Given my irrational fears of getting lost, how could I have risked the worst possible way of getting lost?

My parents and a dozen relatives were waiting for me at the Beirut airport where I flew in from Cairo. Lucine had stayed home to complete the preparations for dinner, when we would be joined by more guests later on. It was a joyful and poignant reunion. My parents and Lucine had aged. I too had changed. I had left four years earlier as a fresh graduate from medical school and was now returning as a professor. I had been trapped in a marriage my family never warmed up to and was now getting free of its shackles. The prodigal son had returned home; the shipwrecked sailor was now standing on solid ground again.

There were many questions and fragmented conversations, but mostly expressions of affection. My relatives hugged me and pinched my cheeks, as if I were still a little boy. My parents stood aside; there would be plenty of time for us to talk. They looked happy and vindicated. I had returned to their bosom. Lucine seemed rejuvenated at the prospect of taking care of me again. Everyone looked more cheerful than when I had left Beirut. The Lebanese civil war had ended and peace and prosperity returned to the country. There was also peace in my heart and the promise of starting a new life that would make me forget the frustrations and turmoil of the past four years and forgive those who had caused it—most of all myself.

8

Home Again

IT TOOK ME A FEW DAYS to find my bearings. My parents had refurbished my old bedroom with new furniture and curtains with a Lebanese version of Mondrian motifs. My mother had chosen them with the help of younger women relatives with European tastes, even though she had a better taste herself. My father had the oriental, squat toilet replaced with a modern, Western one. The kitchen was the same; since I wouldn't be using the kitchen, my father had seen no point in fixing it. I was happy that the rest of the apartment had not been tampered with, since I had come to appreciate it even more after four years of absence. I was touched that my parents had gone to so much trouble.

After I settled in, we talked about my appointment at AUB and my grant. They were pleased and proud of me. It had been years since we had had a conversation like this. However, there was no discussion of Sylvia or my marital situation. That topic was too loaded and painful to get into at this time. I was now back home, and that was all that mattered. None of my relatives said anything about Sylvia either. They had never warmed up to her or to her family and were hardly perturbed to see my marriage coming to an end.

The fact that my parents expected me to be living with them may seem strange. I was almost thirty years old. Was it not time for me to be living on my own? Actually, *not* moving in with my parents would have struck people as strange. Why would I want to live apart from them when I did not have my own family? Didn't I like them? It would have been different of course had Sylvia been with me.

What I needed most right away was a car. My parents had thought of buying me a Peugeot on a cousin's advice. I was relieved that they had not. Cars were an important part of one's public persona in Beirut, and a Peugeot wouldn't do for me. I went back to the agency where we had bought my little Fiat years earlier. There was a virtually new white Jaguar with a 2.4-liter engine that somebody had just traded in for a more powerful model. It whimpered rather than roared but it was a beautiful car, with blue leather seats and walnut burl instrument panel. It was reasonably priced and I could buy it with a small loan. The Jaguar did wonders for my image, which I feared had been tarnished by my marital woes. But the car had a temperamental engine that required frequent trips to the garage. It was like having a capricious and high-maintenance mistress, but the satisfactions it provided made it worthwhile. Even my father was impressed when several people complimented him on my good taste. Appearances counted for half of reality in Beirut.

~

A few days later, I went to AUB to see John Racy. We were happy to reconnect. I told him about the year I spent at the NIMH, and he told me about his work at AUB. He had conducted a survey of psychiatric practice in Arab countries and published a book based on it. I would be sharing his office for seeing outpatients, and he was most helpful in making me feel welcome in my new position. I next met with Dr. Fuad Sabra, a neurologist who was head of the Department of Medicine, where John and I had our appointments as assistant professors. There was no department of psychiatry yet. I was disappointed with my salary of $5,500 a year, especially since most of it was paid for by my NIMH grant. But that was the university salary for Lebanese professors, and there was not much I could do about it. On the other hand, my teaching and clinical responsibilities would be minimal and I could devote the bulk of my time to my research on my own schedule. Fuad Sabra hoped that I would see a full load of patients and generate revenue for the department, but I was not about to do that. I was happy, however, to see a few patients and to help John teach the psychiatry course for medical students.

No one had given any thought to providing me with a research office. When I met with Dean Joseph Macdonald, he was vague about the prospects. There simply was no space in the medical school or the hospital. He suggested that I take over the medical student lounge, but I couldn't possibly do that. It took some persistence and veiled threats to report the situation to the NIH before an office was made available in Fisk Hall on the main campus, in the building that also housed the English Department. I put a handwritten sign on the door that read "Psychiatric Research," and someone from English added the words "Center for." It was comforting to know that someone would be watching over my grammar.

For a start, I would take a stab at Lyman Wynne's Rorschach study project with families who had a schizophrenic offspring. Two such families had been referred to me for treatment, and I could turn them into research subjects. One of the families was Palestinian and consisted of elderly parents with three sons and a daughter. Only the middle son worked and supported the family. The youngest son, Fuad, was the patient. He was in his late twenties and had been in and out of mental hospitals after repeated suicide attempts. The psychiatrist who referred him to me was Dr. Henri Ayoub, head of the Psychiatric Hospital of the Cross (*Deir el Salib*). It was run by a Catholic order and staffed by French-trained Lebanese psychiatrists—essentially the French counterpart of the hospital at Asfouriyeh.

Dr. Ayoub had tried to "psychoanalyze" Fuad but had finally given up. Fuad was not the only end-of-the-rope patient he would refer to me. However, given Dr. Ayoub's position, I agreed to take his referrals. The fact that he thought I might succeed where he had failed could also be taken as a compliment. Fuad had been diagnosed as a schizophrenic, but his primary symptom was a debilitating depression. He had neither worked nor had any friends for years. I had had another schizophrenic patient at the NIMH who complained of being besieged by "too much world." By contrast, Fuad had "too little world" in his life. He morosely shuffled through it under a heavy dark cloud and was so relentlessly pessimistic that just seeing him from a distance ruined my mood for the day.

The second was the family of an Armenian baker who lived in the working-class neighborhood of Bourj Hamoud. The elderly parents still ran the bakery. The younger son helped, but the older one, Sarkis, who was the patient, was the mainstay. When he was in his mid-twenties, Sarkis had become withdrawn and uncommunicative, showing unmistakable symptoms of schizophrenia. However, as long as he silently labored at his work, his parents learned to live with his illness. When he stopped working they took him to a physician, who gave him cortisone shots, which was very odd, and that made his condition worse. The physician then told the parents to get in touch with me. I had just come from America and would know the latest methods of curing him.

I met with each family and explained to them that I would treat the patients free of charge in exchange for the family's participation in my research project. They didn't quite understand the research part, but I was a professor at AUB and was offering them free services so they would put up with the research, whatever that meant. We had a few "family therapy" sessions and I arranged to meet with each family member for the Rorschach tests. The Rorschach ink blots were baffling. They would look at a card, look at me, then look back at the card again. What on earth was this about? I brushed aside their skepticism and pressed them to tell me what they thought the cards looked like. They finally caught on and came up with: A bee? A bat? A splattered piece of dog shit? The prospect of eliciting their cognitive styles looked hopeless. I felt sorry for putting these poor people through this insane exercise. It was especially hard with Sarkis's parents. They looked at me ruefully: Was this the sort of thing Armenians did to other Armenians? How was this going to help their son? I evaded their questions, but at the end I couldn't bring myself to even try to translate the protocols into English. They made no sense. I wrote to Lyman Wynne and told him this was not working out and he readily accepted it. He probably had had meager hopes that anything useful could come out of it anyway.

At the next family session, I announced that the research project was over and from here on we would work together to make their sons get better. Shortly thereafter, Sarkis went back to work. His parents were delighted. They insisted that I come to the bakery to sample the freshly baked bread, and have people to see the miracle worker who had cured their boy. The fresh bread was delicious, and after I shook a few hands, I decided to declare victory while I was ahead. It was much harder to make headway in Fuad's case. The group sessions helped reduce the tensions in the dysfunctional family, and I continued to treat Fuad for the next three years. I didn't cure his depression, but he stopped trying to kill himself (at least, not as often).

The outpatients I saw at the AUB clinic made me aware of the disconnect between the psychiatry I had learned at Rochester and the psychiatry I was now practicing in Beirut. My Arabic had gotten even rustier. Therefore, most of the patients I saw were American expatriates or spoke English. The idea of psychotherapy was strange in a culture where people openly talked about their problems to others all the time. They did not need to pay a psychiatrist to listen to them. The other psychiatrists in practice mostly prescribed drugs. I was still biased against using drugs instead of psychotherapy, which was particularly naïve in that context. Yet I thought this was true not only for patients but for people in general. When the husband of one of my cousins told me that he was having occasional headaches for which he took aspirin, I told him that that wasn't a good idea. It would be better to find the cause of the headache and treat it. That made no sense to him, but he was too polite to say so.

John Racy and I were hired as consultants by the Aramco oil company and we took turns flying to the hospital in Dhahran in Saudi Arabia. The patients we saw were mostly Saudis who complained of physical symptoms for which no medical cause could be found. One man with insomnia insisted that he could never, ever sleep. I didn't think such a thing was possible and he gave up on me.[1]

Even though I knew it was going to be hot in Saudi Arabia, the first time I stepped off the plane I thought I had walked into an oven. As the summer temperature soared to 125° Fahrenheit, I couldn't walk on the cement sidewalk in shoes with leather soles and had to run from one lawn to another. Nonetheless, Saudi Arabia had its attractions. The most interesting part of those trips were the excursions to see the red sand dunes in the desert stretching out to the horizon. We were awakened on morning by the bells of a passing camel caravan. Another unexpected benefit was the tax-free shopping at the market of Khobar, the port town of Dhahran. The Egyptian physician who took me there said had I come ten years earlier, we would have gone to visit one of the local dignitaries and a servant would have passed around a tray of Rolex watches, like candy, for guests to pick from. Those days were gone, but I could still buy a Rolex Oyster Perpetual watch for $100. He was very proud of his Rolex and said the watch could be taken down to a depth 2,000 feet. I asked if he had ever done that. Was I kidding? He took his Rolex off even when washing his hands. But didn't he just tell me that the watch was waterproof? "Yes" he said, "that's true, but why tempt the devil?" I bought a Rolex anyway and kept it for three decades. It never kept good time and I spent hundreds of dollars having it serviced. Finally, I sold it to an antique watch dealer for many times what I paid for it. There must have been people who were dumber than I was who would buy it. But I must confess, I occasionally do miss it, specially after my daughter told me she always associated the Rolex with me since her childhood.

Another purchase at Khobar turned out to be one of the best deals that ever came my way. I paid another $100 for a magnificent carved heavy wooden chest with elegant brass decorations. It had come from the Malabar coast of India and was used as a bridal chest. Since the dark and heavy wood it was made of weighed a ton, I couldn't take it to Lebanon, where duty on furniture was assessed by weight. However, by that time we were planning to move to the United States in year or so. Could I have it shipped there? My companion on that shopping visit was an American male nurse. He said shipping it to the United States would be no problem. The oil company paid for the transportation of the goods of employees returning home. When a family he knew was leaving for the States they would bring the chest with them and I could claim it from them. It might take a year or two, but I would eventually get the chest. I decided to take a chance and bought it, with little hope of ever seeing it again. A year or so later, after we had moved to Stanford, I had a phone call from someone who told me he had my chest from Saudi Arabia. I asked where in Texas was he calling from, since many of these people were from Texas. But I was wrong. He was calling from a motel in Palo Alto, ten minutes away from where we lived. I don't remember how I managed to get the Malabar chest home but it became one of our most cherished possessions.

~

For my next project, I had to find something more feasible to do. I stumbled on the idea of conducting an epidemiological study of mental illness in Lebanon. It may have been inspired by John's survey of psychiatry in the Arab world. The project was hardly exciting but nevertheless necessary to establish a baseline for the prevalence and distribution of mental illness in the country—essentially Beirut, where all the twenty-one psychiatrists practiced. The idea was quite simple. Each psychiatrist would fill out a one-page questionnaire on each patient he saw for the first time. The project would span a period of six months. To help set it up I went to Denmark and Norway and met with eminent epidemiologists who were conducting long-term epidemiological studies based on the excellent statistical data available to them through public registries.

One serious problem I faced was in computing rates of mental illness for the general population and its component groups such as women and men. To do this, I needed current statistical information on the population of Lebanon, but no such data existed. The last census was taken over forty years earlier. The government would not conduct another census because its results would negate the fiction that the Maronites were the largest religious group, thereby justifying the president having to be a Maronite too. (This is still the case another forty years later.)

John Racy was very helpful in putting me in touch with those psychiatrists in Beirut whom I did not know, but I still had to convince them to take part in the project. I would pay each participant a small fee for each completed questionnaire. It wasn't much, but it only took a few minutes to do the task. All but one of the psychiatrists took part in the study. The one exception changed his mind when one of his patients who had heard about the project expressed concerns over confidentiality, even though the surveys were anonymous.

I published a number of papers in psychiatric journals based on our general results and then several more on more specific aspects, such as social class differences in the prevalence of mental illness, with my sociologist colleague Charles Churchill (who had actually taught me as an undergraduate). I presented a paper at the 1964 annual meeting of the American Psychiatric Association in New York. The discussant said it was the best study of its kind from the developing world (actually there were not that many studies to choose from). Beyond that, the study didn't get much more of a hearing. A well-known British

Figure 8.1 John Racy.

epidemiologist spread the word that my survey was not based on a random sample. It was a "service study" that counted patients who had seen a psychiatrist and not everyone with psychiatric problems in the population at large. That was true, but the same could be said for virtually all other studies of this kind; my study at least had the merit of being comprehensive by encompassing much of the country. It was like the dog who danced on stage. The point was not how well it danced, but that it could dance at all.

Once I got settled in our research office, I hired my first research assistant. Adele Takieddine had a master's degree in psychology from AUB and came from a well-known Druze family. She was married to a Druze lawyer, a wonderful man who later became a judge (and who was assassinated during the civil war because he was advocating peace and reconciliation). We could now start on another of Lyman Wynne's research ideas, the value orientation study in four different villages: a Druze village and a Shiite village in south Lebanon, Anjar (the only Armenian village in Lebanon), and its neighboring Sunni village Majd-el-Anjar.[2] I translated the questionnaire into Armenian, and Adele into Arabic. It then took most of one summer to gather the data. Adele and her husband rented a house in south Lebanon and did the work in two of the four villages. We had local assistants to do the same in Anjar and in Majd-el-Anjar. When we were finally done, I sent the results to Lyman. I didn't hear from him for a while. He had several statisticians analyze the results and subject the data to every conceivable statistical test. Nothing intelligible came of it. No wonder. It turned out that the assistant in Majd-el-Anjar had simply completed all of the questionnaires himself by randomly checking the answers. It was simpler than going around and tracking down the subjects and getting their responses. The whole thing was a colossal waste of time. To his credit, Lyman didn't complain that none of the projects he wanted me to carry out in Lebanon were panning out and I didn't complain that I was being sent on one wild goose chase after another. Nor did Lyman ever object to my undertaking research projects of my own devising and, on the contrary, tried to help me every way he could. Moreover, by being in charge of research projects, rather than merely a participant as I had been at the NIMH, I gained greater respect for psychiatric research. I realized how difficult it was to come up with significant results, let alone achieve a major breakthrough.

~

It was deeply satisfying to reestablish my ties with my aging parents and Lucine now as a more mature, independent adult. For the first time, I was able to help care for them. Even though my salary was modest, a dollar was worth over three Lebanese pounds and went a long way. I gave my father a third of my salary to help him build a four-floor apartment building that he could then rent out. I also gave some of my salary to my mother for household needs, so that I would not be living at home for free. When my parents had health concerns, my contacts at the hospital made it easier to deal with them.

I was happy to be back living at home. However, there was a constant stream of guests who came in the evenings, and I had to sit with them for long hours as a member of the family. I wanted to have a place of my own when I wanted to be alone or entertain my own guests. I found a small house, whose top floor was for rent, in the resort town of Beit Meri, which was located on a ridge with a commanding view of Beirut, the sea, and the surrounding mountains. My father couldn't understand why I would want to live at a mountain resort out of season. My mother suspected that I wanted to have a place where I could entertain women friends, and my father acquiesced. Nonetheless, I continued to have most of my dinners at home and slept in my old room whenever I stayed out late in Beirut, so I spent plenty of time with my parents and Lucine and I almost never cooked for myself.

I decided to have a house-warming party at Beit Meri and invited my colleagues. I wasn't sure how many would come, but to my surprise most of them drove up from Beirut. I first had the unrealistic idea of preparing the food myself, but my mother and Lucine took over and set up such a lavish buffet that so delighted the guests that they kept talking about it for weeks. It looked as if I still needed my mother and Lucine in my life, and I was grateful for it.

I was also happy to reconnect with my cousins, especially Elvira and Nora, and their families. On my father's side, I was closest to my cousin Jirair (who used to help me with math problems in high school). My relatives were proud of me. I nonetheless chafed at some of the demands they made, such as consulting me about their health problems even when they were about to see their own physician. But it was a small price to pay, especially since I too occasionally asked for their help. Such close links among extended families in countries like Lebanon are a mixed blessing. In a family whose members are doing well, these connections make useful alliances. However, when most of them are not doing well, they become a drag on the few who are.

I continued the no-emotional-commitment-to-women policy established in Washington. I was now better established, more self-assured and experienced (and the Jaguar helped). I became involved with an American woman from Houston called Ashley, whom I met at the ski resort at the Cedars. She wasn't there to ski and I was the sort of man she was looking for. Ashley worked in a neighboring Arab country for an American company and kept a small apartment in Beirut, where we spent occasional weekends together. She was enterprising, outgoing, and refreshingly uninhibited. Although only in her mid-twenties, she had already been married and divorced. She offered to marry me if I moved to the United States. Her independent yearly income was five times my salary at AUB. However, we had little in common beneath our skins, and she returned home soon after.

My most significant friendship in Beirut was to be with an American couple—Janet and Strother Purdy. Janet was a vivacious woman in her late twenties with a distinctive, long blond braid. I met her when she came up to me at a reception and five minutes later asked if I would like to go for a picnic the next day. The following morning, I went to the address she gave me and saw a tall man in the street loading something on a blue Volkswagen microbus. He paid no attention to me. A few minutes later, Janet came bouncing down the stairs, pointed to the tall man, and said, "Meet my husband, Strode." *Husband?* There was no mention of a husband the day before. Then another woman showed up and Janet introduced her as her sister-in-law. It became clear during the picnic that Janet was trying to fix me up with her sister-in-law, but it didn't work.

Strode Purdy taught English at AUB. He came from a prominent New York family, while Janet claimed Jesse James as an ancestor. The Purdys had lived in India before coming to Lebanon and were avid travelers. Strode was like a nineteenth-century English gentleman who pored over old Baedeker guidebooks and fraying maps looking for ancient sites and read extensively before going anywhere. Janet followed him for the sheer excitement. We went skiing together at the Cedars. And since I spoke Arabic and knew the culture, I became a good travel companion for them for their excursions. Strode opened my eyes to the history of ancient ruins in the region and sparked a lifelong interest in archeology and the classical world. We went to famous sites like Petra in Jordan—the Nabataeans capital, carved into the rose-colored sandstone—and Palmyra in Syria with its magnificent colonnaded piazza, as well as obscure places like Dura Europus on the

Figure 8.2 With Strode and Janet Purdy on our way to skiing at the Cedars.

Euphrates, which I had never heard of (and which looked like a pile of rubble).

On one of these trips, when crossing the northern Syrian desert, we almost got stuck going through a lava field strewn with rocks the size of melons. The microbus was no Land Rover, but it held up valiantly. The car didn't have a gas gauge, so Strode had to keep track of the miles we traveled between refueling. I lived in constant fear that we would run out gas in the desert although Strode never seemed troubled by that prospect. We finally got to Deir Zor, which for me had grim associations with the Armenian genocide. As we were driving through the town, Janet stuck out her head, covered in a flamboyant *keffiyeh*, through the opening in the roof and waved to the people in the crowded street. A band of teenage boys excitedly ran along the slowly moving van and began to rock it from side to side. I yelled at the men sitting at the sidewalk smoking the *nargileh*, and they stopped the boys before the car came perilously close to toppling over.

We continued our journey through the desert until we stopped some distance from an encampment of black goat-hair tents. Janet, despite my attempts to stop her, gestured to the Bedouin guard in front of the largest tent. The man, armed to the teeth with bandoleers across his chest, walked up to see who we were. I apologized for the crazy American woman and he went back to the tent and returned to say that the paramount chief of the largest tribe in the area wished to see us in his tent. Janet took off after the guard and we followed them. We were ushered into the main tent, where the chief held audience with some twenty men sitting cross-legged all around. A beautiful young girl with spectacular eyes enhanced by kohl sat at his feet. She must have been her father's favorite child. There were no women. I expressed my profuse regrets for the intrusion, but the chief was very gracious. A black servant brought in coffee. He first sipped from it ceremoniously (to show that the coffee was not poisoned) and then served in small cups the intoxicatingly aromatic, bitter desert brew. The chief asked who we were and said he knew about the American University in Beirut when I told him that Strode and I taught there. He also turned out to be a member of the Syrian parliament by virtue of being the head of his tribe. He said he had been a member of a Syrian delegation to the Soviet Union but hastened to assure me that he was not a Communist. I said the thought would have never crossed my mind. Finally he genially got up to bid us goodbye. On the last leg of our drive across the vast desert, we saw the tiny figure of a man waving frantically in the distance. Strode turned the car around and raced in that direction thinking someone was stranded and dying of thirst. When we reached the man, he greeted us calmly and asked if we had a light for his cigarette.

~

One Sunday afternoon, I got a call from an American colleague's wife. She wanted me to meet a young woman. Daphne was from Boston, a junior at Radcliffe, and currently studying at the Sorbonne in Paris. She was in Beirut for a few days visiting her aunt, who was married to a senior executive of an American company. The aunt wanted an escort to take Daphne around during her visit. I was given a date and a time to show up at a fashionable address on the Corniche. Precisely at seven, I rang the doorbell and Daphne opened the door. We stood transfixed. She was a classical American beauty with flowing ash-blond hair, pale blue eyes, and prominent, aristocratic cheekbones; a living image of Flora in Botticelli's *Primavera*, with her seductive distant stare and sensuous parted lips. Neither of us moved until she said, "I'll get my coat."

I took her to Les Caves du Roy, the most enchanting nightclub in Beirut. Built underneath a luxury hotel, it conveyed the illusion of a cave with an elegant and exquisitely designed interior. Its softly lit niches were filled with objets d'art and its recessed stained-glass windows enhanced its medieval ambience. The bands that played there came from the Riviera; even the Italian bartender ("Aldo") was world famous. We had dinner, and before we got to the dessert and dancing, Daphne leaned across the table and wanted me to kiss her.[3]

The following evening, we were invited to dinner with Daphne's aunt and uncle. Our host was the president of the third-largest bank in New York. He was visiting Beirut with his wife and daughter, Judy. The dinner was at the Paon Rouge in the Phoenicia Hotel, another legendary nightclub. I rang the door to the apartment and this time Daphne's aunt opened it. I introduced myself and she took me to their living room. The aunt was an elegantly dressed woman in her fifties who exuded a sense of supreme self-confidence. Her husband was a tall, handsome man in a dark Brooks Brothers' suit, a white buttoned-down shirt and understated tie—straight out of central casting—but the money and the pedigree appeared to be on his wife's side. Daphne's aunt thanked me for escorting her niece while looking me over with an inquisitive smile wondering if that is all that I was doing. Her husband was less intrigued by me and more taken by the prospect of having dinner with the head of the third-largest bank in New York. Daphne joined us shortly. She looked stunning.

We met our hosts at the nightclub for a fine dinner. Judy was a somewhat slimmer version of her heavyset mother and totally eclipsed by Daphne. Her date was a nice-looking young American, recruited locally, who looked ill at ease. Following dinner, Daphne and I got up to dance, but Judy's date was so intimidated by my furious twist that he never made it to the dance floor. Judy's father tried to have her dance with him, but she refused. I whispered to Daphne whether I should ask Judy to dance with me. She said I didn't have to. I felt awkward. I was brought up with the notion that being indebted put one in an inferior position ("being under"). I had now eaten the New York banker's bread and the least I could have done was dance with his daughter. So I asked Daphne again and this time she consented. But it was too little, too late and Judy looked disappointed; in some ways, so was I since she was so obviously attracted to me.

Daphne had to return to Paris by the end of the week. I was completely hooked and hoped she was, too. I told her I would come to Paris to see her as soon as I could take some time off. She tried to dissuade me. She would be finishing up her classes at the Sorbonne and getting ready to return to Boston in a few weeks. She didn't know how much time she could spend with me. I said I would come anyway. I must have sensed that this might be the end of our relationship since I arranged for several more stops in Europe as part of my first summer vacation.

I arrived in Paris during a spring thunderstorm. I checked into a small hotel recommended by Frommer's *Europe on $5 a Day*. The rain kept pouring down and I stayed in my room; it would be another two days before I could meet with Daphne. I didn't feel like going anywhere or seeing anything. It hardly mattered if I was in Paris or in Peoria. I lay down on my bed feeling miserable. I had bought a small bottle of brandy on the plane. I got up and took a few swigs. In a few minutes, my mood lifted and I felt calm. I was astonished and alarmed at the effect of the mere sip of brandy. If alcohol could provide such relief when I felt miserable, I could have easily become an alcoholic if I felt miserable in life. My calmer mood made it easier to reflect on what I was doing in Paris. After the heady days in Beirut were over, I wasn't sure if Daphne wanted to see me again, and I had my own doubts about seeing her. I knew that underneath her poise, and exquisitely polished exterior, Daphne was a conflicted and confused twenty-year-old junior in college struggling to get a grip on her life. She had been traumatized by her parents' contentious divorce and felt alienated from them. Although I was not about to play the psychiatrist with her, I cared enough not to turn a blind eye to her problems.

Our dinner and the night we spent together were anticlimactic. We ate at an exclusive private club at Daphne's suggestion (where the spectacular bill wrecked my $5-a-day budget). In the middle of the night, as we slept in my room at the hotel, a banging on the door awakened us. It was the elderly hotel manager. She was waving a paper and gesticulating excitedly that French law required that all hotel guests be registered. I groaned. Daphne filled the form and we went back to sleep. In the morning, when she left my hotel, we were affectionate but had little to say to each other. We both knew that we wouldn't see each other again.

The experience left me worn out and annoyed with myself. I had again gone impetuously after an unattainable and problematic object. What sort of future did I think Daphne and I could have together? There were chasms separating our lives. Even if she had been within my reach, I was far from sure I would want to spend the rest of my life with her. Why, then, pursue her relentlessly? In a perverse way, my experience with Daphne brought back echoes of my relationship with Sylvia. Would I never learn and grow up? I didn't hear from Daphne for a long time. She then called me at home from the Beirut airport one evening when she was on her way to India. I was caught off guard and my reaction was guarded. She apologized and hung up. Several decades later, I was watching a program on public television about terminally ill cancer patients meeting at a retreat in the Santa Cruz mountains. There was a middle-aged woman who caught my attention. Her face ravaged by cancer, she still looked hauntingly beautiful and strangely familiar. Who could she be? When she turned her face, I saw the expression in her eyes and I knew. It was Daphne.

~

From Paris, I flew to Copenhagen. I was in my modernist period and I was keen on Scandinavian design and teak furniture. I couldn't afford to buy any of it, but that didn't stop me from going from one store to another. At one of them, a young woman who worked there spent a lot of time telling me about the beautiful objects in the store. Her name was Ingrid. I was fascinated by her esthetic sensibility and told her I wanted to buy something affordable and easily portable. She suggested a set of steel cutlery designed by Arne Jacobsen, the Danish architect. I had never heard of him. The pieces, especially the knives with their tiny blades at the tips, looked as if they had come from outer space. I bought it anyway. (It was the cutlery that was to be used five years later in Stanley Kubrick's science fiction film *2001: A Space Odyssey*).

Ingrid had been very kind to me and I invited her to dinner. She then took me to her room in an attic and we went straight to bed. It seemed like the natural thing to do and we felt no need to discuss it. I woke up late in the night when Ingrid was fast sleep. There was a small window in the low slanting ceiling that opened to the sky. I stood up on the bed and looked through the skylight over the roofs of the slumbering city. As the cool Nordic breeze washed over my face, I thought about where I was and what I was doing. Ten years ago, I had refused to meet with Talin, whom I would have died for, to uphold an ethical principle. A year ago I had left Sylvia. Two days ago, I had parted from Daphne. Today I had met a woman in the afternoon, talked with her about Scandinavian design, and was now in bed with her. Did this betray an erosion of my moral fiber? Had I fallen from grace? Such qualms used to bother me, but no longer. I was never unfaithful to Sylvia while we were together. I had played the game by the rules but gotten nothing but grief for my pains. The rules had now changed. As I inhaled the fresh night air, I felt a profound sense of peace with myself. Ingrid stirred and drowsily told me to shut the window and get back into bed.

Stockholm was next. Everything was so clean, so neat and tidy. I stopped at a kiosk in the Djurgården park. The blonde young woman behind the counter with clear smooth skin and a lovely smile handed me a delicious drink of fresh cold milk in a finely designed and sparklingly clean tumbler. It didn't have the excitement of a glass of Bordeaux but it was more soothing. The same was true for the slanting light that lingered into the evening. Unlike the blistering sun of Lebanon, the gentle rays of the Nordic sun lightly caressed my skin.

I had read in my guidebook that one should not leave Scandinavia without going to a sauna. So I went to a public sauna and decided to first have a massage. The masseuse was a powerfully built woman, who grabbed me by the abdomen and almost lifted me off the table and then kneaded me like a lump of dough. I finally got off and wobbled into the steamy hot sauna with a towel around my waist. A dozen naked men sat sweating on the benches and looked at me with peculiar smiles. I was too naïve to realize I was in a gay bath. One of them came up to me and asked if I had been in a sauna before. I said I had not and I had no idea what to do. He was happy to show me and we went through the whole ritual, ending up with the cold shower, which I thought was a bit crazy. We left together and he asked me what my plans were for the evening. I told him I had read about a place on Kungsgatan, where one could go dancing without bringing along a partner, and hoped to go there. He said he knew exactly where it was and would take me there. But could we first go to his room so he could change? We went to a dorm at the university where he was attending a conference. When we got there, he poured out for me a large glass of whisky with a trembling hand. My guard went up as I realized he was gay. I was more disappointed than upset. I told him I had no problem with his being attracted to men, but I did mind his bringing me to his room under false pretenses. He had wasted a lot of my time. By the time I got to Kungsgatan, all the women would have gone home. He was profusely apologetic. We would leave right away. On the way, he told me that he was a Finnish dentist who would have liked to have been married and have children but he didn't have that choice. There was an unhappily married woman who wanted to sleep with him, but he couldn't bring himself to do it. Unfortunately, she was in Helsinki; otherwise, he would have introduced me to her. We got to our destination and stood for a moment in the street. I asked him if he would consider coming up to the ballroom with me. He shook his head. I went upstairs feeling rather sad and he went on his way.

~

I had one last task in Stockholm. Back in Beirut, Janet had been talking endlessly about her Finnish friend Stina, keeping up a drum roll of *Stina*, *Stina*, *Stina*. When I went out to a nightclub with the Purdys on a double date, Janet would lean into my ear and whisper, "Your date is very nice, but wait till you meet Stina." Stina was tall, slender, and lovely. Her delicate nose had the sweeping curve of a ski ramp. Stina was going to be in Stockholm during my visit and Janet insisted I contact her. I agreed in order to get Janet off by back. However, when I checked into my hotel there was a note from Stina that started with "My *séjour* in Stockholm is over." It went on to explain that she had gone home but if I came to Helsinki she would be happy to show me around. There was a telephone number and I called her. I used to pride myself that I could tell what people looked like from the way they sounded on the phone. Hearing Stina's voice, I concluded that she was a heavyset woman like a Wagnerian soprano and her flesh quivered when she laughed, which she did throughout our conversation. (Stina told me later that the reason she kept laughing was because she had just put on a recording of Shostakovich at the radio station where she worked and now she was talking to Katchadourian—one Soviet composer on top of the another.)

When I got back to Beirut, I told Janet I was not amused by her attempt to fix me up with a Wagnerian soprano. She swore that Stina didn't look like that at all and I would see for myself when she came to Beirut. Stina was coming to Beirut? Yes, she was. She had enjoyed living in Peru and she found the prospect of spending some time in Lebanon intriguing. However, Stina had to have a job to be able to afford living in Beirut. I told Janet that foreigners couldn't get a work permit in Lebanon unless the employer showed that no Lebanese could do the job. How was she going to get Stina a job? She said she couldn't but I could. And why would I do that? Because Stina was the most fabulous and amazing woman I had met and I would be forever grateful to Janet for introducing her to me. It sounded like another of Janet's wild schemes. However, on further reflection, the idea of giving Stina a job made sense. The psychiatrists in Beirut were giving a hard time to the young Armenian woman who was collecting the questionnaires from them. I thought the psychiatrists would take a European woman more seriously and since Stina didn't know anyone in Lebanon there would be less concern about confidentiality. I wrote a job description for which only one human being on earth could qualify. The dean was puzzled by its third-decimal specificity. But since it was my grant money, he approved it anyway. I wrote a letter to Miss Lindfors in Finland offering her the job, and she graciously accepted.

~

Stina Lindfors was born in Helsinki to an upper-middle-class family that was part of the 6 percent Swedish-speaking linguistic minority in Finland: the former elite during the seven hundred years that Finland was part of Sweden. Her father, Jarl ("Lale"), was a forester by training who had moved up to a senior administrative position in the

Finnish forest industry. Her mother, Runa ("Nunni"), was trained as a physiotherapist and was a pioneer in the women's physical education movement. She taught at a high school (which Stina attended) in Brändö, a suburb of Helsinki. Nunni was also a self-taught gifted painter who had never had the opportunity to attain the professional recognition she deserved. As a young woman, she had a friend who became a famous artist; their work could hardly be distinguished from each other's. Stina's older sister, Maj, was a well-known fashion designer married to Ola Kuhlefelt, a business executive from a socially prominent family.

Stina had always wanted to become a journalist. While in high school, she had published articles and had been editor of the youth page of *Hufvudstadsbladet*, the largest Swedish-language daily. However, since there was no program in journalism at the University of Helsinki, she ended up studying English literature and political science. Stina spent her junior year at the University of Wisconsin, at Madison, as an International Brittingham Scholar. It was there that she met Janet, who was an undergraduate married to Strode, getting his doctorate in English. The Purdys and Stina became fast friends. They traveled together in Europe and she became godmother to the Purdys' daughter Alexandra. After graduating from the University of Helsinki, Stina spent close to two years working in the slums of Lima for a Swedish organization that supported the work of the French Franciscan priest Abbé Pierre. After she returned from Peru, she obtained a position at the Finnish Broadcasting Corporation as an announcer and freelance producer.

Stina arrived at the Beirut airport at 9:50 p.m. on January 18, 1964. I met her alone because Janet didn't want her presence to dilute Stina's impact on me. Stina was indeed tall, slender, and lovely. We got into my car and I hoped she would notice it was a Jaguar. Instead, she politely asked me if it was a Rover. A *Rover*? No, it was not a Rover. That would be like confusing a racehorse with a mule. This was not an auspicious beginning. I delivered Stina to the Purdys with some foreboding. We had a glass of champagne and I left. *Ten days* later, we were engaged at the Caves du Roy. My engagement present to her was a thirteenth-century Byzantine gold coin on a chain that my father had given to my mother when they were engaged.

Stina and I have often wondered about our precipitous engagement. What were we thinking? What was it that transpired during those ten days that led us to such a momentous decision? After being married for forty-six years, neither of us has the slightest idea. I won't go into the details of what it was about Stina's charms, intelligence, and character that turned my head so as not to embarrass her any more than I already may have. Suffice it to say that there were two ways in which Stina stood apart from the women I had known before. She was not only intelligent but also an intellectual. I was not even aware of the distinction before I met her. To this day, she remains for me the model of the humanist without the intellectual pretensions that sometimes go with it. More importantly, Stina was imbued with an elemental sense of goodness. She was not religious in a conventional sense but was a truer Christian than many Christians I had known whose pious hallelujahs rose to high heavens with noisy gongs and clanging cymbals. She actually lived in a Peruvian slum amid people who made their living sifting through mounds of garbage located within smelling distance of where she slept. She learned their language, took care of their children, shared their joys and sorrows, and ate their stew made with the meat of guinea pigs. She didn't do this in response to divine command or ideological fervor. Nor did she want to curry favor or mollify her conscience. She did it because it was the right thing to do. I couldn't have done it; I don't know anyone else who could, but she did it and won over not only my heart and mind but my soul.

Our families were baffled. When Stina wrote to her parents, "He is Armenian. He is the most wonderful Armenian in the world. He dances and skis like a god. I'm going to marry him," it didn't sound like their level-headed daughter, although her sister Maj said, perhaps with a tinge of envy, "I knew she was going to do something like this." Stina's father had to look up "Armenians" in the *Nordic Family Encyclopedia* to find out who these people were. As for my parents, Finland, Iceland were up there somewhere where people sat frozen to the ground. My parents did have the opportunity to meet and come to love Stina before we got married. Her parents and sister laid eyes on me only two days before our wedding. (We told our children to *never*, *ever* do that to us, and they have not.)

~

Figure 8.3 Courting Stina at Les Caves du Roy.

Stina started working with me on the epidemiological study and was remarkably successful in getting the questionnaires completed by the psychiatrists. They liked her, and one of them invited her to go on a trip to Egypt with him. She then became an integral part of the project that formed the centerpiece of our research in Lebanon, a comparative study of culture and psychopathology in two villages—Anjar, an Armenian and Christian village, and Majd-el-Anjar, an Arab and Muslim village—a few miles up the Beirut–Damascus highway. The central question we would address was, How do cultural factors affect the prevalence and forms of mental illness? The basic idea was part of Lyman Wynne's interest in ascertaining the role of familial and cultural factors in the genesis of schizophrenia. However, I expanded it to include all forms of mental illness as an extension of the epidemiological study. In the fall of 1963 Herb and Judith Williams, whom I had met a year earlier in San Francisco, came to Beirut with their three young children. Soon after, Lyman Wynne also came to Lebanon with his family to spend his sabbatical year working with us. I found an apartment for them in Beirut and helped them get settled. I was glad to do what I could for them after all that Lyman had done for me. However, getting Lyman and Herb to work together was not easy. They had very different characters. Where Lyman was driven, Herb was relaxed; the harder Lyman pushed, the more Herb went limp. But somehow they managed to get along.

The research plan was to have Stina work with me in Anjar, where I would do the psychiatric survey and she would do the ethnography with Herb's help. In Majd-el-Anjar, Lyman was supposed to do the psychiatric component and Herb the ethnographic and cultural analysis. It was a highly unrealistic expectation. Lyman didn't speak a word of Arabic, and Herb was less than fluent. So nothing came out of that part of the project. The Anjar part was far more feasible, but it still took us two years to complete it. Stina learned Armenian (on top of the six languages she already spoke), taught the children Finnish songs, and became a beloved figure in the village as *Oriort* ("Miss") *Stina*. She and Herb carried out a census of Anjar and wrote a short ethnography of the village. I conducted the psychiatric survey and saw patients in the clinic. Stina and I lived in the village for a while. Our assistants Antranig and Rahel Kendirjian were enormously helpful. They were important sources of information and helped open doors for us. We became good friends, came to know their children, and years later we helped with the college education of their grandchildren in Beirut. Similarly, Herb, Judith, and then their three children became our, and our children's, lifelong friends. We spent Thanksgiving with the family over two decades when they lived in San Francisco. Herb died when he was only fifty as a result of a mismatched transfusion during an operation. Judith then moved to Providence and then New York, and we kept in touch with her and then with her daughters, Megan and Annie.

The Anjar study was the last psychiatric research project of my career. All my subsequent research had no connection with psychiatry. The most salient finding of this work was the way Armenian culture shaped the personality of the people of Anjar and, by extension, Armenians in general. The relentless emphasis on education, order, cleanliness, and material success shaped their character and made Anjar a spectacular success. In a few short decades, it was transformed from a muddy refugee tent camp to a thriving village and then a small town. By contrast, Majd-el-Anjar hardly changed. One example of the villagers' passivity was the way they dealt with the regional government. The villagers of Majd-el-Anjar got their water from a nearby source, which the women carried in jars balanced on their heads. Finally, the government embarked on a project to have water brought to the center of the village. However, the pipes provided for it were one segment short. So the women continued to carry the water over their heads and the men cursed the government under their breath; but no one did anything further about it. That would have never happened in Anjar. The Armenians would have complained, pleaded, cajoled, and bullied the authorities. They would have begged and bribed the officials, bought the missing pipe themselves, and, if all else failed, stolen it, but they would have gotten the pipe in place. All this drive and initiative, however, came at a price. The Armenian virtues that helped Anjar move ahead of its Arab neighbors could also turn into compulsive, rigid, and intolerant traits that sapped their character.

Stina and I wrote a book-length account of our study in Anjar that was to be published as part of an anthropological series. I sent the manuscript for the final review to a husband-and-wife team of sociologists. They suggested that we combine the Anjar study with a similar study they had conducted elsewhere for a comparative account. I withdrew the manuscript for this more promising prospect. Then the wife died and the husband fell apart. Our book never got published. It is also a source of abiding regret for me that Lyman and I never published anything together.[4]

~

The prospect of marrying Stina prompted me to initiate divorce proceedings. Since Sylvia and I were married in the Armenian

Evangelical Church, we were subject to its jurisdiction. I asked Nora's husband, Noubar, who was a well-established attorney, to file for divorce. Once a week, for the next several months, Noubar would come to our house for dinner from his office and then we would head for the trial at the Ecclesiastical Tribunal. Sylvia was represented by a lawyer in the United States who corresponded with the court. The actual trial consisted of arguments presented by Noubar to the three judges—the minister of the church (who had known me since my adolescent days), a shoe merchant whom Noubar knew well, and a gynecologist who rarely said anything. I was never called on to speak, and the proceedings turned into tedious drudgery for me.

The fact that there were none of the usual grounds for divorce complicated the case. Incompatibility did not constitute sufficient grounds. Had Sylvia not contested the divorce, it would have been less complicated. But she dug in her heels, and since her father was a prominent figure in the Protestant Church, the court felt further constrained. I had no idea what Sylvia wanted out of all this. Did she think that her contesting the divorce would lead to my going back to her? Noubar pressed the dubious claim that by refusing to accompany me back to Beirut, Sylvia had reneged on her marital obligations. The strategy was based on the assumption that Sylvia would never agree to come to Beirut, but this entailed some risk. What if Sylvia agreed to come back? Since she did not, Noubar was able to bring the court around. When it finally looked like I would get the divorce, Stina and I set the date of our wedding for August 30, 1964.

The verdict didn't come until mid-June. Sylvia and I were granted a divorce, but there was a catch: neither of us was permitted to remarry for three years. I was incredulous: Was this to prevent us from making another hasty mistake? To serve as an object lesson to others? Our punishment? Ordinarily, I wouldn't have cared, but now we had a wedding scheduled. We were stumped until help came from a most unexpected corner. I was told that Sylvia's father was unhappy that the verdict seemed to place the blame equally on both me and Sylvia rather than just on me, where it belonged. He saw it as a slight to his daughter's honor and wanted me to be declared the sole guilty party. So they appealed the decision.

The Appellate Court took up the case. Time was of the essence. The Lutheran church in Finland had to post the banns a month before the date of our marriage. If we missed the deadline, we would have had to postpone the wedding, and it would be a huge embarrassment, especially for Stina's family. With *one day* to spare, the Appellate Court announced its decision: the divorce decree was voided and instead the marriage was annulled. I couldn't fathom the reasoning behind the new decision. What mattered, however, was that it lifted the prohibition on remarriage. I was fined fifty Lebanese pounds—a slap on the wrist—which presumably made me the guilty party and mollified Sylvia's father. It was a huge relief. I took the certificate of annulment straight to the airport and sent it to Helsinki as express air freight. For what it cost, I could have shipped a motorcycle, but it was well worth every penny. The document reached Finland in the nick of time.

In less than a week before our wedding, Stina and I flew to Stockholm and took the overnight ferry to Helsinki. Her parents and sister met us with bouquets of flowers in hand. I stayed at a downtown hotel. Since we were not yet married, I couldn't share Stina's bedroom (even though we had been living together in Lebanon). The wedding was only a few days away, and we frantically started to get ready. Stina's lovely white gown designed by Maj had to have a final fitting, and the same was true for the formal white-tie suit that had been rented for me.

Our wedding took place in a wonderful old stone church. Looking radiant, Stina walked down the aisle on her father's arm. There was no one present from my side. My parents were now old and the prospect of their coming to Helsinki never came up (my father had never flown in his life). The best man was Maj's husband, Ola. Our friend Tim Andrews, who taught at AUB and at Helsinki University (where he had been Stina's professor of American literature), was to have read a passage from the Bible in English. He never showed up. Tim was the most accident-prone man on earth. On his way from his island house his outboard motor had run out of gas in a storm and, when he was refueling the engine, it was flooded by a wave. Tim barely managed to row himself back to his island. The ceremony was in Swedish and I understood nothing. When prompted by the priest, I said yes. After the wedding, there was a reception at Maj's house. Most of the guests

Figure 8.4 Stina's parents are in the center holding our daughter, Nina. To their right is Maj and her husband, Ola, with their son Kaj in front and their oldest son, Jan, standing behind. To my left is their youngest son, Mats, with Stina on my right.

spoke fluent English, but the older generation, like Stina's father, spoke German as their first foreign language. Lale, who was a formal man, raised his champagne glass for a congratulatory speech, which he ended with, "Dear Herant, we are so pleased that Stina married *even* you." I thought, that was fine, as long as she married me. Actually, it was a carryover from his German. What Lale had meant to say was that Stina had married *me* and not someone else.

I had often told Stina how much I had enjoyed the turning of the leaves during the fall at Rochester. She told me that the *ruska* in Lapland was equally beautiful when the ground cover of blueberries and a myriad other bushes changed color. So we decided to go to Lapland for our honeymoon. Nunni set up an elaborate travel plan and drew a wonderful map to the ski lodge they jointly owned with a group of friends. It required that we fly to Oulu to spend the night, take the train to the town of Kemi, then go on a six-hour bus ride to the village of Akaslompolo, and finally walk to the cabin in the dark. It was in the middle of absolutely nowhere. We were met by a smiling old man who offered us a pair of trout strung together. That was to be our dinner. He didn't tell us that the fish had been heavily salted and needed to be soaked in water before cooking. They were inedible and we went to bed hungry. First thing in the morning, we walked to a store in a small neighboring village and got some food. The *ruska* was wonderful, and after a few days of hiking we returned to Helsinki.

On a gorgeous day we went out to Stina's family's summer place on an island in the archipelago, an hour away from the mainland. Stina had often talked to me lovingly about the island, but I had no idea what to expect. It was nothing like the Lebanese mountain resorts that I knew. But it was love at first sight. The times I would spend on the island over the next four decades would be some of the happiest moments of my life. On our return to the mainland in the afternoon, fog moved in, blanketing the sea in a thick soup. The old fisherman, Walter, who was taking us back in his finely maintained wooden boat, peered anxiously through the fog to see where he was going and ran aground on some shoals. He had prided himself on being the only person around who had never gotten stuck on them and so got quite upset. He cursed as he stepped out on the underwater rock and somehow managed to push his boat off the rock, and we continued the journey with his limping engine. Lale offered to defray the cost of repairing the bent propeller, and that calmed him down. Staying off the rocks would be one of the main preoccupations of my own life during the summers to come.

~

We returned to Beirut for a round of receptions. Everyone in my family liked Stina. The fact that she had learned Armenian endeared her to them; she was the only non-Armenian they knew who spoke their language. My mother and Stina developed a special bond. Once my mother took Stina under her wing, no one dared ruffle a strand of her hair. The bond between them was based not solely on their shared love for me and their mutual love for each other but perhaps also on my mother's own past. We didn't know it then, but my mother would have been a foreign bride had she married Ramzi, so she could easily identify with Stina who herself was now a foreign bride.

By the mid-1960s tourists were pouring into Beirut from Arab countries and all over the world. It was a great place to live. If Beirut ever was "the Paris of the East," that was the time. We had the best of all possible worlds: the pine-studded mountains and the deep blue sea, fabulous ancient ruins and superb modern hotels and restaurants, the patina of the old mingling with the shine of the new. We lived in a cosmopolitan world where we dined and danced, rode horses, skied, and swam in a bouillabaisse of American, European, Arab, and Armenian friends. As the tourist brochures claimed, one could ski and swim on the same day in the spring (even though the snow would be slush and the water chilly).

The main ski resort was at the Cedars. To get there, one drove north from Beirut on the coastal highway for about two hours and then took off to climb up the winding road through spectacular scenery. The Qadisha Valley was like a cake from which a huge slice had been taken off, exposing its geological layers. The village of Bcharre was perched at the vertiginous top of a ridge. (Bcharre was the birthplace of Khalil Gibran, whose writings were popular in America some years ago; probably the only Lebanese author that most Americans have heard about, if they've heard about any.) The final leg consisted of a succession of hairpin turns that finally reached the small grove of cedars, the remnant of extensive forests that covered the mountains in the distant past. Soon after, one came to the base of the mountain face that provided the open skiing slopes.[5] To its south towered Qurnet el-Souda, the highest peak in the Mount Lebanon range (at 3,088 meters, or 10,131 feet). Skiing was sometimes just an excuse to go to the Cedars. The trip up the chairlift revealed spectacular vistas. The resort had a lively après-ski life that was worth the trip. Women and men dressed up fashionably for skiing as well as for dancing in the evening. Whatever the latest craze, it would have reached there as soon as it became known in Europe. The twist gave way to the hully-gully at the same time that it did on the Riviera, although the Lebanese version ("Habibi twist") lingered on. Yet there was also a kind of self-indulgence that was distinctively Lebanese. When you skied to the bottom of the run, little boys would run up to unbuckle your boots for a tip to save you from bending down to do it yourself.

The Lebanese knew how to live well and how to make money to pay for it. Christian and Muslim elites had more in common with each other than they had with their own common folk. People lived side by side in peace. Saudis in flowing white robes and men in custom-made suits, conservative Muslim women in veils and French-speaking Christian women in miniskirts and high heels, crossed paths without blinking eyes. Tolerance—live and let live—was the order of the day. As long as there was money, anything and everything was possible.

Stina and I lived in this exciting cultural bubble. We were rarely apart, day or night. My house at Beit Meri into which we moved, had a breathtaking view of Beirut spread at our feet like a colorful carpet with the turquoise waters of the Mediterranean glistening in the brilliant sun to the clear horizon. At night, sixteen clusters of light from the villages on the mountains surrounding us shone in the dark under the starry sky. Stina was happy to live in these surroundings. But she also had a good deal of adjusting to do to the culture that we were part of. Her low-key, self-effacing Finnish-Scandinavian manners clashed with the ostentatious materialism of Lebanese Armenians. Whereas Finns played down their social status and material advantages, Armenians played them up. Finns avoided appearing "having too much"; Armenians dreaded appearing "having too little." Finns were sparing with praise, even when deserved; Armenians lavished it, even when undeserved. These generalizations may not have applied to all Finns or all Armenians (they never do). But they did reflect our own cultural experience.

Stina never felt to be in her element in Beirut society, although she gamely went along to please me. People wondered about the value of the jewelry she wore reluctantly. She cringed when a woman was called "nice," which simply meant attractive and rich. She went to the fanciest hairdressers and dressmakers with my encouragement. As part and parcel of that culture, I was also part of the problem as I tried to push her into the local mold. It was not until we had lived on a more neutral ground in the United States that we were able to bridge the cultural gap between us as Stina became more Armenian and I became more Finnish in our own ways.

~

The fact that I was able to adapt to a Scandinavian culture so readily was reflective of the ambivalent relationship I had with my own culture. Although most of my formative years during childhood and adolescence were spent in Beirut, I never became an integral part of its native society. I'm not sure why. An important part of the problem, or perhaps one of its symptoms, was my relationship with the Arabic language. Language had always been important for me beyond its utilitarian value as a means of communication. I referred earlier to my intimate relationship with Armenian and the unwarranted sense of mastery I had over it as a teenager. I was keenly aware of the power of language in influencing others, especially after it became central to my teaching and administrative career. There was also the sheer pleasure I derived from using words to express precisely what I wanted to say and how I said it.

I could hardly do this in Arabic, which I spoke worse than a semiliterate laborer. The fact that English was the language of instruction in college made my ignorance of Arabic less of a liability, but I still felt hampered by it. I was like an expatriate in Lebanon even though there was no country I could call my own. My mother and Lucine spoke Arabic poorly. My father was somewhat better at it. He could also read the Arabic script because the Ottoman Turkish he knew was written in the same script. (Turkey adopted the Latin alphabet in 1928.) The Armenian schools I attended neglected the serious teaching of Arabic. But beyond these considerations, I actively resisted learning Arabic because it was not my language, and I didn't want it to be my language. I not only found Arabic difficult but disliked its harsh guttural sounds and was specially irritated by the loud, undulating, repetitive music that blared around me.

I came to realize years later that Arabic didn't have to sound that way. When visiting Isfahan, we stood at the entrance of the bazaar assaulted by the cacophony of sounds and noises that poured out of it. The shouting of people, the braying of donkeys, the banging of metal, and the screeching of wheels compounded each other. Suddenly everything fell silent. It was the hour of the noon prayer. The reedy voice of the muezzin flowed out of the minaret in the exquisite sounds of the classical Arabic of the Qur'an. The chanting of the name of God (*Allahu… Akbar*) was like a soothing balm on troubled souls. It was impossible not to stop and bow one's head. The very same words booming out of loudspeakers on minarets in Beirut made me grind my teeth.

I would have loved Arabic had I learned it well. It is a highly evocative language with extraordinary imagery and range of expression. It's no wonder that Arabs become intoxicated with it. Speeches go on endlessly because people revel in the sounds of the words and abandon themselves to its lilting strains as if listening to music. Arabic can be as ornate as the most intricate arabesque in Islamic art. It can also evoke an image or express in a few choice words the essence of an idea. In what other language can one hear a man say, "My wife is such a light sleeper that she is awakened by the farting of a flea"?

The four years I spent in the United States had increased my distance from Lebanese culture. It changed me to a point that I had trouble navigating through it. When I joined the medical faculty at AUB, I needed to get my medical license from the Ministry of Health. There were brokers who took care of such things. Against my father's advice, I decided to do it myself just as I would have done in the United States. I found out, with some difficulty, what I needed and put together a package of a dozen items (identity card, photograph, photocopies of certificates and diplomas, fiscal stamps, and so on).

I went to the Ministry of Health in downtown Beirut, where it was hard as hell to park. A man with thinning hair and a bulging belly was sitting at a desk reading the paper, with an empty cup of Arabic coffee sitting desolately in front of him. He didn't look up from the paper, conveying a message that he was not going to lift a finger without having his palm greased. But on principle, I was not about to bribe him. I got his attention and he said, "Yes?" (Meaning, why are you bothering me?) I told him I wanted to apply for my medical license. He shook his head gravely and said, "It's very difficult. You need *many* things." I said I had everything that was required. But how was that possible? He had worked there for thirty years and nobody ever had everything. Would he please tell me what was needed so I could give it to him? He sighed and as he listed each item, I put it on his table. When we were done, he declared, "God is merciful! I have never seen such a thing before." I thanked him. When should I come back for my license? His face clouded over. He lowered his voice down to a conspiratorial whisper. "The director is a little tired today. Maybe he'll feel better tomorrow." It took three more visits until I wore him down.

Having learned my lesson, when I had to have my driver's license renewed, I had a broker do it. It only took one visit. I found the broker inside the building and gave him my license. As people crowded around the counter (standing in line would have been very strange) the broker would go behind the clerk's desk and place a stack of papers in front of him with a tip prorated on the number of licenses to be renewed. If anyone protested, the clerk would make sure that that person would wait the longest. As he gave my license back to me, I noticed that the broker looked downcast. "Sayyed Mustafa," I said. "You don't look well. Are you sick?" He looked at the floor. "Don't ask, doctor." I told him I had to know. "If you insist," he said with a resigned look, "I will tell you what happened. You may find it hard to believe me, but I swear on my father's grave that it's true. You can see for yourself what I have to do to earn a living in this place. So for years, I have been buying a lottery ticket each month with the hope that God will smile on me one of these days. I would win a few pounds now and then, but nothing much. Last month, I bought a lottery ticket from the same grocer and put it in my pocket. When I went home, I looked at it and couldn't believe my eyes. It was 55555. Now, you tell me, how can a number like that win anything? So I took the ticket back and got another one instead. That damned ticket I gave up won the big prize. I could have really used the money, but what bothers me most is that God finally opened the door for me and I slammed it in his face. He will never smile on me again."

Having to deal with the wheeling and dealing that I kept running into irritated me until I realized that this is the way people conducted their business in Lebanon. There was no subterfuge. Everyone knew what was going on and did more or less the same thing. I was also reminded of how important social status was to people's sense of self-esteem. I ran into a relative that I had not seen for years. He was a successful and self-important businessman and he complained bitterly about how it had been necessary to hire armed guards during the civil war, and how he even had to arm himself. I asked what sort of weapon did he get. He said he had a Kalashnikov. A *Kalashnikov*? Wasn't that a military assault rifle? Yes, it was. Where did he learn to use it? Actually, he didn't know how to use it. Why have it, then? "Well," he said, "what else could I have done?" How about getting a pistol? "A pistol?" he asked, incredulously. "You expect a man of my position to go around brandishing a pistol?"

One of the people in Beirut I remember most fondly was my shoeshine man. He was an Armenian in his sixties with a great sense of dignity despite his grinding poverty. I always referred to him as Baron (Mr.) Garabed and he called me *Docdor*. The fact that he was shining my shoes instead of my shining his shoes was not due to any disparity in our merits, but life circumstances beyond our control. Moreover, he was by far the *best* shoeshine man in town. Neither I nor any of my colleagues in the medical school could make that claim for anything we did.

Baron Garabed sat in front of El Faisal's restaurant opposite the Main Gate. Every couple of days, I stopped by to have my shoes polished. I wore fine English and Italian shoes, and he took special pleasure in shining them. While at it, he also gave me a trenchant analysis of the political situation in the world. He saw international conflicts as personal contests between national leaders. ("De Gaulle called the Lebanese prime minister, who is a jerk as you well know, and told him, 'You cut that crap or I will personally kick your sorry ass.'") I thought paying Baron Garabed each time he shined my shoes was unworthy of our friendship, so I proposed paying him a monthly fee instead. He agreed. Then on, every time he saw me across the street he would call out, "*Docdor*, your shoes!" and would not relent until I let him shine my shoes, which he may have already polished earlier that day. ("Let me just give a quick brush-up. A man like you shouldn't walk around with dirty shoes.")

One day, I invited Baron Garabed to join me for lunch at Faisal's. He was reticent at first but then straightened up his cap and we went in. The old waiter wasn't pleased at the prospect of serving the shoeshine man, but he dared not offend me. My next attempt to be kind to my friend was more quixotic. As Christmas approached, I wanted to give Baron Garabed a present—not just money. He deserved to have the best of *something* in his life that money could buy. The logical thing would be a pair of shoes from Hashem, the exclusive store where I bought my own shoes. When I told him this, he protested. "Please, *Docdor*," he said, "don't do this to me. My wife and I don't have a decent mattress to sleep on. Our roof is leaking. What will I do with shoes that cost a fortune? The old slippers I wear are good enough for me. Give me the money instead." I said no. I couldn't lift him out of poverty but I could buy him a pair of shoes like the ones I wore. With great reluctance, he accompanied me to the shoe store. The salesman took one look at him and then looked at me. "This is my friend," I said, "and he will pick out whichever pair of shoes he likes." To kneel in front of the shabbily dressed Armenian with fingers stained with shoe polish to fit him with a pair of shoes was hard on the snobbish salesman. He did that only for gentlemen, but he had no choice. I was one of their best customers, and his boss was glaring at him. We walked out with a pair of English shoes that Baron Garabed picked out after much contemplation. They had the thickest soles and would last forever. However, I never saw him wear them. They just sat next to his shoeshine box like a trophy. To this day, I can't make up my mind if what I did was kind and generous or self-indulgent and thoughtless.

~

A year after Stina and I got married, we began to think of emigrating to the United States. I'm not entirely sure why. It wasn't Stina's idea, nor was it a concession to her. However, she didn't object to it. She had lived in the United States and could readily conceive of doing so again. All things considered, it would be easier for her to live in America than in Lebanon. One other hand, we had an easy and pleasant life in Beirut. Stina found her work at AUB interesting and continued with her freelance journalism for Finnish papers and radio. At the medical school, I was my own boss. However, I was concerned that once my research grant ran out, I would have to earn my keep by doing whatever was required of me. John Racy had become frustrated by AUB's failure to make good on its promise to start a department of psychiatry, so he decided to go back to Rochester. I doubted if I could do any better. I feared that I'd be frustrated by the lack of challenge and intellectual stimulation. I also felt a lingering insecurity about the political instability of the region. Lebanon looked like a magnificent house of cards. I didn't want to put down deeper roots and then live through another Iskenderun.

Leaving my parents and Lucine was going to be hard, but they were brave and encouraged us to go. They were also concerned about the political future of Lebanon. They were too old to move with us, so going to America entailed a lifelong separation. The next question was where to settle in the United States. I could have gone back to Rochester, but I felt so close to John Romano that it would have been like going into my father's business. I might have returned to the NIMH, but a career in research as a civil servant was not for me. Then I remembered David Hamburg's comment about Stanford when we were parting company in Berkeley. I found the Stanford catalogue at the AUB library and leafed through the pictures. It looked like a nice place. I wrote to David Hamburg reminding him of what he had said about my thinking about Stanford should I get tired of Beirut. I was now tired. I mailed the letter and we left for a vacation in Iran. It was a fascinating journey as we went from the Caspian Sea to Shiraz visiting on the way the mosques in Isfahan with their magnificent tiles and the fabulous ruins of Persepolis. The shops in Teheran were amazing. I bought Bronze Age swords and arrowheads, and exquisite in-laid boxes for a few dollars. In no other country in the region had I encountered such a variety of people and places. During a long bus ride, men took turns reciting epic poems and the guides could quote the poetry of Hafiz and Saadi.

On our return, my father told me that a Dr. Hamburg from America had been trying to get in touch with me. He wanted us to meet in Vienna. The contact information was garbled and I was not able to contact David. I inquired at the Austrian embassy about international psychiatric conferences that David might be attending, but there were none. (It turned out he was at a Wenner-Gren Foundation conference on primatology, which I could not have guessed.) I wrote to him at Stanford explaining what had happened. He wrote back saying he was sorry we had been unable to meet. He wanted to tell me in person that he was delighted to offer me a faculty position as an assistant professor of psychiatry at Stanford. It was as simple as that; nothing like that could possibly happen today.

I initially thought I could start at Stanford the following autumn, but the Anjar project was not finished, so my starting date was moved to the fall of 1966. Meanwhile, I was to present a paper at the annual meeting of the American Psychiatric Association in New York, and David invited me to fly out to Stanford to visit the department. I was delighted to see the Stanford campus and struck by its similarities to AUB—the red-tiled roofs, stone buildings (sandstone at Stanford, limestone at AUB), and lush vegetation. The most surprising similarity was between the Stanford hospital building and the Phoenicia Hotel in Beirut—both of them in Edward Stone's signature architectural style.

David and Betty Hamburg received me most cordially. We had dinner at their house and renewed our earlier ties from Washington. Their children, Peggy and Eric, had grown up quite a bit and hardly remembered me from four years earlier. I stayed at a motel on the El Camino Real, close to what was to become the Stanford Shopping Center. Over the years I would drive past it and note the successive changes it went through paralleling the transformations in my own life. At the departmental meeting I attended, I gave a presentation on my Anjar project. It was well received. I had heard about some of the well-known members of the faculty and tried to match their names with their faces but got it wrong in every case. Betty Hamburg drove me to some of the exclusive residential areas in Woodside and Portola Valley. I assumed we would have a house somewhere there, and on returning to Beirut I told Stina she could look forward to living in very fine surroundings.

The excitement of going to a great university in a wonderful setting was mingled with the apprehension of facing the unknown. Parting from my parents and Lucine was very hard; I can imagine how much harder it must have been for them. (I got a taste of that after my daughter settled in New York and my son on Maui.) How often would we be able to see each other? How would they manage on their own, especially as they got older and their health failed? The fact that I was now happily remarried and seemed destined for a bright future in a secure land of plenty gave them much comfort and made things easier for us as well. The next, and most important, phase in our lives was about to begin.

~

Stanford would irrevocably change our lives. It is startling to think that it was a serendipitous dinner in Berkeley with David Hamburg that led to our coming to Stanford four years later. Could such a momentous development be due to pure chance? Virtually everyone I have spoken with about this has a story to tell where chance had played a critical role in their taking a fork in the road that changed their lives. While there is no doubt that if I had not met David Hamburg for dinner at Berkeley, I would have never come to Stanford, does that mean that I could have become a coal miner in West Virginia instead? Obviously not. The chances are that I would have gone to some other academic institution, but the fact that I ended up at Stanford and not somewhere still made a huge difference.

What does it mean when we ascribe an event to chance? Defined broadly, everything could be attributed to chance, from the origin of the universe to the birth of an individual. And even if events are random, and choices are made without method or conscious deliberation, each outcome is unique and the chance of its happening again in exactly the same way is virtually nil. The only way someone could have an identical life experience to mine, or be my mirror image, would be in a hypothetical parallel universe (whatever that means).

By the same token, everything is also predetermined. Each event has its antecedents—nothing just comes out of the blue. In that context, chance is a shorthand for probability. If there is a high probability that it is going to rain, and it rains, then we do not ascribe it to chance. If there is only a low probability of raining and it rains, then we attribute it to chance. If the outcome is positive, we call it fortuitous or serendipitous. If it is not, we call bad luck.[6] In either case, chance refers to the unpredictable element in an event or the absence of some assignable cause for it. Or as Tolstoy put it in *War and Peace*, we ascribe to chance what we cannot explain. It is a veil for our ignorance.

This is the secular view. In a monotheistic context, nothing happens by chance; everything is ordained or determined by the will of God—not a sparrow flies nor a leaf falls without it. In the classical world, the Greek goddess Tyche (or her Roman counterpart, Fortuna) personified luck. (It is interesting that it should be a goddess and not a god that determines the unpredictable in human lives.) However, the Greeks did not think that what happened in our lives was exclusively the result of what the gods willed. It was also due to our own choices—a paradoxical idea of double determination, which the Greeks never fully resolved in their tragedies like *Oedipus Rex*.

As I see it, we are born into a playing field that reflects our biological and social heritage. These are the givens over which we have no control. However, what happens in the playing field is subject to choice as well as chance. Of the two, choice, which entails the exercise of free will by which we make things happen, appears the more active element and chance the more passive. But chance is not something that just happens to us. Chance merely opens up opportunities to which we respond, or fail to respond. There are also certain requirements for chance to act on. To win the lottery, you have to buy a ticket. By the same token, choice does not necessarily reflect our actions as a fully free agent. The Greek idea of dual determination has its parallel in our current secular models of human behavior whereby our biological heritage interacts with our social upbringing—an intermingling of nature and nurture. This seems to leave no room for free will, but we must assume nonetheless that there is such a thing so that we can hold people morally and legally accountable for their actions.

Applying these considerations to my own life, having dinner with David Hamburg at Berkeley was a chance event that was necessary but insufficient for me to come to Stanford. There were further actions necessary on both of our parts in order for me to end up where I am. Opportunities are also time sensitive. You have to get hold of them at the right moment. The ancient Greeks had a god called Kairos, who embodied the spirit of opportunity. He was represented by a man bald except for a tuft of hair on his forehead. As he ran by, one had to grasp the tuft of hair from the front; there was simply no way to get hold of his bald pate from the back.

9

The Joy of Teaching

Early in August of 1966, we left Beirut for Finland on our way to Stanford. The departure this time was far less traumatic than the previous time I left for the United States. Actually I remember little of who came to bid farewell at the airport. The fact that I was now leaving under better circumstances than eight years earlier made it easier. I was now happily remarried and heading for a new career. There was far less for my parents to worry about. And our separation did not loom as large. We could come back and visit them every year.

Since we did not need to be at Stanford until the beginning of September, we stopped in Finland on our way. We spent some time on the island with Stina's parents and then had a memorable weekend at Orisberg, the eighteenth-century ancestral mansion in northern Finland of our good friends, the Björkenheims. We had met Benni and Babben in Beirut, where he was the Finnish consul and worked for the Finnish Paper Mill Association. Benni had also worked in Egypt and Syria, so he knew the region well. We had wonderful times together in Beirut. Over the years, we continued to see the Björkenheims during our summer visits to Finland, and our enduring friendship provided a cherished sense of continuity to our lives. Through them we met renowned Finnish visitors like the designer Timo Sarpaneva. No less noteworthy were the five striking Finnish models who put on a fashion show (with Stina as the compère) at the Paon Rouge nightclub. It coincided with my father's eightieth birthday, and we invited my parents and a few close relatives to have dinner and attend the show. (I remember my father at the time as a very old man.)

From Helsinki we left for Orisberg by train and arrived late in the afternoon, when the slanting rays of the sun bathed in light the mustard-colored walls of the Björkenheim mansion. It was the work of the German architect Carl Ludvig Engel. His distinctive neoclassical buildings, dominated by the monumental Lutheran Cathedral from the 1840s, surround Senate Square in the heart of Helsinki. The stately Björkenheim mansion was in a magnificent setting next to a small artificial lake with a small island which one of Benni's ancestors had put in place overnight as a surprise gift to his wife.

In the evening, we gathered in the living room that had seen so many memorable occasions. The vodka flowed freely. We feasted on delicious wild mushrooms pickled in the Russian style, before sitting down to a sumptuous dinner. By the time we went to bed, the early summer sun was rising. With his exuberant character, Benni could turn an ordinary gathering into a party, and this was a special occasion for him to shine. The guests included Benni's mother, Hélène, and Maire Gullichsen, two remarkable women who had been part of Finland's history for over half a century. Gullichsen was cofounder with Alvar Aalto of the design company Artek, which still carries Aalto's signature furniture and glassware. Aalto had built for her an exquisite country house, Villa Mairea, which we visited on another trip. Wealth, breeding, intelligence, beauty, and a passion for the arts had endowed Maire Gullichsen with a special aura. Even late in life, she exuded a sensuous, earthy quality. Benni said that the weekend was the last of the special gatherings to be held there, and we were glad and honored to be part of it.

From Finland, we left for Barcelona to celebrate our second wedding anniversary. I reserved a table at a fine restaurant where waiters in tuxedos fluttered around like a flock of crows. I was bewildered by the vast menu and asked the waiter to recommend a distinctive dish of the region. He said he had the perfect choice for me. The dish came. It was pitch black—squid cooked in its ink. It tasted as strange as it looked. I ate it grudgingly, thinking of all the other wonderful dishes I could have had instead. Stina and I had our picture taken for the second year in a row as part of a plan to have our photographs taken on each wedding anniversary; it was a good idea but we didn't manage to keep it up.

Barcelona had fascinating buildings. Unfortunately, I knew next to nothing about its architecture, including the Art Nouveau buildings by Antoni Gaudi. We then went on to Madrid to attend the World Congress of Psychiatry and had dinner with Irv and Marilyn Yalom from Stanford, who were also attending the psychiatric congress. It was interesting to hear about their experience of working and living at Stanford. They were to become good friends over the years. I remember nothing of the congress, but I do remember the great Prado Museum and a pathetic bullfight with hundreds of psychiatrists cheering an aging matador killing a tired old bull. It was time to pack our bags and head for our new homeland.

Our port of entry was Los Angeles. Years later, when visiting the immigration museum on Ellis Island, I was especially taken by the exhibit of immigrant luggage piled on carts. I wondered which piece from the many lands would have been mine. I settled on a bag made from a colorful kilim. Had I been an earlier immigrant coming through Ellis Island, they would have probably changed my name to something shorter and more American—Harry Dorian? But L.A. was not Ellis Island. We were let through in a prosaic manner and the immigration officers left my name alone.

We took the connecting flight to San Francisco and a taxi to Peter and Eva Rosenbaum's house in Menlo Park. Peter was an assistant professor of psychiatry, and they had invited us—total strangers—to stay with them until we found a place of our own. It was a kind gesture that made easier the frustrating search for a place to rent. After a fruitless week of searching, I told Stina we needed to take a break. The day after, on a warm Sunday afternoon, as I lay down on the bed still too discouraged to go house-hunting, Stina showed me a notice for a two-bedroom house on Amherst Street, in Palo Alto, at the edge of the Stanford campus. At $250 a month, it was more expensive than comparable-sized units and we thought it might be worth the price. As we walked down its corridor with its grass-cloth walls, Stina and I looked at each other and knew that this was it. It was a nicely designed house with its living room overlooking the patio, which gave onto a well-tended garden. It became our home for the next three years.

~

I knew very little about the department of psychiatry I had joined. David Hamburg and I never discussed what I was going to do. I had assumed that Stanford would be some other version of the Department of Psychiatry at Rochester, but it was nothing like it. The faculty consisted of a heavily research-oriented multidisciplinary group. Many of the senior members of the faculty were not psychiatrists but neuroscientists and biochemists; the clinicians were mostly the younger assistant professors, like Peter and Irv. The theoretical orientation of the department was heavily biological and didn't quite fit with the psychiatry I had learned at Rochester or practiced in Beirut. I felt like the clove of garlic in the salad David had tossed together. When I asked David years later why he had hired me, he said he remembered no specific reason for it. I looked like someone he would want to have around. He had done this occasionally and had never regretted it.

I had written to David from Beirut that I hoped to start my psychoanalytic training on coming to Stanford since I had been unable to do it at Rochester. He advised me to wait until I had settled in. It was a shrewd move. He must have known that once I realized what it would take to make several trips a week to San Francisco to see a training analyst, I'd think better of it, and I did. Had David said that I could not go into psychoanalytic training, we would have had a problem. I actually never heard David tell anyone not to do anything. If he didn't want it done, it somehow never got done. My cross-cultural research interest in Lebanon came to a dead end. No one in the department had the slightest interest in it. However, the research and the publications that came from it turned out to be important when I came up for tenure later on.

I was assigned as a tutor to several residents. I knew how to do that and rapidly gained a following. Soon after, David appointed me coordinator of the Residency Admissions Committee. It was my first administrative position and involved sifting through the applicant pool and preparing a short list for the committee to choose from. As with so much else I had done and would do in the future, my approach was largely improvisational ("innovative" may sound better). I was a good judge of character and I knew who the best residents were. The task was to find the candidates who were most like them. It was a method based on analogy, not analysis. To expedite the process, I relied on a few shortcuts. I would first look at the applicant's photograph (required at the time) and some key qualifications to weed out the losers. For example, one candidate had sent in a picture taken in a tuxedo. That showed bad judgment and disqualified him. This may now look appallingly arbitrary and discriminatory. However, forty years ago there was virtually no diversity in the applicant pool and no candidates to discriminate against, even if one were misguided enough to do it.

David was keen on my getting involved more broadly in the university at large, and that suited me fine. I replaced him on a university committee on student activism, and another committee on student health. The sexual revolution of the 1960s was in full swing and there was a lot of concern over the risk of pregnancy and "venereal disease" among students. I suggested that someone should perhaps teach a course on sexuality. The committee members thought that it was a great idea: I should do it myself. I protested that I couldn't possibly teach such a course since I knew next to nothing about the subject. Robert Sears, who was dean of the School of Humanities and Sciences, said, "Don't worry about it. During the thirty years that I've been at Stanford, ignorance has never stopped anyone from teaching anything." The ensuing laughter drowned out whatever further objections I tried to raise.

I went to David to get his advice and he encouraged me to develop such a course. Had he not supported it, I would have dropped the idea and there would have been no sex course—which was to play such an important part in my life. I was in fact quite ignorant about sexual matters. No one at home or in school had ever taught me anything about the subject. I had learned about the reproductive system and sexually transmitted diseases in medical school but nothing about sexual behavior. During my psychiatric training, what I learned about sexuality was purely from a psychoanalytic perspective and hardly relevant to a course for undergraduates. However, having put my foot in it, I thought I could go to the library, find some books and learn what I needed to know. But there were no books in the library except for the two Kinsey volumes and a few outdated books by authors like Havelock Ellis.

What was I to do? I asked around and was told to go to the Kinsey Institute at Indiana University, which I had not heard about. I wrote to the director, Paul Gebhard, the last surviving member of Kinsey's collaborators, and he cordially invited me to come for a visit. I flew to Bloomington and spent several days with Gebhard and his staff, who were most helpful. I was amazed at the wealth of books, photographs, and films on sex that had been gathered. It was fascinating to be immersed in this material, but after several days, I felt I had had enough. Some of it was downright sickening. In one book on sadism, there was an old photograph from China that showed a criminal being flayed alive as he looked away with an expression of resigned agony. The caption explained that before criminals were subjected to such punishment, they were drugged. That did little to allay my horror, and I could not get that image out of my mind for a long time.

The curator of the film library was the gruff former police chief of Bloomington. Most of the films had been confiscated under obscenity laws and sent to him by other police chiefs. He bluntly asked what I was looking for. People wasted his time when they didn't reveal their interests and made him go through film after film, until they saw what they wanted (for the last visitor it was nude women with garters and stockings). I assured him that I had no special interests and I would be happy to see whatever films he wished to show me. Many of these films were from Argentina made in the 1930s. They were refreshingly candid and genuinely erotic, unlike the mind-numbing pornography now flooding the Internet. I then went back to the library and found the syllabus of a course someone was teaching at the university, a successor to the course originated by Kinsey that launched his revolutionary career in sex research. It gave me something to start with.

By the start of the following academic year, I had cobbled together a course called Human Sexuality: Undergraduate Special 112. It was first offered in the spring of 1968 and attracted 68 students. The following year the number went up to 420, and in the third year we moved to Memorial Auditorium to accommodate the 1,022 students who flocked to the course. At its peak, enrollment reached 1,200, accounting for one out of four undergraduates. "Hum-sex" was now

Figure 9.1 Lecturing to the Human Sexuality class.

the most popular course at Stanford. All told, well over 20,000 students would eventually take the course, thereby making it the most heavily subscribed class in the history of the university.[1]

From its outset the course had a broad-gauged approach. It not only covered the biological, behavioral, and relational aspects of sex but also included lectures on erotic art, film, and literature by guest speakers. Two of the residents who were my advisees served as section leaders. After enrollments got larger, we could no longer provide discussion sections and the course became an exclusively lecture class. When Donald Lunde, one of the two section leaders, joined the faculty, I made him co-instructor of the course.

There was no textbook. During its first year, my transcribed lectures served as the reading for the course. The following year, we used a recently published book, which proved inadequate for our needs. Don Lunde then suggested that we write our own textbook. I readily agreed and wrote an outline for the book organized along the format of the lectures. My assumption was that I would write most of the chapters. But Don said he wanted to write half of them so he could get half of the royalties. I hadn't thought about royalties, and agreed to his request. (It proved to be a mistake.) I let Don write the more medically oriented chapters on reproduction and sexually transmitted diseases since he had graduated from medical school more recently and was more up to date on these topics.

At David Hamburg's suggestion, we contacted David Boynton at Holt Rinehart and Winston in New York. Our proposal for a college textbook was readily accepted and the first edition of *Fundamentals of Human Sexuality* appeared in 1972. It included a guest chapter on erotic art by Lorenz Eitner, chair of the Art Department, and chapters on erotic literature and film by my friend Strother Purdy, then teaching at Marquette. David Hamburg wrote the introduction. It was an instant success. The first edition sold 69,000 copies, and the second edition 133,000 copies. The book went through five editions and was translated into French, Spanish, Portuguese, and later, Chinese. It was widely adopted by burgeoning college courses on sexuality and served as the basic model for sexuality textbooks by others that came on the market over the next decade. Although intended to be used as a textbook, *Fundamentals* reached a wider audience and became a selection of the Literary Guild Book Club, selling several hundred thousand copies.

Although we started by my being Don's tutor, we were able to develop a more collegial relationship. Our collaboration worked fairly well with regard to the textbook, but not the course. Lecturing to an audience of over a thousand students turned it into a performance. I knew how to tell a story and was in my element—Don was not. Instead of developing his own style of lecturing, he tried to imitate and upstage me. It didn't work. Student evaluations gave me rave reviews, but savaged him. He ruefully complained that he wasn't a bad teacher but that I was *hors concours* and made him look bad. I took no pleasure in his plight. Nonetheless, I had to protect the course, so I cut back his involvement.

Nevertheless, our collaboration on the textbook was supposed to continue. When preparing the third edition, I submitted my revised chapters and left for Finland for the summer. By this time, Don and I were hardly communicating with each other. Soon after, I got a frantic call from David Boynton, who said Don had informed him that he was not going to revise his chapters and was pulling out of the third edition. He was cutting off his nose to spite my face. I couldn't possibly take on revising Don's chapters on such short notice, especially while sitting on an island in Finland. Boynton pleaded that his job was on the line. Our textbook was their second-best seller, and its loss would be very costly to the publisher. I agreed to have Boynton hire John and Janice Baldwin, who taught a sex course at the University of California at Santa Barbara, to revise Don's chapters. The three guest chapters were dropped over my protests. The third edition appeared in 1980. It was a rushed job and nowhere as good as the previous editions. Moreover, it now had to compete against several new textbooks that had come on the market. They followed the general format of our textbook, but their lighter tone and sex-is-wonderful message were more attuned to the times. Our sales plummeted, and we lost the market. I felt betrayed by Don and wanted to end my collaboration with him. We negotiated a settlement, whereby I would be the sole author of subsequent editions. Soon after, he left Stanford to go into private practice and we lost touch. My falling out with Don Lunde was the only serious conflict I would ever have with a colleague.

~

David Hamburg's encouragement and support of my initiating the human sexuality class and his facilitating the publication of the textbook were but two of his critically important influences on my career. He not only brought me to Stanford, but made sure that I stayed there. His unstinting support continued over the ensuing decades. David was not only a colleague and mentor but also became a devoted personal friend. In spite of his enormous accomplishments, he retained his soft-spoken and unassuming character. Whatever else we talked about, he never failed to ask me about my family. I felt the same way about his wife, Betty, and their children, Peggy and Eric. I got to know Eric better when he was a Stanford undergraduate. He then went on to law school and became a filmmaker. Peggy went to medical school and eventually became commissioner of the Food and Drug Administration in the Obama administration. Betty Hamburg had her own distinguished career as a child psychiatrist and president of the Grant Foundation.

Like John Romano before him, David Hamburg's influence on my life and career was enormous as a mentor, role model, and friend. Therefore, his own life story is significant for me. David was the grandchild of immigrants. He was educated at Indiana University and received his psychiatric training at Yale and his psychoanalytic training at the Chicago Institute. In 1958 Hamburg became chief of the adult psychiatry branch at the National Institutes of Health, where he first hired me. Three years later, he left for Stanford to head the Department of Psychiatry and Behavioral Sciences, where he hired me for the second time. After he left Stanford, he became president of the Institute of Medicine of the National Academy of Sciences, director of health policy research and education at Harvard University, president and chairman of the board of the American Association for the Advancement of Science, and finally president of the Carnegie Corporation (actually a foundation) in New York in 1982. Hamburg served at the highest levels of national and international bodies, including President Clinton's Committee of Advisors on Science and Technology, and chaired parallel committees at the United Nations and European Union on the prevention of genocide. David received numerous honorary degrees, and other honors, including the Presidential Medal of Freedom (the highest civilian award of the United States), which was given to him by President Clinton.

These publicly recognized accomplishments, no matter how outstanding, do not do justice to David's tremendous influence in shaping the lives of individuals like me. Perhaps the single most important way in which David served as a role model for me was to show that a psychiatrist could be more than a clinician who treated patients and engaged in psychiatric research. While I harbored no aspirations to emulate the scale of his accomplishments, I was emboldened to step out of the narrow confines of psychiatry and go into university administration and foundation work. The nature of David Hamburg's influence on me was in some ways quite different from that of John Romano, yet there were also striking similarities between them. They were both influential in my gaining greater intellectual breadth and avoiding parochialism. The multidisciplinary perspective I pursued in my teaching and writing was derived from them. Their styles of leadership differed, but both of them encouraged and helped me in

Figure 9.2 David Hamburg.

taking on administrative responsibilities. After David and I had both retired, it was especially gratifying for me to collaborate more closely in his work in the prevention of genocide. It brought us even closer together than our various engagements had done earlier.[2]

~

I continued to teach the human sexuality course for two more decades. The huge enrollments had subsided by the end of the 1970s to relatively smaller classes of several hundred students. We moved to the Art Department's Annenberg Auditorium, which became my favorite venue for teaching the course. I much preferred its rows of seats rising up from the stage rather than lecturing from the stage looking over the heads of students. With the entire wall behind me serving as a huge screen, I could use two projectors to simultaneously show the numerous slides I used in my lectures.

There were several likely reasons for the declining enrollments in the sexuality class. As times changed, the topic of sex lost much of its novelty and notoriety. Schools began to teach the subject, so students now came to college knowing more about everything they always wanted to know about sex but were afraid to ask (the title of a highly popular book at the time that was followed by a Woody Allen movie with the same title). The decline in attendance wasn't due to a loss in the quality of the course. On the contrary, the relatively smaller classes in Annenberg were more substantive than those in Memorial Auditorium. The popularity of the class obviously owed much to the subject matter and novelty and shock value compounded its interest. ("Mom, guess what class I'm taking?") It was also one of the few academic experiences that students had in common. The bandwagon effect added to its popularity. Yet, these considerations don't do justice to what the class meant to students in terms of its substance and how it was taught. One year, when I was on sabbatical leave, a lecturer in psychology taught the sexuality class using my textbook. She had forty, not four hundred, students.

Despite the hype, human sexuality was a serious course. As some students discovered to their dismay, it was not a shortcut to an A. The final grades were curved to conform to the grade distribution of the Human Biology Core Course. One plaintive student complained, "You can't give me a C in sex—you'll ruin my reputation on campus." I told him I wasn't grading his performance, only his knowledge. I worked hard, constantly refining my lectures. An early pitfall was packing too much material into each hour (the bane of novice teachers). I would begin to talk at a stately pace, but as I fell behind, I accelerated my pace and finally got frantic in rushing to a finish. Since I didn't write out my lectures, I couldn't easily tell where I was at any given time. The slides helped to maintain my pace, but they could also add to the problem if there were too many of them or if they carried too much text. I came to limit the slides to fifty per lecture—a slide a minute—and gauged my progress by predetermining what slide I should be on midway in the lecture. Finally, I learned to start and end a lecture precisely on time, like a proverbial Swiss train. Another early problem was my habit of gesturing with my arms when I talked. It was part of my Middle Eastern heritage, but it proved distracting. To control it, I tried to lecture with my arms held stiff at my sides. Once I had gained enough control, I could relax and keep my gestures within bounds, thus making them part of my distinctive style.

After thirty years of teaching the course, when I was sixty, I noted in my diary at the start of a course: "[I had] a wonderful time working on my first lecture. Spent about three hours on it even though I had given that lecture so many times. That is what it takes." In my earlier years of teaching, I played up to the audience like a performer on stage. I gradually came to compete more and more against myself, to see if I could exceed my own standards. Teaching became my main source of professional satisfaction. Few things left me more exhilarated than giving a particularly good lecture. Yet, I was also aware that the performance aspect of teaching was not its central purpose. I noted in my diary, "There is a certain pleasure in a flawless performance. I suppose that is what drives artists. But I am keenly aware that words as such can be empty like tinkling brass."

~

I expanded the course to include panel discussions on homosexuality, transsexuality, sexual morality, and other topics of special interest. Over the years, Dr. Donald Laub, one of the pioneers in the treatment of transsexuality, brought over a hundred of his patients to speak to the class. It was an important way for students to confront their own gender identity. After one of the panels, I went on stage to thank the participants. A statuesque black woman asked me, "Dr. Katchadourian, do you remember me?" I said I was sorry I didn't. (I didn't tell her I would have surely remembered her if I had seen her before.) She said, "I don't blame you. When I was in your class, I was a man and on the football team." Don Laub told me that several of his patients had come to him after they heard him lecture. Another dramatic encounter involved a young male-to-female transsexual who was also strikingly beautiful. When the class was over, half a dozen of the students surrounded her on the stage showering her with complements and questions. One of them asked her when she had had her surgery. She smiled sweetly at him and said, "I haven't yet had my surgery." The vision of what was hanging between the legs of this enchanting creature made the young man go weak in the knees. I also heard the most egregious example of sexual discrimination from another of these panelists. He was an airplane mechanic who went back to his old job after his male-to-female transformation. He was rehired, but at one third less pay because she was now a woman despite the fact that the very same hands would be doing the very same job.

In the 1980s, the specter of AIDS added a sense of urgency to the course. Until then, syphilis and gonorrhea were conditions readily controlled by antibiotics, and other sexually transmitted diseases were mostly a nuisance. AIDS made sex a matter of life and death. I dealt with the subject using neither scare tactics nor head-in-the-sand denial. I focused on the risks involved and the importance of taking responsibility—for oneself and for one's partner—without preaching on what was right to do and not right to do.

There is a thin line between bragging and false modesty. I have thousands of laudatory comments in course evaluations about the class and my role as a teacher.[3] What appealed to students was not only the subject matter but also the manner in which it was presented. I was conventional in how I looked, talked, and acted, but I spoke about topics that others did not. There was an authoritative and formal aspect to me that made my students sit up and pay attention. Yet I also made them sit back, relax, and laugh. I spoke about sex openly and unabashedly, without being a freewheeling advocate for it, or a censorious critic, thereby making talking about sex safe and respectable. I showed sexually explicit images, but only when they served a clear didactic purpose, not for entertainment. I showed pornographic materials when discussing pornography, but not otherwise. I worked hard to gain the students' respect, trust, and affection, but I didn't pander to them. Another aspect of the course that appealed to students was the objectivity and fairness with which they perceived my approach to sensitive subjects. At the end of my lecture on abortion, I would ask for a show of hands to indicate how many could tell whether I was pro-life or pro-choice. A few dozen hands went up. Then I asked them if they thought my personal position was consistent with their own. The same hands went up again. They were not detecting a bias but hoping that I was on their side.

One of the few criticisms leveled at me was that it was not enough for me to be objective: I had to stand up and be counted in support of liberal positions in favor of gay and lesbian rights, or conservative beliefs in heterosexuality and virginity before marriage. In response to such criticisms, I pointed out to the difference between a lecture and a sermon. There was nothing wrong with sermons, but they had to be delivered from pulpits, not lecterns in classrooms. I didn't claim to be free of all bias, but what bias I might have had was never intentional, to the best of my knowledge. Perhaps most importantly, what came across to my students was that I cared about them; their health and happiness mattered to me. Decades later, former students, now in middle age, would continue to tell me about the positive impact the course had had on their lives.

~

It would take many pages to tell the many stories and experiences I had teaching the sex course. During the flamboyant 1970s, Memorial Auditorium provided an irresistible venue for eccentrics, exhibitionists, and political dissidents to parade their wares in front of a massive, captive audience. A tall, long-haired, disheveled young man stood in front of the class playing the harmonica until a few seconds before I started my lecture; I never had to tell him to stop and never found out who he was. A couple in formal wear sat in the front row sipping champagne as they might at their wedding. When "streaking" was the rage, we had the inevitable display of buttocks from fraternity brothers. It was a perfect setting for their initiation rituals. One student faked being ill and "threw up" spaghetti and red sauce while uttering sounds of choking; by the time I got to his seat, he had run out. (I got a scathing letter from the head of the Art Department for letting one of *my* students soil *his* carpet.) Others ran down the aisle in various forms of outlandish dress and undress.

It was not, however, all fun and games. Since my class met Monday mornings at 9 o'clock, it became a target for antiwar protestors. Disrupting a class with a quarter of the undergraduate body in attendance meant that business was not going to go on as usual. They never managed to shut down the class, but they did manage to disrupt it time and again. They lay down on the steps leading into the auditorium and students had to step over them to get in. I knew that, sooner or later, we would have a visit from the "guerilla theater." Indeed, half way into one of my lectures, I heard the distant sound of a flute and could tell the theater troupe was advancing down the stage. I pretended not to notice, until their leader touched me on the elbow. "Excuse me," he said in a mocking tone. "Yes?" I asked, feigning surprise. He said they wanted to put on their skit. But, I said, I was in the middle of giving a lecture. "You can give us ten minutes or argue for twenty," he said defiantly, eying the audience. The tension in the hall was palpable—you could hear a pin drop. I brought my face even closer to his and said, "Look, I don't know how it's going to be after the revolution, but right now, I'm in charge." "However," I added, "let's ask the students to decide what they want: watch your skit or have me go on with the lecture." He was silent. I turned to the class for a vote. There were a few loud voices in favor of the theater but a massive "Aye" for going on with the lecture. I resumed my lecture. The group retreated to the wings, but they didn't go away. After the class, their leader came back to me and said in a hurt tone, "Did you have to humiliate me like that in front of the class?" "Come on," I told him, "you interrupted my class and tried to make an ass out of me and you're complaining of my humiliating you?" "Let me tell you," I added, "I'm on your side; I also oppose the war. But disrupting my class isn't going to end it."

I didn't set myself up as a personal advisor or therapist for my students, but that didn't prevent some from seeking my help. One young woman told me that she woke up some mornings with the pillow between her legs feeling relaxed and at peace with herself. Only after she heard the lecture on masturbation did she realize that is what she had been doing. During a lecture on conception and pregnancy, I mentioned that there was a sperm bank in Los Angeles that specialized in donors with high intelligence and achievement, including some Nobel Laureates. At the end of the class, one of the students came up to me with a question: How did they get the sperm out of the Nobel Laureates? Was there some sort of machine? I asked why they would need a machine. "You mean," he said looking astonished, "they..." and made a suggestive gesture. When I confirmed that that was indeed what they did, he said, "I'll be damned," and walked away. He couldn't imagine a Nobel Laureate doing what he did regularly.

One freshman was particularly distraught. In high school, after he learned that he was admitted to Stanford, he made a list of what he must do during his first quarter. High on the list was "getting laid." Soon after he got to Stanford, he began to look for a prospective partner. Finally, he lured a young woman at the dorm to his room. While he considered how best to broach the subject, she unceremoniously unbuckled his belt, unzipped his pants, and said, "Let's do it." He thought he had died and gone to heaven. But to his horror, he couldn't get an erection. When his partner's gentle, then increasingly energetic, and ultimately frantic ministrations failed to have any effect, he was mortified. He wanted to know if he was going to be impotent for life. Other students were concerned over disparities in the sizes of their genitalia: Was the penis too small, the vagina too tight? "The penis is not a foot and the vagina is not a shoe," I would tell them, "so stop fretting about it."

~

I was deeply touched by students with severe disabilities. Jane was born blind and her father, a professor of electrical engineering, invented a laptop typewriter that allowed her to take notes in class. At the end of one lecture, I saw a young woman walking slowly down the aisle toward me. She had highly unusual facial features and a serene smile. As she got closer, I realized she was blind. Her name was Martha. She was holding hands with a handsome young man, whom she introduced as her boyfriend. I invited them to join me for a cup of coffee. I wondered how well she could follow my lectures without seeing the slides. I asked her if I should look for anatomical models of the reproductive system at the medical school. She said, "Why? What's the problem?" The problem seemed obvious to me, but I couldn't tell her that, so I changed the subject. We were in the spring quarter, and I asked about their summer plans. They were going to Paris with her boyfriend's parents. I thought why go to Paris if you couldn't see anything there. But I realized there was more to Paris than "seeing" it. You could also hear, touch, smell, and taste Paris—you could *be* in Paris.

Melanie had lost her hearing in early childhood. She approached me after class and told me she couldn't follow the lecture after I had the lights turned off. She had to read my lips. She spoke in such a normal tone of voice that I had not realized that she was deaf. I had a light installed on the lectern and explained to the students why I did that so they wouldn't think I was being vain. I had student volunteers for a panel on sexual morality. When Tom took his place on stage, I noticed that he had no hands—his arms ended up in stumps where the wrists would have been. I invited him to lunch at the Faculty Club. Then I kicked myself. How was he going to eat? Would I have to feed him? By this time I had learned not to offer help unless asked. As we got in line for the buffet, Tom had me insert the fork under a wrist band and he served himself. When we sat down, he asked me to cut the meat on his plate and then he could eat on his own. He used the same device to write with a pen. How did it go when he went out on dates? Well, some girls freaked out, but others adapted themselves to the unusual circumstances, and a few might have actually enjoyed the novelty.

The most difficult of these encounters was with Dorothy. At the start of my seminar on academic and career choice, when I was making some introductory remarks, there was a commotion at the door. Someone was wheeling in a young woman who had jerky and agitated movements. As she got closer, I could recognize the spasticity, ataxia, and athetoid movements of her limbs. It was disconcerting when she first tried to say her name. With flailing arms, contorted neck, and an expression of supreme effort in her reddened face, the word exploded out of her mouth. I stopped for a moment to welcome her to the class and carried on. My brain was racing. How was this girl going to handle the class? Then I realized that that was not the problem; she had been in classes for many years and done well enough to get into Stanford. The problem was how was *I* going to handle her in my class? How were the other students going to deal with her? Dorothy sat next to me and I noticed the eager, pleading look in her eyes. She wanted to speak, but did not want to interrupt the class. I asked her a question the way I would have asked any other student. With supreme effort, she uttered several sentences, which were no different than what could have been said by another student irrespective of the manner in which they were said. That broke the ice, and we gradually figured out how to involve Dorothy in class discussions to give her the opportunity to

Figure 9.3 A decade of change: On the left, psychiatric resident at Rochester in 1959; on the right, Assistant Professor of Psychiatry at Stanford in 1969. (Photograph by Kaius Hedenström.)

express herself while allowing for her physical limitations. The woman who had initially brought her in was her mother, who had essentially devoted her life to Dorothy's care—that's what it took. Then on, an attendant would bring her to class and help her in the dorm. The paper that Dorothy submitted for the class was beautifully written. She was an English major and wanted to be a writer. She planned to spend the summer studying literature at St. Andrews in Scotland.

I taught the sexuality course for the last time in the spring of 2002. After my final lecture, on May 30, the students gave me a standing ovation. Then, in walked Russ Fernald, my friend and the director of Human Biology, with Linda Barghi, the program administrator, several other staff members, former teaching assistants, and my wife, Stina. Two huge cakes decorated with reproductive motifs were brought in. A couple appeared on stage to dance an Argentinean tango. Students came down to shake my hand and hug me. It was hard not to be moved. The sexuality course was in some ways a by-product of the decade of the 1960s. It was a turbulent period for which I was ill-prepared but to which I managed to adapt myself. I never used LSD or smoked pot. Nor did I embrace the revolutionary ideology and politics of the period. Mercifully, our children were too young to get caught up in the pervasive madness whose epicenter was San Francisco. But it was impossible not to be affected by it. Like so many others in academia, my appearance, dress, and demeanor changed, mimicking the transformations of the younger generation. I was the last person in the Department of Psychiatry to grow a beard, but also the only one who never shaved it off.

~

The undergraduate sexuality course was the cornerstone of my teaching career. It established my reputation as a teacher and set the model for the other courses I developed. The visibility it provided was an important factor in my going into university administration. For a few years, I also taught a course on sexual dysfunction at the medical school. It attracted many students, but little acknowledgment and support from the Department of Psychiatry (now without David Hamburg) or the dean's office. It contributed nothing to my salary; so as my workload expanded, that course was the first to go.

I was more successful teaching medical students elsewhere. I was invited by the Albany Medical School to help them set up a course in sexuality for medical students. That led to my teaching a two-day course, which proved highly successful. I taught it for almost ten years, until it fell victim to budget cuts. During one of my visits to Albany, I thought of contacting Sylvia. I had heard that she was living in Albany. I found her phone number in the telephone directory (she had taken back her maiden name) and called her. We had not spoken over the three decades since our divorce. She was surprised, but agreed to have dinner with me. We met at a restaurant close to my hotel. Neither of us had changed that much over the years. We spent a pleasant and congenial evening. Over several hours, we talked about various topics but avoided sensitive subjects. We had both done well in our careers. She had not remarried. I made a tentative attempt to get into questions about the past, but it went nowhere. It was refreshing that we could at least laugh together and nothing unpleasant marred our meeting. As we left the restaurant, I offered to accompany her to her car. She said I didn't have to do that, but I told her having walked out on her the least I could now do was walk her to her car. She laughed and we parted amicably.

Encouraged by this experience, I tried to see her again on my next visit to Albany the following year, but she had a cold and we couldn't meet. A year later, we set up another time to have dinner at the same restaurant. However, when I stopped in New York on my way, I found a letter awaiting me at the hotel. It was an angry and accusatory note from Sylvia informing me that she no longer wished to see me. She had agreed to have dinner two years earlier so we could ask each other's forgiveness. Having looked me in the eye and satisfied herself that she had forgiven me, she had no further need to see me. This was so inconsistent with the pleasant memory of our previous meeting that I was thoroughly baffled and decided not to contact her again. It looked like she had reverted to her old self.

A few years later, I got a message from Sylvia. She wanted to meet with me. I told her that I was no longer coming to teach at Albany and expressed my regrets. She wrote wanting to know if my refusal to see her was due to her "unfortunate letter." I explained that it was not. She then asked if it was because of what had happened at the lawyer's office at Rochester. I didn't respond to that. We exchanged several more emails, in which she provided some observations and explanations about our past. Her tone was generally candid and conciliatory, lapsing into being critical occasionally. We agreed that our perspectives on the past were too disparate to be reconciled. However, even if we couldn't reach a common understanding we could still forgive each other. Sylvia then sent to me a cordial note about the natural pearl necklace I had given to her. It was an unusual and lovely piece that had belonged to my mother and meant a great deal to me. She said she had worn it on many occasions and been complimented for it, but now wanted me to have it. I thanked her and said I would give the necklace to my daughter. I considered it a peace offering and accepted it as such.

~

After I stopped teaching my class at Albany, there were new opportunities for me to return to teaching courses on sexuality The Stanford Medical School asked me to give a few lectures to the second-year medical students on sexual dysfunction, as I had done many years before. They were well received by the students (the associate dean in charge said I had "scored 10 on the Richter scale"). The lectures were expanded to nine hours. However, when the curriculum was being revised under a new dean, my lectures were cut back to two hours. I quit.

When the State of California instituted a requirement that clinical psychologists and social workers attend a day-long course in sexuality as a condition for licensing, the head of a school of psychology in southern California approached me to develop such a course. The first set of lectures were to be given at the Claremont resort hotel in Berkeley over a weekend. I drove there in the morning, to be met by a harried staff member. The lectures had to be moved to a local high school auditorium to accommodate an unforeseen overflow of participants. Several hundred had registered, but twice as many had showed up. The lectures were moved to a run-down high school in a poor section of town. I stopped at the bathroom—there was no soap, no towels, and no toilet paper.

I walked on stage to face a tightly packed audience of psychologists and social workers seething with anger. They had come from all corners of the state and paid quite a bit of money. This was not what they had expected. I stood quietly sizing up the situation and told them I was just as unhappy as they were with the location but we would do the best we could. By the end of the first lecture, they seemed to have forgotten where they were. It ended happily, with the organizers walking away with $40,000. I tried to renegotiate my fee, with mixed success. Nonetheless, the experience reconfirmed my confidence that I could do the job whatever the circumstances. I did several more of these presentations until the pool of potential participants dried up.

Forty years ago, it would have been easy for me to go full time into the field of sexuality. Sexology was reemerging as a field.[4] Masters and Johnson had seized the initiative with their studies in the physiology of sexual function and their method of sex therapy. Passing through St. Louis, I met with Virginia Johnson at their clinic. William Masters was out of town, but he called me during the visit and was highly complimentary about my textbook and contributions to the field. I could have done something similar at Stanford with respect to teaching sexuality. I was the only person offering a large sexuality class at a nationally renowned university. Given the strong support I had at Stanford, I could have expanded the course into a wider teaching or research program.

I considered starting a survey of sexual behavior of the students in my class, which over the years would have added up to a huge sample and provided a perspective on changes in sexual attitudes and behavior. I didn't do it because it would have meant more work when my plate was already full. I also couldn't have kept Don Lunde out of it. With our relationship beginning to fray, I didn't want to be further involved with him. In retrospect, it was foolish to let such considerations deter me from taking advantage of a unique opportunity.

I also avoided joining the budding associations for teachers and practitioners in sexuality. When attending a sexuality congress in New Delhi, John Bancroft from Edinburgh (who became head of the Kinsey Institute) commented on my reticence to get involved with sexologists. "You shouldn't hold yourself back so much," he told me. "A lot of people at the conference know and respect your book which they have on their shelves. They'd like to know you personally." At another meeting, a psychologist told me, "I've read everything you've written, but I've never met you before. People think you're like Betty Crocker—a well-known label with no actual person behind it."

I became more aware of my reputation after a chance encounter with the famous Dr. Ruth in, of all places, Dubai. I had been asked to give lectures on Islam to the World Presidents' Organization and Dr. Ruth was there as a celebrity speaker. While standing in the buffet line, she looked up at my name tag and said, "There's a very famous person with the same name in my field." I said, I supposed that would be me. Her eyes opened wide and she called out in a loud voice, "This man is a giant on whose shoulders we've been standing," and so on. The cameraman who always traveled with her got it all on film.

I am still unclear about why I didn't take the fork in the road that would have led to a career change at that point in my life. I may have been turned off by the flakiness and flamboyance of some of the people in the sex field. Perhaps I was embarrassed by being labeled a sex expert, even though that is how a lot of people came to think of me anyway. At any rate, it would have meant having a very different career; whether a better or a worse one is hard to tell. Yet another "what-if" for which there is no ready answer.

~

The sexuality class that started as an Undergraduate Special was adopted a few years later by the Human Biology Program as one its general offerings (as Human Biology 10). That led to my closer association with the program and to its eventually becoming my academic base until my retirement. The Human Biology program was established in the mid-1960s, about the time I came to Stanford. The initiative came from the Ford Foundation based on its concern that undergraduate education was becoming too specialized and increasingly removed from the broader educational needs of students. The Ford Foundation approached David Hamburg and Paul Ehrlich with an offer of a million dollars for Stanford to establish an interdisciplinary program that would bring together the biological and social sciences. They were joined by a stellar group of faculty, including the Nobel Laureate Joshua Lederberg; Donald Kennedy, future president of Stanford; Al Hastorf, future provost; and several other senior professors and department chairs. That got the program off to a great start. The Ford Foundation made a similar offer to Oxford University, but it was turned down; probably the same would have happened at Ivy League schools, which were not as receptive to this sort of educational innovation. Actually, the proposal might not have been accepted even at Stanford had it not been backed by a pride of lions.

There has been an innate resistance to such innovative programs at universities. Like medieval guilds and trade unions, they are organized into departments based on academic disciplines. That is where knowledge is generated through research and scholarship, which provide the basis for teaching undergraduates and for the professional training of graduate students. Occasionally two fields come to overlap and spin off into a new field (such as biochemistry). Such natural pairings are well accepted. However, hybrids like Human Biology are suspect. The faculty in Human Biology had their primary base in their own departments. They came together in Human Biology because of their interest in interdisciplinary subjects and in teaching undergraduates with similar interests. That is why the program attracted some of the best teachers at the university.

Human Biology became an instant success and remained one of the most popular majors at Stanford. It appeals to students who find departmental majors—such as in biology or psychology—too confining and sometimes too forbidding. Human Biology provides greater breadth, albeit at the cost of relatively less depth than departmental majors in a given field. There are also students who come to college uncertain about what to study or what career to pursue. Human Biology gives them a wider berth by exposing them to various areas and allowing them time to refine their interests. Human Biology provided me with an ideal academic base. Even though I rose through the ranks to become a tenured professor of psychiatry, I lost my moorings in the department. Had it not been for Human Biology, I would have had nowhere else to go, since I didn't fit into any of the departments of the School of Humanities and Sciences. The program also allowed me to develop courses—like human sexuality—that were multidisciplinary and that addressed the personal concerns of students. No department I can think of would have sponsored such a course at the time.

The academic anchor of Human Biology was the Core Course, typically taken during the sophomore year by majors in the program. It was taught by faculty members from biology, the medical school, and the behavioral sciences. When Donald Kennedy was director of Human Biology, he and I developed a new model for the Core Course based on the human life cycle. It was never fully implemented because Don went to Washington to head the Federal Drug Administration and I joined the Stanford administration as vice provost and dean.

I was well liked by my colleagues in Human Biology and became an integral part of the program. I initiated and taught more new courses in the program than anyone else. Human Biology made me feel part of Stanford more than did the Department of Psychiatry. However, while my colleagues in Human Biology had a departmental base and clear professional identities, I ceased to be part of a departmental fabric. I roamed like a masterless samurai. That provided me with a greater sense of independence but also made me feel rootless. Some faculty colleagues may have wondered why a psychiatrist was teaching undergraduates and siphoning off so many students from their classes. (One faculty colleague whose class enrollments dwindled to the point where finally no one showed up thought it particularly unfair that I should have so many students.) I hoped that the pubic recognition awarded to me by students so generously would not generate envy. Each class of graduating Stanford seniors chose a faculty member to speak to them and their parents on Class Day before Commencement. It was deeply gratifying for me to be selected Outstanding Professor and Class Day Speaker seven times and to receive the Teaching Award of the Associated Students of Stanford University.

Despite all the accolades I received, I was troubled by lingering doubts over legitimacy or being "real." Whereas my colleagues taught subjects based on their specialty and their areas of research, there was a disconnect in my case. I was no longer teaching in my specialty and the courses I taught were not directly related to it. Just as I had felt I was using someone else's language when speaking English, I now felt that I was teaching courses in someone else's subject. And unlike my colleagues who drew from their deep knowledge of their field, I often had to teach myself the subject before I could teach it to students. Perhaps that made me a more effective teacher since undergraduates needed a swimming pool to learn how to swim rather than be thrown into the ocean. Moreover, most of the courses I taught were my own creation and not borrowed from anyone. This was particularly true for human sexuality. No one else was teaching it, nor presumably would have, if I hadn't. So my qualms were groundless and I deserved some credit at least for being an innovator. Besides, as a colleague pointed out, I knew more about the subject then anyone else. That was true in the aggregate, even though there were a few who knew more about some of its specific aspects than I did.

These doubts spilled over into my professional identity and role. Looking at it in a musical setting, I couldn't settle in my mind if I wanted to be a composer/scholar who produced original work; a virtuoso performer/teacher who played the music; or a conductor/administrator who led the orchestra. I wanted to do all three and couldn't decide in which role I could have made the greatest contribution or derived the most satisfaction if I had pursued it single-mindedly. Perhaps this was yet another manifestation of my need to be an outsider and unwillingness to be contained in a professional mold any more than a personal mold.

~

Several additional courses that I developed followed the same basic format of the human sexuality class. They were multidisciplinary, drew from the biomedical and behavioral sciences and to a lesser extent the humanities, and addressed topics close to students' personal concerns. These included my course on adult development, the course on academic and career choice, and finally my seminar on guilt and shame. Further afield, I taught twice in the Great Works in Western Culture course, a successor to the famous Western Civilization program at Stanford. This was a humanities course taken by freshmen and based on eight books. At the outset, I knew Freud's book well and Darwin's book somewhat less well. I had a hazy idea about two others and no idea about the rest. Yet, the second time I taught the course, I received the highest evaluation and perfect scores from every student in the class, but only grudging recognition from the humanist who coordinated the program.

While I was fascinated by the books we studied, teaching with humanists turned out to be a mixed blessing. The other instructors were lecturers who were hired to teach in the course but were not members of the academic faculty, as I was. Moreover, I was a psychiatrist and not a humanist. I felt frustrated by the discussions we had in preparation for the classes. Very little of it had anything to do with how best to teach the students. Instead, my humanist colleagues tried to outdo each other in expressing even the simplest thoughts in the most convoluted language possible. They seemed to be speaking in code in order to exclude me from the conversation. One of them said it was doubtful if human behavior could be studied at all which was, of course, what I did. In hindsight, I wonder if any of this was true or simply reflective of my own anxiety about grazing in someone else's field.

The selection of the authors fed into the touchy debate of what constituted the intellectual canon. Some rejected works by "dead white males." Their opposites were suspicious that authors were being chosen on the basis of political correctness rather than merit. I leaned toward the conventional side of this issue but came to appreciate the importance of studying authors that students could identify with. One of the authors was the Argentinean Jorge Luis Borges. Assuming that the students knew little about South American history, I made some general comments about it. At the end of the class, one of the students whose parents were from Bolivia came up to thank me. She was happy to hear the name of one of *her* heroes, like Simón Bolívar, even if it was no more than a passing mention.

These smaller seminars added an important dimension to my teaching experience. They allowed me to get to know my students on a much more personal basis than the large lecture classes would allow. My success in teaching these other courses was also a vindication that my popularity with students was not the result of my showmanship or peddling sex (although no one had ever told me that to my face). My close relationships with advisees extended these dimensions even deeper. I became more than an accomplished lecturer; I was a mentor and for some students, I became like a substitute parent. One young woman who had a distant and troubled relationship with her father told me tearfully in her senior year that I had "filled a void" in her life. At a Human Biology graduation, a student left a rose and an anonymous note in my mailbox that read, "Thank you for making my four years here so special and wonderful. You are an amazing professor."

Some of my former students, like Beth Dungan, Grace Yu, and Elaine Cheung, who also worked for me as teaching assistants, became good friends and I kept in touch with them after they were married and settled in their careers. The student that I have known longest is Mary Hufty, who became a physician (and Lucy's doctor). I met Mary in 1967, the year after I came to Stanford and was teaching my first freshman seminar on identity. It was limited to sixteen students but seventeen of them showed up. Mary came up to me and said she was not on the class list but would very much like to attend the seminar. She smiled and I caved in.

It has given me much satisfaction over the years to watch former students grow into full adulthood and get settled in their careers and personal lives. They have enriched my life and provided it with the perspectives of successive younger generations. The greatest satisfaction in my career came from touching the lives of young people

Figure 9.4 Seminar on guilt and shame.

even when I was not aware of it. I still have former students who tell me, "You don't remember me, but years ago you told me something that changed my life." When I ask what it was that I had said, it sounds embarrassingly like fortune-cookie wisdom. ("Never give up.") I used to make light of these comments until I realized their significance, and now I thank them and say how touched I am.

One evening, Stina and I ran into a young mother with a toddler in tow at a restaurant in Palo Alto. She came up to me and said, "Doctor Katchadourian, you are responsible for this child." My wife looked at me and I looked at the woman. I was absolutely sure I had never seen her before (but could something have happened on some dark and rainy night that I did not remember?). As I looked baffled, she said, "My husband and I were married when we were students in your class. I was trying to get pregnant but couldn't. Then we heard your lecture on conception and I promptly got pregnant."

There was yet another teaching experience that could have led to a significant shift in my career. I collaborated in teaching a course on family law with a Stanford law professor. Subsequently, I was offered a half-time position in the law school. I didn't give it serious consideration at the time, but in retrospect it could have been yet another interesting twist in my circuitous career with longer-term consequences.

The last course which I initiated, and which I ended up teaching for a decade, was a seminar on multidisciplinary perspectives on guilt and shame. I covered the psychological and relational aspects and invited colleagues to deal with topics outside my areas of competence, such as guilt in religious and philosophical traditions and the law. It was one of the most rewarding teaching experiences I ever had. In addition to giving me an opportunity to present a complex subject from a variety of perspectives, the course had a direct relevance to my own struggles with guilt when I was younger. I wasn't sure if students would be interested in such a heavy topic, but the seminar was oversubscribed every year it was taught. I learned an enormous amount from my colleagues and my students and the seminar provided the basis for a book on guilt, which I wrote after my retirement. Three decades of teaching had not dulled my appetite for teaching, and I had not lost my touch.

A final extension of my teaching, and one that I still maintain, was in connection with the Stanford Alumni Association. It began with talks that I gave to alumni clubs across the country. Later, these turned into lectures in the Stanford Travel Study Program. I took these tasks seriously and put in the same sort of effort and care that I put into my regular courses. The fact that I was addressing fellow travelers, who were closer to my own age, on topics that had nothing to do with my professional background made it a special challenge. The twenty-five or so trips that Stina and I went on were wonderful opportunities that took us to many out-of-the-way places we would never have gone to otherwise. For these services, I received the Richard W. Lyman Award of the Stanford Alumni Association.[5]

While I have done a variety of things in my nonlinear career, my primary sense of professional identity has always been that of a teacher. Teaching is what I did best and what gave me the greatest satisfaction. Ultimately, it mattered little what subject I taught and who my students were; the older I became, the younger the students I enjoyed teaching. Were I qualified, I might have preferred to teach elementary school students rather than medical students. I always reminded myself that if I can't explain something to a child, then I don't understand it myself.

There is a six-page entry in my diary from my junior year in college on "How to give a talk." It is both amusing and prescient. It starts with, "I never write down my talks, but that does not mean that I do not prepare them." The planning process started with invocations of God's help ("Put your thoughts in my mind and your words in my mouth"). Next, I began to think about the topic in general terms and made a preliminary outline. Then I read about the subject until I knew enough about it (but no more than that). And I collected suitable quotations. The "rehearsals" that followed involved talking to myself, often walking outdoors. ("I was roaming at the seashore last night and gave a talk to the sea and to its waves from a cliff.") A few days before the talk, I would rehearse it in front of a mirror. I memorized the quotations and never wrote out the actual text; the words would come on their own.

I made a careful distinction between what I called the "background" and the "real subject." I compared it to a window display with a black velvet cloth with a sole object at its center. I noted that the talk should deal mostly with the context and then deliver the essence in a short and intense burst. Talking about the key message at greater length would dissipate its impact. ("If you drag out a bit the secondary subjects, it is not serious. But if you drag out the heart of your talk, it will be deadly.") I wrote this almost sixty years ago when I was nineteen years old, but I could just as well have written it yesterday.

~

The first decade at Stanford that marked my transition from the medical school to the university at large was an equally transformative period for our family. We had settled down at Stanford for the long haul, but our initial period of adjustment wasn't easy. As I became immersed in trying to find my niche and venturing in novel and uncertain directions, Stina felt stranded at home. The sunny climate that delighted her at first came to feel monotonous and oppressive. She resorted to turning the portable sprinklers against the kitchen window to simulate rain. I recall coming home to find her crying: she said she missed the birds on the island. I told her there were plenty of birds right there in our backyard. "Yes," she said, "but they're the *wrong* birds." I thought to myself, we're not going to make it here—but we did.

Our life in Lebanon had been exciting but imbued with a feeling of impermanence; life in California was secure and stable but felt static and dull. Stina missed her family and the cultural and intellectual life of Finland. We had no relatives nearby, and virtually all of our friends were from Stanford; they were primarily part of my world, not hers. When Stina learned that Stanford had a good Latin American Studies department, it turned out to be a godsend. She was already fluent in Spanish and could work for a master's degree. It became an academic anchor for her and introduced her to new friends. She also resumed freelancing for the Finnish radio and newspapers and became involved with organizations like Amnesty International and Physicians for Social Responsibility. She was strongly opposed to the war in Vietnam and brought me around to the same position. Actually, I was never for the war, but simply didn't care because it wasn't my problem.

During this period of uncertainty before we had put down deeper roots at Stanford, we were saved from making a move that would have been disastrous. My former chief resident, Elmer Gardner, had established an ambitious program in community psychiatry at Temple University. It was based in one of the poorest black communities in Philadelphia at a time when such programs were flourishing with the shutting down of state hospitals. Elmer asked me what it would take to get me to join him in Philadelphia. It was a time when I was beginning to feel detached from the Department of Psychiatry and acutely aware of Stina's feeling homesick. I told Elmer that if I could get a full year's salary with only six months in Philadelphia, I would seriously consider it. The idea was to spend the balance of the year in Finland, where I would engage in some sort of research and Stina would be happier. We didn't give much thought to how we would sustain such an arrangement as the children got older and had to go to school. It was a harebrained idea, but it was accepted. However, before we could act on it, Elmer and his department chairman had a falling out and he left to go into private practice elsewhere. Had I made the move, we would have been stranded in Philadelphia in a job I had no interest in. Our lives would have been derailed, with consequences I would rather not think about.

~

Our daughter Nina's birth on February 19, 1968, was an amazing experience. I had attended deliveries before and had delivered babies myself, but this was different. The nurses bundled up the baby and gave her to me. Covered with blood and gore, she was a mess—but to me she looked like the most enchanting sight in the world. I put my finger in her hand and she closed her tiny fingers around it. My head told me it was just a reflex, but my heart knew it was because I was her *daddy*. I couldn't believe how cute she was and marveled at the light of intelligence and creativity shining in her lovely eyes (even if she could hardly open them). Suddenly, my life felt full in a way it had never felt before.

The following summer, we took Nina to Beirut. Wearing my baptismal gown, she was christened in the Apostolic Church. She was terrified by the priest, but my parents and Lucine were ecstatic. There is a photograph of my father at our balcony bouncing Nina in his arms. I don't recall seeing him so happy. We then went on to Finland and presented Nina to Stina's parents. They already had three grandsons, but she was their first granddaughter and was to develop a special bond with them.

We realized that our house on Amherst Street was going to be inadequate as Nina grew up and if we decided to have a second child. I was now eligible for university-supported housing, and a few plots were left in the new Frenchman's Hill development. We had a house built and moved into 956 Mears Court in the fall of 1970, a few months before our son, Kai, was born. Over the years, the furniture we bought in Finland, the oriental carpets and artifacts from my parents' house, and what we had picked up during our travels over the years, turned our house into an attractive space where we were to live for thirty-five years. It was interesting enough to be featured in a Finnish interior design magazine, where the special lighting made the house look even more striking than it actually was.

The news of Kai's birth on December 18 caused much jubilation in Beirut. When Nina was born, we received many congratulatory letters and telegrams, but now they poured in—this one was a *boy*! My father had announced that his *vakil* (representative) had now arrived and was pleased that Kai's middle name was Aram. Unfortunately, he didn't live to hold his grandson in his arms. Like his sister, Kai was a perfectly formed baby from the top of his head to the tip of his toes. My mother said he had had my great-grandmother's striking green eyes and Nina had my mother's warm brown eyes. They enriched our lives immeasurably and were to continue to do so in the years to come (even though I continued to miss their childhood).

~

In 1971, shortly before Easter, I got a call from my godfather, Khatchig, in Beirut: my father had suffered a stroke. He had been undergoing radiotherapy for cancer of the prostate, but this was a new development. If I wanted to see him, I had to come right away. I took the next available flight to Beirut. Full of foreboding, I found my father in bed with his eyes wide open but oblivious to his surroundings. He looked at me, perplexed, and didn't seem to recognize me. The next day, I sat by his bed with the bright spring sun pouring in from the windows. I had given up trying to communicate with him. On an impulse, I took out of my wallet a photograph of Nina and Kai and held it close to his face. He looked at it and tears began to roll down his face. I was stunned. I said, "Baba, it's Nina and Kai." He looked bewildered and the momentary connection was lost as the light of recognition went out of his eyes.

During the hours I spent by my father's bedside, I reflected on our lives with an aching sense of regret for the distance that had kept us apart. I had thought of him as a demanding and forbidding father in my youth and feared that he would crush me without my mother's protective mantle. He now lay helpless and lost to the world. His influence on my life had been inhibiting and paralyzing at times. All this, however, was now behind me as a deep sense of love and gratitude welled up in me. My father was an honorable man with tremendous integrity. He was a devoted father who had been unable to freely express his affection toward me, just as I had been unable to express the love for him that now suffused me. I continue to regret deeply that I was not able to take care of my father at the end of his life as I was able to look after my mother and Lucine. Their accounts of my father's undergoing radiotherapy for his cancer, the ordeal of taking him back and forth from the hospital when he was in pain and incontinent, were heartrending. And I was not there. I realized there was not much I could have done from continents away. So what I felt was not guilt but a profound sense of regret.

I had an acute sense of the burdens my father must have carried in his life, particularly the loss of Iskenderun and the thinly veiled resentment that my mother and I harbored over it. He must have sensed it, and that must have added to the pain of his loss. How galling it must have been for him to see people less able than he was become wealthy while he toiled into his seventies to support his family. The tight bond that my mother had formed with me, perhaps to compensate for the loss of the love of her life, had perhaps led to my keeping him at arm's length. I was never disrespectful to him, but shut him out of my life when I was growing up. It was a great pity that the closeness to my mother may have been at the cost of my distance from my father.

My regrets over not having known my father better have lingered over the years in the various objects he passed on to me. The rugs, the brass and copper braziers, my father's kilim money belt, his gold Zenith pocket watch with its exquisite decorations, as well as more mundane objects are a legacy from a long-lost world. Yet I know so little about them. Objects without a history are like orphans. Experts can identify them and tell me about their origins and value but no expert can tell me about their personal history. Where did my father get them? When did he buy them and what did he pay for them? What did they mean to him?

As my father and I both got older, the tension between us dissipated. There was more openness and a greater sense of closeness with the reversal of our roles, but that could hardly make up for lost time. As a child, I had been dependent on my father; as an old man, he became dependent on me. Yet dependence is not intimacy. I envied those who were able to form bonds of friendship with their fathers that went beyond kinship. My father and I were so different from each other and our relationship so loaded that it was impossible to think that we could be friends, or would have had anything to do with each other had we not been father and son.

These times of reflection at my father's bedside were frequently interrupted by the stream of relatives that came to visit my father even when they could no longer communicate with him. The burden of hospitality fell on my mother and Lucine, who had to cater to their guests while caring for my father singlehandedly. I suspect that the obligation our relatives felt to visit my father may have been burdensome to them as well. They seemed all trapped in a well-intentioned cultural tradition that came at a cost to everybody concerned. I could no longer stand it, and for a few hours every day I would slip away to the poolside of the Hotel Phoenicia to read Solzhenitsyn's *Cancer Ward*.

A week passed. My father's condition remained unchanged and there was no telling how long he would last. One evening, when the crowd of guests was particularly dense, the wife of one of my cousins, who came from a very different cultural background, took me aside and told me in a whisper that it was customary to withhold food and water from the dying person to hasten the end. It was euthanasia without the benefit of the name. I was horrified at the thought even though it might have been the merciful thing to do. Another cousin asked me why we were keeping my father at home when there were hospitals to die in. Yet my mother and Lucine carried on the burden heroically and lovingly.

I was in a dilemma. I wanted to stay with my father until he died. But I had colleagues covering my courses, and they couldn't do that indefinitely. My wife and young children needed me back home. With a heavy heart, I hugged my father's insensate body for the last time. I bade a tearful farewell to my mother and Lucine and headed back to Stanford. By the time I reached home, my father had died.

10

Would-Be Administrator

Teaching in Human Biology was my primary activity during the early 1970s. Concurrently, a succession of assignments outside of the medical school gradually led to my effectively moving out of the Department of Psychiatry. Amid the turmoil that engulfed Stanford during the Vietnam War, a committee was established to advise the president on campus unrest. I succeeded David Hamburg on the committee in 1969, when Kenneth S. Pitzer had recently become president and Richard W. Lyman was provost.[1] My membership on the committee helped to bring me further to the attention of the university administration. Consequently, when the university made an attempt to douse the fires by establishing an Office of the Ombudsman, they turned to me.

The office had originated in Sweden in the eighteenth century and a Swedish ombudsman now reported to parliament. On my way to Finland for the summer, I went to Stockholm to meet him. He was a distinguished retired judge, who kindly took me to lunch at his club and explained to me the functions of his office. He told me a story that exemplified it. Two sisters living in a remote region in northern Sweden had contacted him with the complaint that their father had been improperly buried. They had appealed to local authorities, but got no satisfaction. The ombudsman asked the local magistrate to have the body exhumed. Lo and behold, one of the legs was askew. The news was picked up by the national press. The public reaction was that if one could get satisfaction for an odd problem like that, then there was hope for a remedy to more compelling injustices as well.

During the year and a half between 1969 and 1970, when I was ombudsman, about two hundred persons came to see me. Two thirds were students, a fifth staff members, and the remaining few from the faculty. Most complaints involved relatively minor issues, but some cases were more contentious. I prevailed to have one student's grade changed, which was not easy. A more serious and eventually tragic case involved a graduate student. He had been working on his dissertation in mathematics for ten years, until his graduate fellowship was terminated. He didn't have a case, but he looked so forlorn and pathetic that I sent him to Lincoln Moses, the dean of graduate studies. More out of charity than anything else, the dean gave him some more money, but it hardly helped. He left Stanford, only to come back a year later and beat his faculty advisor to death with a hammer; the very man who had helped him more than anyone else. No one, including me, had suspected how disturbed he was. I was subpoenaed by the defense attorney to testify during his trial, but I couldn't support the claim that he had been unfairly treated. He was given a shockingly lenient sentence, and I worried that he would come after me after his release, but nothing further was heard of him.

At the end of my term, I submitted a report on my experiences as ombudsman. I pointed out the modest but significant gains we had achieved as well as the cost; at $320 per case (in 1970 dollars), it was an expensive operation. However, once in place, such services were hard to eliminate. My successor was a contentious person who made an uncalled-for public criticism of the highly respected provost. The office didn't have much teeth, but it still could do damage.[2]

Next, I was delighted to be appointed a University Fellow. The program had been established at the instigation of law professor and vice provost Herbert Packer, with a grant from the Ford Foundation to bring together senior administrators with younger tenured faculty who had the potential for becoming "university statesmen." It was meant to narrow the barrier between faculty and administration. The fellowship provided full salary for three years to a few select faculty members who would spend half their time in some form of service to the institution and the balance as they saw fit.

This was an unparalleled opportunity for me to become more deeply involved in the intellectual life of the university, but I failed to take full advantage of it. I continued to teach my human sexuality class, work on the textbook, and develop plans for a similar course on the human life cycle. I dredged up the residual data from my research in Lebanon and wrote a number of additional papers. In retrospect, I should have taken courses in art, classics, and literature to compensate for the gaps in my undergraduate education.

Several years later, I was appointed director of the university fellows program by president Donald Kennedy and continued to run it when Gerhard Casper was president and Condoleezza Rice provost. I reported to Condi Rice and enjoyed working with her. She was unfailingly supportive and she impressed me with her competence. (Years later, I would have more to do with her when she joined the board of the Hewlett Foundation, of which I also was a member.) On one occasion, I had to introduce Condi before a talk. I felt at a loss. What could I say about her that people didn't already know? I started with a story about a man who stood at the entrance of the jewelry market in Beirut. He had a finely made mahogany box suspended from his neck in which he had dozens of small bottles of concentrated scent which he sold as perfume. (The audience was wondering where I was going with this.) Condi reminded me of this man. She spoke in pithy sentences with a lot of meaning like vials of perfume. Condi was pleased and told me she had been introduced many times before but never like this. I was proud of her when she became the national security advisor and then secretary of state but watched with puzzlement and then dismay as she too became mired in the war in Iraq—even though compared with some of her colleagues in the cabinet, she was often the voice of reason and restraint in the Bush administration.[3]

~

In the fall of 1975, and at the end of my ninth year at Stanford, I was finally able to take a year of sabbatical leave. I had been developing a course on the human life cycle on the model of the human sexuality class and had planned to use my sabbatical working on a textbook for the course. We could have spent the year virtually anywhere, but we settled on England since it would be easier for the children's schooling. We considered going to Oxford but were dissuaded by ill-informed advice. I then saw a notice for a house exchange posted by a professor of chemistry at the University of London. His family lived in the small town of Shiplake, outside of Henley-on-Thames, and it sounded like a fun place for us to live in. It was an hour's drive from Oxford, where I could use the Bodleian Library, and an hour by train to London, making visits easy.

Our choice of Henley turned out to be a mixed blessing. The children enjoyed living there and developed a hilarious British accent. However, Henley was dull and going to Oxford turned out to be more time-consuming than I had thought. The house we lived in required a lot of maintenance. The au-pair girl we brought with us got homesick and left after a few months. My life-cycle book was set on an unrealistic time schedule. And before I actually could start working on it, I was invited to take part in a conference on adulthood organized by Erik Erikson and to write a chapter for a book that Erikson would edit. It was an important project and it took me until Christmas to finish it.[4] I then managed to write the section on the biology of adolescence, which got expanded and published as a separate book.[5]

The only person we knew in the area was Sir Geoffrey Vickers, who had moved to a nearby town after his retirement. I had been introduced to him by my friend Erich Lindemann, who had come to the United States from Germany and had an outstanding career as chief of psychiatry at the Massachusetts General Hospital in Boston. He had made important contributions to the development of social psychiatry and was a pioneer in studies of grief. After Erich retired from Harvard, David Hamburg brought him to Stanford. We became good friends. I had weekly meetings with him until he developed cancer and moved back to Boston. Our conversations ranged over a wide array of subjects, which gradually veered away from psychiatry. He was an unusually kind and thoughtful person who left his mark on me.

Sir Geoffrey was a distinguished public figure, "educated in the classics and trained in the law," as he put it. He had been knighted for his services as head of economic intelligence during World War II, when he worked closely with the three chiefs of the British armed forces. He was a scholar and the author of books on institutional learning. Stina and I became good friends with him (and for five-year-old Kai and seven-year-old Nina, he was "Geoffrey"). He visited us on the island in Finland, which he called a "healing place." His exquisite command of English made everyday conversations sound like literature. When I shared with him my uncertainties over my career, he told me what eventually mattered were the lives I touched, and it seemed like I had touched quite a few.

As our stay in England was coming to an end, I became increasingly preoccupied with what I would do on my return to Stanford. I took long walks in the nearby copper beech forest to ruminate about the future. I had been away from the Department of Psychiatry for five years and did not relish the prospect of returning to it when David Hamburg was no longer there. His successor was having problems running the department, and a revolt was brewing among the faculty. I was surprised to receive a phone call from the dean of the medical school informing me that the head of psychiatry had resigned and my colleagues in the department wanted me to take over the chair. I was flattered, but the dean sounded less than enthusiastic. I suspected that he preferred to have the hospital director, who was a psychiatrist, to take over and reestablish order in the department. Besides, I wasn't sure I wanted to go back to the department, let alone be in charge of it. I declined his offer—another path not taken.

In the early spring, I heard that the dean of Undergraduate Studies was ending his five-year term. I didn't know much about the position, but it sounded as if it might be a great opportunity for me even if the chances of my getting the post seemed remote. We left England for Finland on a cargo ship in late May to spend the summer. My friend Lincoln Moses, the dean of Graduate Studies and his wife, Mary-Lou, came to visit us on the island. Lincoln was an avid birdwatcher who hardly put down his binoculars during their stay. At an opportune moment, I asked him about the undergraduate dean's position and hinted at my interest. He thought it might be an interesting prospect. I felt encouraged but didn't want to get my hopes up.

We returned to Stanford in the fall and I tried to get resettled in the Department of Psychiatry. I had nowhere else to go. I was happy to resume teaching my sexuality course and began to see some outpatients. The new department head knew that I had been the faculty's preferred candidate for his position, but he was gracious enough to welcome me back. But that didn't make my longer-term prospects at Stanford look any more certain. Then, in the spring of 1976, I got a message from Gordon Wright of the History Department. He was chairing the search committee for the new position of vice provost and dean of Undergraduate Studies and wanted to meet with me.

~

The Office of the Dean of Undergraduate Studies (ODUS) was created in 1970 on the recommendation of the Study of Education at Stanford (SES) chaired by Herbert Packer. The study was a comprehensive attempt to review and improve education at Stanford. The new position was intended to be like a counterpart to the position of the dean of the college in other select institutions. The dean was to act as a senior administrator with responsibilities for ongoing review of undergraduate education and educational innovation.[6] The responsibilities included undergraduate advising and oversight of undergraduate courses and programs outside academic departments. One of these was the Freshman Seminars program. The other two—SWOPSI (Stanford Workshops on Political and Social Issues) and SCIRE (Student Center for Innovation in Research and Education)—were more politically oriented and led by activist staff. SWOPSI had been initiated by students to provide opportunities to examine social problems and become actively engaged in their solutions. Although the university administration was lukewarm in its support of these programs and faculty engagement was limited, SWOPSI offered, at its peak, some 100 workshops, attended by nearly 2,000 students. It was part of the wave sweeping across campuses in the 1960s to get students more engaged in the "real world" and make their education more "relevant." The undergraduate dean responsible for their oversight had to walk a fine line between supporting their legitimate educational efforts and moderating their zeal in reforming the university.

The undergraduate dean's position was a thankless job. It wielded no influence in the two key areas that mattered most: the appointment and promotion of faculty and the allocation of budgetary resources, including faculty salaries. The deans of the schools and the provost controlled these activities. Gordon Wright pointed out that the next dean of Undergraduate Studies would also be a vice provost for Undergraduate Education, thereby giving the office more clout. Would

I be interested? Yes I would be—there was no need to be coy about it. I met with President Richard Lyman to express my appreciation for his confidence in me, but I also raised my concerns over how Halsey Royden, the dean of Humanities and Sciences (H&S), would react to my intruding into his turf with these expanded responsibilities. He said I was right to be concerned, but that shouldn't stop me from accepting the position. I met with Provost William Miller, who was warmly supportive. And I talked with James Gibbs, the incumbent dean, to get a better sense of the prospects and the problems of the office. Finally, I went to the Halsey Royden and asked him if he would advise me to take the position. He said he would, and I decided to take him at his word.

I was pleased to have my office located in Building 10, in the Inner Quad, where the president and the provost had their offices. The symbolism would not be lost on people. I appointed three associate deans, whose positions would parallel those of the cognizant deans with line responsibility in H&S. I was fortunate to have Alexander Fetter from Physics to be my associate dean for the natural sciences, David Abernethy from Political Science for the social sciences, and Diane Middlebrook from English (and later on, Marsh McCall from Classics) for the humanities. I would myself work directly with the deans of Engineering and Earth Sciences. The key concern in Engineering was to ensure that students would obtain a broad liberal education despite their heavy load of engineering courses—a concern shared by Dean Bill Kays, with whom I developed a cordial relationship. The same was true for the much smaller school of Earth Sciences.

The assistant dean of Undergraduate Studies, who had oversight over SWOPSI and SCIRE, had agreed to step down. He had been a thorn in the flesh of my predecessor and I was relieved to see him go. I would appoint someone who was less contentious and would represent me to the programs rather than be the advocate of the programs to me. The academic departments tolerated these programs but were hardly enthusiastic about their course offerings. If the departments were like shops in the bazaar, the smaller programs I would be in charge of would be the street vendors. They served a useful purpose in providing academic opportunities for socially and politically activist students, but they tended to be seen by the faculty as academically marginal.

In hindsight, I didn't fully appreciate at the time the significance of the contributions these extra-departmental programs made in addressing political, economic, and social issues that deserved to be part of the university's offerings (and eventually made their way into the regular curriculum). Part of the reason was my own political bias. I came from a conservative background and saw left-wing politics as being disruptive of the social order of the university. The excesses of politically activist and antiwar students—the sit-ins, class disruptions, and broken windows—alarmed me, and I didn't want to aid and abet their mischief. That led me to tar with the same brush the more responsible activist programs. More importantly, I saw these reformist programs as a distraction from my efforts to improve the quality of undergraduate teaching and academic advising at the mainstream departmental level. I feared that the association of my office with these more marginal entities would detract further from its already ambiguous academic legitimacy. If I was going to have a significant impact on undergraduate education, I had to find my way into the shops in the bazaar themselves rather than lead the street vendors in an assault on the "establishment."[7]

As a member of the medical faculty, I was an odd choice for the position of dean of Undergraduate Studies. It is to Stanford's credit that it was open-minded enough to take that chance—I doubt if it could have happened elsewhere. Two experiences from my past helped to set my agenda. My administrative style would be modeled after John Romano's. I would listen to my staff, but I would have the last word. My research in psychiatric epidemiology came in handy as I set out to establish a baseline to determine how undergraduates made academic choices and how satisfied they were with their education at Stanford. To that end, I established the Senior Survey based on questionnaires that seniors completed before graduation. It asked both general and specific questions about their undergraduate education, including their encounters with courses, departments, and members of the faculty. I wanted them to name names, not indulge in generalities. Graduating seniors turned out to be generally well-satisfied with their Stanford experience. However, they had problems with residential education and the freshman advising system. Moreover, two renowned departments attracted a good deal of criticism with respect to their introductory undergraduate courses that were required for premedical students. Many of these students didn't want to be in these courses any more than the departments wanted them there. Otherwise, these departments took good care of their own majors.

To ascertain trends in academic choices and grading practices, the registrar, Sally Mahoney, agreed to tap into the academic records. There were some surprising findings. (In one small department, close to 90 percent of the grades were A's.) I took these findings to the school deans. I now had information about their departments that they didn't have, and this knowledge earned me a place at the table. I proposed setting up meetings with each department chair to review their undergraduate program, and I wanted the cognizant deans of the schools to accompany me and my associate deans on these visits. No one had done this before but the deans agreed to it. These visits helped establish the legitimacy and relevance of my office as much as anything else I was to do. A few department heads may have found the visits intrusive, but most of the rest welcomed them. They were glad that somebody was finally paying attention to these issues in a systematic way.

Problems with freshman advising were endemic and hardly limited to Stanford. Good freshman advisors needed to combine a grasp of their field (which was easy) with a wider knowledge of the undergraduate curriculum as a whole (more difficult) and be able to relate to students personally and spend time with their advisees (most difficult). Once students declared a major, departments provided good academic guidance for students who knew what they wanted to do. Students who were uncertain about what to major in and procrastinated in declaring a major were in limbo; they needed the most help and got the least. It was frustrating for faculty to advise students when they didn't know what they wanted. And they weren't generally eager to hold their hands while students tried to figure it out. Similar considerations applied to residential education. Stanford didn't have residential colleges like Harvard and Yale, which were a central part of undergraduate life. Instead Stanford had theme houses that served some of the same purposes, but they involved only a part of the student body. Consequently, many undergraduate residences didn't do much more than house and feed their students. The dean of Students and his staff who were in charge of residential education worked hard at it, but they had little leverage on the faculty to provide intellectually stimulating programs in the dorms.

When it came to dealing with the problem departments, the dean of H&S was reluctant to confront them head on. They were some his strongest departments, with nationally eminent faculty, and he was loath to upset them. I appealed to the provost and he agreed to have me take up the matter with the Advisory Board of the Academic Council—the highest-level academic committee that reviewed all faculty appointments and promotions. I met with the Advisory Board and laid out the problems. I asked if they wanted to see the supporting documents I had stacked up on the table. They didn't need to. They put the two departments on notice: all their faculty appointments and promotions from then on would have to include a statement on how the action would improve undergraduate education in their department.

Undergraduates typically took a third of their courses in their major, which provided the element of academic depth. Another third was chosen from a set of approved courses in three areas: the natural sciences, the social sciences, and the humanities. These distribution requirements were intended to add breadth to their education. The rest were electives. A highly structured set of courses had fulfilled distribution requirements at Stanford until the late 1960s, when they were dismantled as part of the wave of liberalization that swept through colleges and universities. Another casualty of this trend was the famous course in Western Civilization ("Western Civ"). Established in 1936 by Edgar Robinson of the History Department, it had been a mainstay of Stanford's liberal education curriculum. During its thirty-four years of its existence, some 40,000 students took the course and many thought it was the best course they ever had at Stanford. By 1969 the course had been reduced to a mere shadow of what it had been.

The approval of courses to fulfill the distribution requirements was one of my responsibilities. It was a meaningless task, since the criteria for course approval had become so lax that close to two thousand courses fulfilled them. It would have been hard for students *not* to fulfill these requirements in their ordinary course of study. I took up the issue with the provost, now Donald Kennedy, and we submitted to the Faculty Senate, through the Committee on

Undergraduate Studies, a proposal to tighten the requirement to serve its intended purpose. The courses would no longer be subsumed under the three broad categories, but in one of seven areas defined by their intellectual content—areas that I had delineated in the senior survey. Simultaneously, Halsey Royden initiated a concerted attempt to revive the Western Civilization course under the rubric of a new Western Culture program. These two efforts came together, and with Provost Kennedy's strong support the Faculty Senate approved a new set of distribution requirements, including the Western Culture course. The revision of the distribution requirements was to be the main contribution of my office to academic policy.

~

Our general surveys of undergraduate education were useful, but we needed more specific information on how students made academic choices and the consequences of those choices. The provost had provided my office with $25,000 in budget-based funds available each year to be used at my discretion.[8] I put the money into establishing a research office and we launched an ambitious four-year longitudinal study based on a 20 percent random sample of the class of 1981. It came to be known as the Cohort Study. Each spring, the participants responded to a detailed questionnaire and were interviewed by a team consisting of twenty-five individuals, including several senior administrators and members of our staffs. In the second year of the study, I appointed John Boli, a sociologist, as director of our research office. He became my primary collaborator and we coauthored a book on the patterns of academic and career choice in the undergraduate years.[9]

It is generally thought that college education should serve two main functions: career preparation and liberal education. On the basis of how students pursued these two objectives, we placed them in four categories. *Careerists* were primarily concerned with career preparation, typically at the expense of liberal education. *Intellectuals* were their polar opposite. *Strivers* went after both objectives, whereas the *Unconnected* seemed to go after neither, making their academic choices seem random. Strivers constituted the largest group, accounting for about 35 percent of students. Intellectuals were the smallest, at 15 percent.

Careerists and the Unconnected each accounted for 25 percent of our sample and presumably the undergraduate student body.

Considering that each category had its advantages and liabilities, we avoided attaching value judgments to these orientations. However, we made no secret of our concern over the careerist ethos that drove the choices of many of Stanford's students. There was nothing wrong in having career aspirations. However, college was the last chance students had to get a broader education that would do more than lead to a good job. The ability to think; to integrate literature, art, and music into one's life; and to develop a social conscience and good citizenship were equally important for leading a fulfilling and purposeful life. It is ironic that I should have become the advocate of liberal education with a strong emphasis on the humanities since I had been the worst kind of careerist premed student in college and had short-changed my chances of getting a good liberal education—all for the sake of getting into medical school, which I would have done anyway. Perhaps it was that very fact that impelled me in this direction.

The implications of our findings for academic policy were considerable. Stanford provided a first-rate college education, but one size did not fit all. Undergraduates had particular needs and liabilities: Careerists had to be induced to broaden their intellectual horizons in order to become more than mere technicians. Intellectuals to focus more on preparing for a career after college so they could at least wield a broom. Strivers to rein themselves in so they wouldn't spread themselves too thin, and the Unconnected to find their bearings sooner rather than later to spare themselves from meandering through life. It was a tall order. I went around giving talks to anyone who would listen: trustees, administrators, faculty, students, alumni, and educators at large. My presentations were well received, but I had no way of knowing how significant and lasting their impact would be.

In 1991, when I was no longer dean, we launched a ten-year follow-up to the Cohort Study. I carried out the interviews and John Boli (then at Emory) analyzed the questionnaires. I had developed a good relationship with the subjects I interviewed during their college years, and it was fascinating to see them again in their early thirties pursuing their professional and personal lives. Our book that was based on the follow-up study was published in 1994.[10] The first volume had elicited good reviews. The second book, which we thought would attract even more interest, did not. There were a few fairly positive reviews, but others were lukewarm. Some objected that the interview reports lacked the objective tone required of psychological reports. We thought the lively and sometimes provocative language we quoted from interviewer comments made the book more interesting. (One interviewer wrote that listening to one student was like "watching paint dry.") Others were unhappy with our focus on "elite" education. Americans pride themselves on living in an egalitarian society and don't like to hear anything that might dispel that fiction. At any rate, the Cohort Study was important for me professionally and personally. It was the most extensive research project I was to ever undertake. Getting to know the remarkable young women and men who participated in the study was my greatest reward.

~

At the end of my five years in the dean's office, there was no indication that I should step down at the end of my term, and I had no wish to leave. Donald Kennedy was now president and Al Hastorf had succeeded him as provost. I had a close and cordial relationship with both of them. However, I needed a break. I had accumulated six months of sabbatical leave, and I decided to take it over the summer and fall quarters. I would have an extended summer in Finland and finish up various writing projects in the fall. I expected to return to my office refreshed and eager to resume my job. However, my reaction on getting back was quite different. My absence seemed to have made little difference (why did I think it could be otherwise?). There were no unfinished tasks for me to complete. The revised distribution requirements were in place. The first phase of the Cohort Study was completed. The excitement and enthusiasm with which I had moved into the dean's position had largely dissipated. My responsibilities no longer seemed sufficient to justify my staying in the job. I couldn't keep battling the dean of H&S and was tired of haggling with the small fry. Nor did I feel up to launching yet another major research project.

I had been elected to the Faculty Senate from the medical school for two successive terms and then became an ex-officio member as dean. All told, I ended up serving in the senate for a total of twelve years. I was

Figure 10.1 Sulking at a Faculty Senate meeting.

an effective and sometimes combative participant in senate discussions. During the turbulent early 1970s, as a politically conservative faculty group was trying to take over the senate leadership, I was approached by the university president to chair the senate to prevent that from happening. For some reason I felt hesitant and consulted a friend. He advised me against it, arguing that it would not be worth my while. This reinforced my own misgivings and I declined. (A few years later that same friend took on the senate chairmanship, which may have helped his becoming a university president elsewhere.)

My doubts about staying on in the dean's position escalated to my wondering about weather or not the position itself made sense. I had quite a bit to show for the past five years. Yet, those achievements had come about through my ability to improvise and mobilize allies and the unstinting support of the provost and the president. Would the next person in my position be able to do the same? Unlike the position of deans of schools, who had clearly defined tasks and line authority, my position had neither. It was a solo performance without an orchestra to conduct. At a more personal level, I felt more like a "would-be" administrator than a "real" one. I was enacting a role on stage and not fulfilling it in real life. My fundamental doubts over being "real" were rekindled with a vengeance.

I realized that my concerns over the structure of the dean's office should be the administration's problem, not mine. If I was tired of the job, I should quit and let someone else deal with the problems of the office. I was mindful of not letting my personal sense of disenchantment dictate the university's administrative structure. I convinced myself that this was not the case. I took up the matter with the provost. Al Hastorf could understand my concerns, but eliminating the dean's office was a big step that required careful thought. Al himself had been dean of H&S and could see the inherent conflict between school deans and an effective dean of Undergraduate Studies. He also knew that leaving the academic fate of undergraduate education entirely in the hands of school deans would not be enough. Jerry Lieberman, who was vice provost for research and a trusted advisor, was brought into the discussion and then Don Kennedy joined in. I had now taken on in earnest the task of persuading the administration to dismantle my office (and unlike Winston Churchill, I was willing to preside over the dissolution of my empire). The fact that these discussions were taking place at a time of budgetary constraints gave my proposal greater impetus. There would be substantial savings that could be more easily achieved by abolishing my office than cutting the budget in more painful ways. I knew that, and I played it up.

The H&S deanery was delighted at my proposal but also puzzled by it. Administrators typically tried to expand their turf; I was doing the contrary. What was in it for me? The dean of Students told me that he had been wrong about me. He thought I was an empire builder but now he knew better. The members of my staff, in particular the directors of programs, had special cause for concern since their jobs could be on the line. There were even people who got upset without having a stake in the matter. As it is said of cemeteries, no one cares about them until you try to move them. Nevertheless, my proposal was taken up by the Faculty Senate, and the dismantling of my office eventually came into effect. As a substitute for an undergraduate dean, the chair of the Committee on Undergraduate Studies was appointed associate dean for Undergraduate Education at H&S. It would no longer be a

Figure 10.2 Marching at commencement with Chancellor Wallace J. Sterling.

university-wide office but it would at least reaffirm Stanford's concern over undergraduate education. My last public function as dean was to march at commencement.

My tenure as dean ended not with a bang but a whimper. There would have been public celebrations if I were retiring from my position at the end of my term. Instead, there was only a gloomy and cheerless farewell party, which only a few attended out of a sense of obligation. The director of SCIRE presented me with a "gift" consisting of an old attaché case cut in half to symbolize my truncated tenure as dean. I thought of telling her that I would use the gift to store the harebrained ideas that were sent my way over the past five years to which she had so richly contributed, but I did not. A smaller gathering was held at the home of the chair of the Committee on Undergraduate Education. I was given a telescope as a parting gift. Had they not consulted Stina, I would have ended up with one of those white elephant armchairs with the university logo. The lesson I learned from all this was that one gets more credit for building things up, which shows progress, while tearing them down seems regressive—even if more useful in a given case.

~

Al Hastorf asked me to continue as vice provost for Undergraduate Education. I appreciated his confidence but accepted the offer with some reluctance. Despite its title, the position had little to do with undergraduate education. I would be one of the provost's four deputies doing what he didn't have the time to do himself. Nonetheless, I was happy to continue working with Al Hastorf in a position of some importance even at the cost of placing myself again in an ill-defined position with uncertain prospects.

The three years I spent as vice provost were relatively uneventful. One of the responsibilities that I shared with the three other vice provosts was to advise the provost on appointments and promotions. We went through over a hundred faculty dossiers ("red folders") and I learned a lot about the background of faculty members (my most welcome contribution to these meetings may have been the Armenian cookies made by Lucy for our coffee breaks). Preparation of the annual university budget was another major task. When Bill Miller was provost, we worked on one occasion until five minutes to midnight. As I was leaving, I told Bill, "Did you know that it's my birthday today?" He said, "Congratulations. Take the rest of the day off."

One of my more burdensome tasks was to act as a faculty grievance officer. A particularly contentious case involved an assistant professor in H&S. The department had endorsed her promotion to associate professor with tenure, but the Appointments and Promotions Committee of H&S and the new dean had turned it down. It took several months to deal with the case. I spent many hours talking with people, listening to the plaintiff at great length, and studying her voluminous file. I interviewed every tenured member of her department and satisfied myself that the candidate had been deemed deserving of tenure on the basis of her substantial record of publications, a good reputation as a teacher, and very positive letters of support from referees outside Stanford.

I reported the results of my investigation to the provost and he requested a meeting with the A&P committee, the H&S dean, and the cognizant associate dean. I presented my findings and wrote down on the blackboard the positive tally of the letters of support. My findings and conclusions were not well received. Some committee members argued such letters of support were pro forma and meaningless. Why then, I argued, did we bother to get them? Stanford required that a candidate for tenure be among the three or four leading scholars in the field. But whether or not the candidate met that criterion depended in part on how one defined the field. In this case, the candidate was a feminist historian working in what was several decades ago a relatively small and developing field. If judged in that narrow context, the candidate would pass with flying colors; if the field was broadened to American history, then she probably would not, and within the entire discipline of history, she certainly would not. I argued that the committee was judging the merits of the candidate's field rather than her academic qualifications for tenure within that field. If the problem was with the field, then the candidate should not have been hired in the first place. The discussion became increasingly heated and contentious. Some of those who argued most vociferously against me were highly respected members of the faculty, and friends of the provost, which put him in an unenviable position. The head of the committee said grudgingly that my ability to make a case was a mark of my competence (in other words, I was a good lawyer), but I was wrong nonetheless.

The provost accepted my recommendation to overrule the dean of H&S, and the candidate got tenure. It was a difficult decision, and Al Hastorf showed impressive courage in reaching it. I respected him for it, but had he rejected my recommendation I would have said nothing. In the end, for all this grief, we received small thanks. A member of the law faculty who had been one of the candidate's key supporters made a public statement that the administration had reversed itself because of the external pressures brought on the case. That was not true. While I did receive numerous letters of support from all over the country, it seemed to be part of an orchestrated campaign. There were threats of withholding contributions to Stanford from people who had never contributed a dime. I treated most of these letters as the academic equivalent of junk mail. There was no basis for the self-serving claim that Stanford had caved in to political pressure.

At the end, I felt hard-pressed to make a fair assessment of my eight-year administrative career. It would have been easier to justify it if it had led to a definitive career change by my becoming president or provost at some other university or college. Some such prospects had presented themselves, but I didn't pursue them. One reason was my unwillingness to give up teaching. Another consideration was the disruption that moving elsewhere would have caused my family. Stina was fully engaged in her life and work, and our children were going through their adolescence and would have most keenly felt the effects of the dislocation. My mother and Lucy were happy in their new surroundings. Whatever the advantages to me, the cost of moving to my family would have been too high.

Consequently I left university administration with mixed feelings. On the positive side, it had saved me from drifting in the Department of Psychiatry. It gave me a taste of what it was like to be part of running a great organization. And perhaps the dismantling of my office was a good thing for the university. However, the experience still left me feeling like a good rider without a horse. Moreover, I felt that the contributions I made in my administrative role had not been sufficiently acknowledged. (My friend Jim Adams called me "the most undecorated man" at Stanford.) Perhaps I should have heeded Jerry Lieberman's advice that the surest way to avoid disappointment was to expect nothing.

These feelings abated when at the commencement of 1993 I received the Lloyd Dinkelspiel Award—the highest honor for service to undergraduate education—from President Donald Kennedy. Even though I felt at the time that it came ten years too late, it still was gratifying to receive it when I did. I had turned sixty, and it helped provide a sense of closure. Another lesson I learned (in addition to the virtue of being patient) is that you don't get recognized simply for having done a good deed; you get your reward because people go to bat for you. Had my friend Craig Heller, the director of Human Biology at the time, not taken the initiative, I would not have received the award. Those he approached for letters of support were surprised; they thought I had already received the award years earlier.[11]

In retrospect, these lingering feelings of discontent that I had at the time seem unjustified and petty. Given the praise and affection

I had received from students, the respect of faculty colleagues and the administration, as well as the opportunities I had been given that no other university would have offered me, what more did I want? A continual outpouring of public adulation to satisfy my relentless need for approval? What would it have taken to eliminate the self-doubt over the value of my contributions and the recognition afforded to them? Even if my feelings of being insufficiently appreciated were only with regard to my administrative roles and not my teaching (I had no problems on that score), I should have realized that the lack of public recognition for administrative services is usually what one should expect. Administrators have to make difficult decisions that do not enhance their popularity. The fact that much of their work is done in private keeps it hidden from public view. Consequently, even university presidents who labor for years in jobs that take over their lives may get small thanks for their exertions. Why should it have been any different in my case?[12]

The matter actually went deeper, to how I valued myself and expected others to validate my worth. On the positive side, I had a good deal to be pleased and proud of, but I will focus instead on the more problematic elements. The first of these was a feeling of discontent with respect to my career as a whole. It reflected a feeling of disconnectedness. The pieces appeared strewn around and failed to cohere into an integrated whole. Everything that I had done seemed to suffer from a lack closure. I felt as if I had been on an arduous journey where I stopped short of getting to the top; or when I reached the top, the view was obscured by fog. Nothing I had done had been a failure but was simply good enough. There was a lack of authenticity, a "would-be" quality rather the certitude of having been a "real" psychiatrist, researcher, scholar, teacher, administrator, or whatever else I might have been.

How do I explain this negative cast of mind? I must allow the possibility that these negative assessments were at least partly justified. Maybe the reason I felt that I never got to the top of the mountain is because I actually never did. I dealt with many challenges, but I was never yoked to a task that would have tested my full capabilities—a job that was big enough for me. If this explanation is insufficient, there are biographical and psychological elements that may shed light on the negative self-image of my professional career. The first of these is my early experience of loss. At the time it occurred, the loss of Iskenderun was a real event with real consequences. By the time I was an adult, that loss no longer had any real bearing on my life. I neither needed nor wanted to be back in the Iskenderun of my childhood. However, the reality of that loss had become superseded by its symbolic significance. Consequently, no matter what else I achieved in life, I could not succeed where my father had failed—either to prevent or to repair the damage of that historic loss. Iskenderun became the Heavenly Jerusalem that would always be beyond my reach.

The second element was connected to my sense of identity. My feeling of disconnectedness and the uncertainty over the value of what I had done could be a reflection of the opacity and fluidity of my sense of who I am. The feeling of being a perpetual foreigner meant that I would also be a foreigner in whatever field I chose as my career. No uniform would fit me. If I didn't know who I "really" was, then I could not know if the work I did was "real." Internal consistency required that if I was to be out of place in one realm, I had to be out of place elsewhere also. Being out of place was not a disability imposed on me; rather, I was where I *wanted* to be. The lack of closure was an active choice. I progressed along a career path, but when I came close to a point of culmination, I backed off. The circle never got closed. This is why I didn't pursue career opportunities that would have locked me in place and put an end to my being out of place. To give up my sense of identity as a foreigner would have meant no longer being myself—the loss of my sense of self altogether.

The third factor was the streak of compulsivity in my character. It colored everything I did. Since nothing was good enough short of being perfect, I could never be satisfied with the view from the top of the mountain even if I had gotten there—fog or no fog. It was not enough to be an accomplished swimmer—I had to *walk* on water. If my father was the significant figure with regard to my struggle with loss, my mother would be the person I needed to satisfy at a level that was simply not possible. I recall one instance when I asked my mother if I was an obedient enough son. She was taken aback by the question. I was not a child anymore, but a teenager—I should have been asserting my growing autonomy, perhaps rebelling, instead of currying her favor like a puppy. And if my mother was indeed the unattainable pinnacle, it was I who put her there.

~

It would take volumes to describe in any detail my personal life during my four decades at Stanford. But even a brief account of the important changes in our lives would provide a more personal context for the events I described above. Nina was ten and Kai was seven when I became dean, and they went through the joys and challenges of growing up in the ensuing decade. Their care and upbringing fell mostly to Stina. I was not an absentee father, but I found it hard to spend as much time with my children as I should have and wanted to. Nonetheless, Stina managed to do it while keeping up with her writing career, although it wasn't easy. She continued to contribute to various papers and magazines and gradually moved into literary translation, focusing on the work of the nineteenth-century Swedish Finnish poet Edith Södergran and then more modern authors. Her translations of Södergran's poems went through three editions, and she wrote a play based on her life. When I couldn't find a publisher for my mother's memoir, Stina wrote the book about it that I referred to earlier. She then wrote another book based on the life and letters of Theresa Huntington, an American missionary in Turkey who worked among Armenians of my mother's generation. These works made Stina better known among Armenian literary circles than I was ever to be.[13]

On our way to England for my sabbatical in the fall of 1975, I had stopped in Lebanon to visit my mother and Lucine. They were at the mountain resort of Mrouj with Elvira, and I spent a bittersweet week with them. After my father's death, they had managed quite well but they seemed forlorn and lonely even though our relatives continued to be very attentive to them. I went down to Beirut a few times and found it changed. People appeared more subdued and in a vaguely apprehensive mood. I could no longer sense the exuberant *joie de vivre* of earlier times. Even walking down the streets didn't feel the same. I wasn't sure if it was me or the country that had changed. Lebanon was still prosperous, but its political stability more uncertain. As other Arab countries reined in the Palestine Liberation Organization (PLO), it gravitated to Lebanon, becoming a virtually autonomous force. There was an incident when Maronite militias shot up a bus full of Palestinians. There was no immediate reaction to the attack, but it turned out to be the prelude to the brutal civil war that exploded shortly thereafter. Lebanon's neighbors, particularly Syria and Israel, backed by larger players, had been meddling in its affairs and vying for dominance. Ultimately this game of power politics destabilized the country, leading to the fratricidal civil war of 1975, which lasted fifteen years, until every faction had fought every other faction. Protracted conflict led to the Syrian army moving into Beka'a Valley. The Israeli invasion of south Lebanon in 1982 was followed by the siege of Beirut. Atrocities during the civil war culminated in the massacres at the Palestinian refugee camps of Sabra and Shatila by Phalangist militias, while Israeli troops stood by. Israeli occupation of south Lebanon led to the rise of the Hezbollah, which, with the support of Iran, was to become the most powerful political entity in the country.

The civil war finally ended when the warring parties became exhausted. The PLO was forced to leave Lebanon (only to return several years later). Israel withdrew its troops in the mid-1980s under a brokered United Nations cease fire. The Syrians remained entrenched longer, until they too were forced to leave Lebanon in the aftermath of the assassination of Prime Minister Rafik Hariri in 1997. When all was said and done, the war had claimed the lives of hundreds of Israeli soldiers, thousands of Lebanese and Palestinian fighters, and as many as 200,000 civilians, if not more. Many more thousands left the country. The infrastructure of much of Beirut was shattered. Lebanon never fully recovered to regain its former position in the region. All for what? The civil war was fought along sectarian lines, but it was hardly a war fought over religious doctrine. Nonetheless, religious loyalties fueled its ferocity. An Armenian Catholic priest told me that at the height of the war he kept a handgun on the altar when saying Mass. He sensed that I was taken aback and hastened to add, "Please don't misunderstand me. It's not that I don't trust God to protect me. But God is busy and may be distracted by more important matters. So when people walk into the church to shoot me, I want to be holding something other than the chalice in my hands."

~

After we returned home from England we followed the civil war in Lebanon with mounting dismay. I called my mother every Sunday and she reassured me that all was well. In reality her house was in the middle of a war zone. When I could no longer get through to the operator, I panicked. When I finally could talk to my mother again, I pleaded with her to take Lucine along and come to stay with us until the situation in Lebanon cooled down. She was reluctant to make the move at her age, but I begged her to do it for our sake. We lost contact again until early December, when my mother called me from Paris. They were on their way to Los Angeles. A Shiite neighbor's son in the militia had taken my mother and Lucine to the airport in an armored personnel carrier. Their connecting flight in Los Angeles was canceled because of an airline strike, so they would spend the night at an airport hotel and fly to San Francisco the next day. l told my mother I would contact her the next morning before they left for the airport.

I called the hotel the next morning, only to be told by the clerk that there was no one staying there by mother's name. How could that be? Where could they have gone? After twenty-three frantic phone calls to plead with and yell at anyone I could talk to, I finally located my mother. She was having coffee in the garden of the hotel and overheard the exasperated clerk saying my name on the phone. The greatly relieved clerk said, "Dr. Katchadourian, here is your mother." It turned out that since the passengers on her flight had checked into the hotel as a group, their names were not registered individually. The rest of the passengers had left for the airport early in the morning, and the clerk had no idea that my mother and Lucine had stayed behind for a later flight to San Francisco. I asked my mother why she had not called me when she hadn't heard from me. She said calmly, "I didn't want to disturb you on a Sunday morning."

We had a joyous reunion. My mother and Lucine settled in with us to wait for the end of the war. They had come with a suitcase each. Half of my mother's suitcase was filled with raisins, nuts, and assorted dried fruits. Did my mother know that Armenians produced all that in vast quantities in Fresno? Yes, she knew, but the nuts and raisins she had brought were the ones I liked. What about the family albums, Lucine' needlework, and other irreplaceable things they had left

Figure 10.3 My mother in her older years. (Photograph by Margo Davis.)

behind? Well, we could get them later. For now, I should enjoy the raisins and the nuts.

Six months later, when it looked like there was no end in sight for the civil war in Lebanon, we rented for my mother and Lucy a two-bedroom unit on Williams Street at the end of College Terrace in Palo Alto, close to our house. The years they lived there were to be among the happiest of their lives. At least one of us visited them every day. Having the grandchildren growing up close by was beyond their fondest dreams. My mother was in her eighties and Lucy in her sixties, but they were rejuvenated and adapted amazingly well to their new life without entirely relinquishing their old ways. This could happen only in America.

Shopping at JJ&F, the family-run supermarket around the corner, was not only convenient but a source of pleasure for my mother and Lucy. They had to get used to the different cuts of meat and prepackaged foods, but they learned fast. I initially would take my mother shopping, and when we got to the meat counter she would say, "Ask the man if the meat is fresh." Or point to a piece of sirloin steak and tell me to "Ask the man to grind it up." When I pointed to the mound of ground beef, she would shake her head; you could never know what had gone into it (in Beirut, it could be cat meat). I was relieved when they became able to do their own shopping and developed a personal relationship with Joe, John, and Frank of JJ&F and then with their equally friendly sons. Lucy never bothered to take a number to wait her turn; she just went straight to the counter and was served first. After cooking *koeftes* with the meat she had bought, she would take some of it back to the store for them to taste. No wonder they all loved her. Yet despite having JJ&F next door, my mother and Lucine never gave up their old custom of putting food away for the winter, buying sacks of onions, flour, and sugar in bulk in the fall.

Even though we were their mainstay, my mother and Lucy did more for us than we were ever to do for them. A constant stream of food found its way to our house. They delighted in babysitting the grandchildren. Lucy knit one sweater after another and crocheted tablecloths and bedspreads for Nina's trousseau and baby clothes for her and Kai's children to be born in the distant future. They had a few, but devoted, friends. Mrs. Elize Manoukian, who knew my mother was from Beirut, used every opportunity to visit them and take them along on her trips to the Stanford Shopping Center. Her untimely death was a great loss. Then there were Helen and Nello Sartoris. We met them through their son, David, who was a Stanford student and who house-sat for us one summer. When I learned that his parents would be visiting him, we invited them to stay in our house while we were in Finland. When we returned in the fall, everything in the house that needed fixing was fixed. Nello had even made a wooden handle for my broken-down electric drill. We were delighted when Helen and Nello moved to Menlo Park to escape the winters in Chicago and to be with David and other members of their expanding family (which eventually numbered thirty children, grandchildren, and great-grandchildren).

Nello and Helen came from a family of coal miners, and Nello had also worked as a welder for General Motors locomotives. They were salt-of-the-earth people who became our devoted friends. Their solicitous kindness and companionship to my mother and Lucy, particularly when we were away during the summers, were a great comfort to us. The four of them played pinochle for nickel stakes, which my mother (who was an accomplished poker player) constantly tried to raise to a dime. Lucy kept a constant stream of coffee and treats coming, being especially attentive to "Mister Nello." Yet there were some things in American culture that my mother could never get used to. Stina came home one day from a brown bag lunch given by the university president's wife. My mother was curious to know what they were served. Stina told her that the guests had brought their own sandwiches. My mother was flabbergasted: the guests brought their *own* sandwiches? *Sandwiches?* At the *president's* house?

One of the consequences of the Lebanese civil war was the exodus of the more educated and better-off people. That brought Elvira's family to Montreal and Nora's to San Francisco. Elvira and Nora were essentially like sisters to me, and their children and grandchildren treated me as a beloved uncle. It gave me great satisfaction to be part of the generational chain of my mother's family. Elvira joined two of her children—Lena and Johnny—in Montreal (another daughter, Mayda, died in Beirut some years ago, as did Johnny more recently). It was a special pleasure to see Elvira's grandchildren come of age and establish their own families. Our dinners at the fabulous restaurants run by Lebanese immigrants became one of the highlights of the year.

Nora and her husband, Noubar Tour-Sarkissian, moved to San Francisco to join their children—Christine ("Tina") and Paul, both of them lawyers, who developed a successful practice with an international clientele. Tina was married to Roger Bernhardt, and Paul to Tania Doumanian—both of them also lawyers. Tina's daughter ("Little Tania") married Philip Wood, from England. Living close by, Nora and I grew especially close. Our families celebrated Christmas, Easter, and other special occasions together. They came to be our closest relatives, and we cherished the bonds between our families. As Nora grew older, she came to look more and more like my mother as well as sharing many of her character traits and taking over as the family matriarch. Elvira's granddaughter, and Lena's daughter, Christine, was another family member who was endowed with my mother's generosity of spirit, strength of character, and competence.

~

With their advancing age, it was only a matter of time before my mother and Lucy would run into serious health problems. Lucy had suffered from a chronic peptic ulcer for years and undergone a radical gastrectomy. However, she was wiry and resilient and always bounced back. My mother was in good health for her age, except for osteoarthritis of the knees, but she coped stoically with the pain. However, new problems began to plague her in her nineties. She developed arterial thrombosis in her left leg and retinal macular degeneration, which blinded her left eye. She had been complaining of visual problems but tended to downplay them. By the time I took her to an ophthalmologist, it was too late. I deeply regretted not having been more alert, but she never complained. I was thankful that this wasn't all happening in Beirut. What would they have done on their own?

In the spring of 1986, my mother had surgery for blockage of the bile duct. The grim-faced surgeon came to the waiting room and told me the blockage was due to pancreatic cancer. She had six months to live. I braced myself to break the news to her, but she didn't seem to want to know. I wouldn't lie to her, but if she would rather not hear what lay ahead, I would let it be. Her condition began to deteriorate. With Lucy attending to her day and night, we decided to keep her at home as long as we could. I wanted to get a hospital bed to make it easier to care for her, but my mother refused; the bed would make the house look like a hospital. I convinced her that we should get it on a trial basis. The bed lasted a day before it was sent back. Eventually she acquiesced with great reluctance, and the bed was brought back this time to stay.

We arranged for a hospice service to attend to her, but the brunt of her care still fell on Lucy. I spent all the time I could with my mother, as did Stina. Elvira came from Montreal for a last visit. I feared that her feelings would get out of hand, but she was a steady and loving presence. Then Nora returned from Lebanon, and she and my mother had another memorable reunion. It was important for both Elvira and Nora to see their aunt for the last time; she had been as much of a mother to them as she had been to me.

I spent many hours at my mother's bedside. We talked about the past, about our days in Iskenderun and Beirut, and most of all about Ramzi. I set aside everything I could to translate her memoirs and promised her I would have it published in one form or another. Her mind remained sharp and her memory intact. She explained to me what I found unclear in the manuscript and added many details. My mother's final days were a wrenching time. As her strength began to fail, she spoke less and the silences grew longer. I sat holding her hand and reflected on her life as I had done with my father a decade earlier. I felt so close to her. Her parchment-thin skin, wrinkled face, and shriveled body were so dear to me. I wanted to savor every minute left of her life. She would occasionally lapse into speaking softly in Turkish—blessing me, wishing for me a long life and boundless joy in the lives of my children. ("May God show you their days. May they care for you as you have cared for me.") The memory of how she had taken care of me during my protracted illness and had always been there, a tower of strength and unconditional love, filled me with an overwhelming sense of gratitude and love for her.

I reflected on what it was that made my mother such a remarkable woman. There was her generosity of spirit. She had a big heart imbued with tremendous good will. You could always count on her. No one came away from her empty-handed. Had she had the resources, she would have been a great philanthropist. Her positive outlook on life persisted despite the catastrophic losses she suffered. My mother was resourceful, brave, and ready to face whatever life dished out. During the Lebanese civil war, men from the militia had come up to her door claiming that a sniper was shooting at them from her balcony. As my mother fiddled with the lock to gain time, she told them there was no one in the house except for her and old Lucine. They looked tired. Would they like to come in to rest and have a cup of coffee? The men were astonished by her reaction. One of them asked, "Madame, aren't you afraid?" "Why should I be afraid?" my mother replied. "With God in heaven and you boys on earth to protect me what is there for me to be afraid of?" The men were abashed and urged my mother to call on them for anything she might need, and left.

My mother's capacity for leadership made a big impression, especially on men. Someone said, "Had Mrs. Efronia been a man, she could have commanded an army." The other replied, "She can command an army now." Her strength of character and self-confidence made her hard on people who lacked the same tenacity and perseverance. She didn't suffer fools lightly; actually, she didn't suffer them at all. My mother was not only intelligent but also full of common sense and foresight. She could see ahead and around corners, which is why so many people sought her advice. She had a broad generational view. "Before you embark on anything important," she would tell me, "always ask the opinion of two persons: one older than you, one younger than you." It is a pity that my mother never had the chance to get a higher education. She had a keen intellect and an inquisitive mind. Her tremendous energy, capacity for work, integrity, and unflinching courage would have qualified her for any profession or public office. Unfortunately, she was a woman who was born at the wrong time and in the wrong place.

My mind kept going back to my mother's tragic loss of her father in infancy and then the loss of Ramzi. How different would her life have been if she had been united with him? I kept wondering how that loss had affected her attachment to me and my own relationship with women. Had she set up the impossible standard against which I had measured them and found them wanting? Was there a residual debt from her loss that I had to pay? How could I make sure not to pass the mortgage on to my own children?

My mother's physician and I agreed that when the time came there would be no heroic attempts to prolong her life, no Red Code. I was awakened one night by Lucy. She was sobbing on the phone. I hung up and rushed to their house. My mother was sitting at the edge of her bed, gasping for breath. Was I supposed to stand there and watch her die? It took me ten seconds before I called an ambulance. The emergency room physician said my mother had suffered a heart attack and would not make it through the night. I stayed by her side. In the morning, another physician, who was visibly pregnant, came by. My mother slowly opened her eyes, and said to her, "Do you want a boy or a girl?" It was time to take her back home.

During her last hospitalization my mother developed a decubitus ulcer in her lower back. The hospice nurse came in the late afternoon to change the dressing. My mother wanted it done sooner in the day and sometimes more than once. I decided to take care of it myself. It made things easier for her, but it was hard for me to see the body that gave birth to me disintegrating. She still looked beautiful, like a ruined Greek temple. As my mother's condition steadily declined, she began to experience severe pain. Her physicians were of little use. The person who was most helpful was the hospice nurse. She had initially struck me as being a little odd, but I came to cherish her. She told me to insist that her physician prescribe a morphine solution that my mother could take orally, as she needed. There was no issue of her getting addicted. It made a dramatic difference. She calmed down and began to sleep through the night. The awesome power of morphine gave her peace.

I was on a committee that met in New York every week. It was important for me to attend the meetings, so I would take the red-eye both ways. I fretted that my mother would die when I was away, but the hospice nurse told me she would tell me a week in advance before that happened. True to her word, I found my mother one morning lying in bed with her face turned to the wall. She would no longer respond to me and only stirred when I mentioned the children. Her time had come and she was ready to go. She died during the night on December 30, 1986, at the age of ninety-three.

We held a memorial service at our house. Nina and Kai came home to be with us. My mother's relatives, friends, physicians, and nurses were all there, along with some of our friends and neighbors who had come to know her. A few of us spoke briefly, struggling to keep our composure. Many left farewell cards. Nina's card has been lost. Kai's note read, "Endless, timeless, enduring, always present, always remembered."

~

The year of my mother's death marked the end of a long and difficult decade for me. I was burdened by multiple responsibilities in teaching, writing, and administration, compounded by the residues of my psychological conflicts and problems. My administrative responsibilities were time-consuming and seemed endless. I spent almost twenty hours a week in committee meetings (much of it wasted). I had undertaken ambitious research projects in undergraduate education. I taught thousands of students and spent much time with many of them outside the classroom, in dorms and other settings. My duties as vice provost presented their own demands. The most thankless of my tasks was the dismantling of my office, which turned into a huge undertaking and left me feeling spent and empty over the pointlessness of it all.

In addition to caring for my mother and Lucy, I had a family to support and a wife and children to attend to. Our family expenses mounted. The royalties from the first two editions of *Fundamentals* initially exceeded my salary and lulled us into the false expectation that we could always count on them. The land we bought and the house and the sauna we built in Finland were paid for by my royalties. The initial outlay was just the beginning. We also had to buy a boat, construct a harbor, build outhouses, replace roofs, and repair the damage from the brutal winters over and over. There was something to be done each and every summer, which I tried to do myself so far as I could, but there was no end to it. Another financial drain was the cost of travel. Between the four of us, we were to fly to Finland and back over a hundred and sixty times.

As our children got older, our expenses escalated further. We had made no provision for their educational costs. We never had a family budget. We spent what we spent, and I had to earn it. As the royalties of the sexuality textbook began to dwindle, I had to scramble to cover the shortfall. I had a higher salary as an administrator, but it wasn't enough. All told, I produced two more editions of the sexuality textbook and wrote five more books and book chapters. I went across the country giving lectures, consulting, and taking on a string of time-consuming projects.

One of these was funded by John D. Rockefeller III and included organizing a conference with the proceedings published as a book, which I edited.[14] When David Hamburg became president of the Carnegie Corporation in 1983, he initiated a major effort to address problems of children and youth.[15] I became involved in several of his projects, including the Middle School Life Science initiative. It was intended to develop a curriculum and a set of textbooks for middle schools derived from our experience in the Human Biology Program.

Craig Heller, director of Human Biology, Mary Kiely, who had worked for David Hamburg at Carnegie, and I collaborated for over a year on this. I spent a lot of time on it and wrote three of the textbooks with a middle school teacher.[16]

David would invite me to various meetings and conferences at the Carnegie Foundation as a speaker or a participant. When I gave talks, people seemed more taken with how I was saying things rather than with what I was saying. On one occasion, I pointed out that historical dates were hard to remember—they were like little birds that perched on the short branches of our memory and then they flew away. At the end of my talk, a woman from the Department of Education came to me and said, "Why can't I put things the way you do?" The birds were probably all she would remember from my talk.

Through these occasions, I met people I would never have met otherwise. One particular event was attended by three governors (including Bill Clinton and Michael Dukakis), four senators, and three former cabinet secretaries. Unlike David, I was not engaged with such public figures in any significant way, but these encounters nevertheless expanded my horizons. (Michael Dukakis said to me, "You know, my father was Greek." He knew my father was Armenian, and that seemed to create a bond between us.) I had never heard of Governor Clinton, but I was greatly impressed by his keen mind; I never knew politicians could be so smart. Many other activities involved the Stanford Alumni Association. I was one of their most popular speakers and gave talks to clubs and conferences across the country. I led over two dozen Stanford Travel Study trips (three of them going around the world with private jet). I wrote two volumes for the Portable Stanford series.[17] Despite my frenetic schedule, I greatly enjoyed the trips I took and the talks I gave; they helped to sustain me through difficult times.

This insane schedule would have been hard enough on me and my family under the best of circumstances. Its burden was compounded by the residues of the obsessive-compulsive problems that bedeviled my adolescence. Although they had largely subsided, they had become woven into my character. That made it more difficult to see them as problems, particularly since they also had positive aspects. They helped me become highly organized and orderly. My streak of perfectionism forced me to set high standards for whatever I did.

Figure 10.4 Standing in front of the relief of an Armenian tribute bearer on the ceremonial staircase of the Apadana Palace in Persepolis on a Stanford Travel Study trip.

It made me think with precision and endowed me with tenacity, perseverance, and an enormous capacity for work. Like a dog with a bone, once I got my teeth into a task I wouldn't let go until it was done, and done well. I was perpetually in the process of "finishing" something. Every task was time-consuming because it wasn't enough for me to be comprehensive—I had to be exhaustive. (Stina would beg me not to start every lecture on a Mediterranean cruise with the founding of the Roman Republic.)

When working on my book about guilt, I learned with some satisfaction that my compulsivity had august company. Some of the most revered figures in Christianity—St. Augustine, Martin Luther, Ignatius Loyola, and John Bunyan—suffered from similar problems. This had a particular bearing on their struggles with pathological guilt, which was also true for me. It is not an accident that I taught a course on guilt for a decade and wrote the most multifaceted book on the subject ever written, to my knowledge. However, the benefits I derived from my compulsivity came at an exorbitant cost, which complicated my life. I was aware of these burdens but couldn't help it. This is who I was, and to change meant becoming someone else. That didn't make it easier to live with myself or to live with me. I had trouble getting rid of obsessive thoughts that got stuck like a needle in the groove. When I got upset with something I said or did, I would keep replaying the incident in my mind in an endless loop. Once embarked on a path, I had to stay on it even if it no longer made sense. (I should have followed what the Buddha said—after you cross the river, get rid of the canoe.) If I got involved in a discussion or argument, I wouldn't let go before driving it into the ground. All encounters and confrontations had to be frontal. I could do nothing by indirection. When I was wrong, I admitted it grudgingly; if you were wrong, you had to admit it too and do it unequivocally.

My perfectionism led me to fret endlessly and needlessly even over trivial matters. A picture hanging slightly askew would bother me as much as a gaping crack in the wall. As a student, when I made a mistake on an exam, I rehashed it endlessly. Now when something was amiss in a lecture, I did the same thing. Of the fifty slides I showed in a lecture, forty-nine would be fine, but the one slide that was not spoiled the rest. Everything had to function like a frictionless machine. I planned the future with the expectation that all would go exactly as planned. I would never be tired, sick, or distracted. There was no allowance for the unpredictable and the unforeseen. I made list after list in an attempt to control my life. If I got the list right, everything would fall in place; if it did not, I made a new list. I also used markers, which determined what tasks would be completed when. This was useful in providing clear starting and end points. But when such markers turned into rigid deadlines, they disturbed my peace of mind.

I seemed to work nonstop. (On November 28, 1988, I wrote in my diary: "First Thanksgiving weekend in the last 15 years that I did not work.") I lived not in the present but in a time warp between the past and a future. I was either regretting the past or fretting about the future. There was no living for the moment. My unrealistic expectations of myself and others couldn't be fulfilled and hence everything seemed doomed to failure, which led to renewed efforts to do better. No sooner did I come close to finishing up a task than I would embark onto the next. If I set out to write ten pages a day and managed to do it, I raised the quota to twelve, then to fifteen. My awareness of these problems became a problem in itself. I was hard on myself for being hard on myself. The remorseless intolerance for failure and relentless need for approval placed enormous burdens on me. The fear of failure was the engine that drove me on relentlessly. No wonder I often felt a persistent, albeit vague, sense of dissatisfaction that led to a lingering moodiness. I felt tired, slept poorly, or woke up in the middle of the night (when I would go to my study to work some more).

The topic of one of my Class Day talks was equanimity, and it was inspired by an address given decades earlier by the great Dr. William Osler. Despite the great stock I put in it, I was easily discombobulated and thrown off-center. ("Judging from yesterday, how brittle is your equanimity," I wrote to myself.) In another talk I cautioned students against trying to do too much in their lives. It was a case of do as I say, not do as I do. Otherwise, the flavor of that talk was fairly typical of these public addresses.[18]

The worst of my fears was being alienated from God. It was not a matter of guilt over wrongdoing, but merely being distant from him. I wanted to submit to God's will utterly and unequivocally. (I would have made a good Muslim; submission to God—*taslim*—being the core of Islam.) My childhood faith sustained me. God was my unmovable anchor. As long as I was attached to him with a stout rope, nothing would faze me. Since my childhood struggle with death, I was no longer afraid of it. My faith protected me from deep existential anxieties, but it didn't help with mundane headaches. My problem was not drowning in a storm at sea, but being seasick.

These stresses and strains couldn't help but have an impact on my relationship with my family. I must have been difficult to live with. (Someone once asked my wife, "What do you get from him, crumbs?") They all worried about me. My mother told me my life was overburdened and I was growing old prematurely. When Nina and Kai were living at home, they would come to my study to hug me and thank me for working so hard for them. They would leave little notes exhorting me to slow down, to take it easy, to let some fun into my life. One Father's Day, Kai gave me a card that said, "Dad, *have fun* once in a while." (When he was younger, he told me, "Dad, don't be so rambunctious about work.") One of Nina's Christmas cards said, "Dad. Once again, you are setting up the year so that you have nothing but work." Stina's exasperation would take too long to describe adequately. I knew all this. I wrote in my diary, on April 21, 1992: "Something is not right. I act fine on the surface but there is an underlying restlessness and tension that I cannot get rid of. It is manifested by irritability, fretting about past events, anhedonia, lack of interest in the future." I then reassured myself, "But I am not overtly depressed."

Notwithstanding all this, I didn't come across to the rest of the world as a dour and driven workaholic (if there is such a thing outside of pop psychology). Instead, I appeared serene, calm, full of equanimity, and contented. A psychoanalyst friend who read the above account was surprised. "I had always seen you as effortlessly good-natured and humorous, without a care in the world," he wrote to me. Another friend and colleague remembered me as "confident, decisive and humorous" and was puzzled by my characterization of myself as "melancholic, troubled and self-doubting." When I asked my daughter Nina whether she was aware of the turmoil in my life when young, she said she remembered me as always calm and collected. Kai said that he knew I was very busy and so every free moment had to be savored. People told me I exuded an Old World charm combined with a New World sensibility.

How can I explain this dichotomy between the negative way I felt about myself and the positive image I presented to the world? Which side was the real me? Were there two Herants that coexisted uneasily or warred within myself? Was my public persona a façade to hide behind and conceal my pain that would have made me vulnerable to the world? Actually, the bleak picture I have painted above did not represent the totality of my life—even during its most stressful and turbulent period. Some of it is an artifact of diaries written during difficult periods. Even during this decade of turmoil, there were many joyful times at work and at home. I loved teaching and enjoyed the attention and perquisites of being an administrator. My family was a source of immense satisfaction. I have wonderful memories of our children's early years. I took them for a family swim on Sundays to Ed Jane's indoor pool in Palo Alto. Nina already knew how to swim and

Kai would cling to my neck when we went to the deep end of the pool. Before going home we had a cold drink from the vending machine, and it felt very special. We went to the ocean to fly kites and run along the edge of the surf. Where my children were concerned, even simple things gave me much pleasure. Stina and I attended soccer matches on Saturdays to watch Nina's Red Ants and Kai's The Demons. I took much delight in watching them star in school plays (Nina in *Rapunzel* and as Eliza Doolittle in *My Fair Lady*; Kai, Prince Charming in *Snow White* and the Tin Man in *The Wizard of Oz*). I got more enjoyment out of watching these plays than I would have from a production of the Royal Shakespeare Company. There were Nina's piano recitals and Kai's drumming at home. Having my mother and Lucy in the audience greatly added to the pleasure of these events. Summers on the island were a whole other world of fun and fulfillment, which I will relate in a subsequent chapter.

As the 1990s rolled along, my life settled down considerably. Life was simpler as a full-time teacher—doing what I did best and loved most. My inner turmoil had largely subsided by the end of the century. "One of the best years that I recall," I noted. "I feel at peace with God, my family, at work and with people." Stina continued with her writing career and became engaged in organizations like the Stanford Center for Research on Women and Gender, where she was an affiliated scholar, and the Global Fund for Women, where she served on the board. Nina graduated from Brown and went on to earn her MFA at the University of California at San Diego. She then went to New York for an independent study program at the Whitney Museum and embarked on a successful career as a conceptual artist and a gifted art teacher. Kai graduated from St. Anthony High School on Maui and became an accomplished competitive windsurfer, tackling mountainous waves, traveling widely to the four corners of the world, and eventually going into the windsurfing business. Nina bought a house in Brooklyn and settled down with her husband, Sina, the founder and highly accomplished editor of *Cabinet* magazine. Kai married Anni in Finland—a wonderful young woman and a talented silversmith whose parents have a summer house on an island next to ours. Watching and helping our children reach maturity and lead happy and successful lives and their love and regard for me and their mother were my deepest sources of satisfaction.[19]

Becoming a university administrator meant I could no longer see patients, effectively ending my career as a psychiatrist. Yet in a more general sense, I never left psychiatry. The skills that I had learned in relating to people and dealing with their problems continued to serve me well for the next four decades. Just as importantly, my knowledge of myself and awareness of my own thoughts and feelings became important assets in my understanding of others. I didn't use these skills to "analyze" or "treat" everyone who came my way. My relationships with students, colleagues, employers, employees, friends, and members of my family entailed a different set of roles with their own expectations and responsibilities. The broad perspectives of figures like Freud and Jung on human history and culture and the views of mentors like Romano and Hamburg on contemporary society greatly expanded my own vision of the world.[20]

At this stage in my life, I assumed that the rest of my career would be a continuation and elaboration of a life of teaching and writing that I had settled into. Little did I know that a whole new phase in my career was to open up to me that would give new direction and deeper meaning to my life and work. It would take me beyond the confines of academia and into the realm of philanthropy.

11

Doing Good

THE CHRISTMAS PARTY at Robert and Helen Glaser's house was one of the highlights of the holiday season. The buffet with its mound of fresh shrimp and fine wines, and the well-dressed women and the men in their bright red vests, made the occasion especially festive. When I came to Stanford, Bob Glaser was dean of the Medical School and Helen was a child psychiatrist. However, I didn't come to know Bob well until some years later, when he was president of the Kaiser Foundation and became my friend and mentor. The guests at the Glaser's Christmas parties were mostly from Stanford, with a sprinkling of others that included Roger and Esther Heyns. Roger was president of the Hewlett Foundation and had been chancellor at Berkeley. We had a number of common interests and would occasionally meet for lunch. Like other American friends, Roger often questioned me about the Middle East—an area of the world that baffled and worried him. Even though I read the same newspapers he did, I could provide a historical and cultural context to help understand the troublesome region.

Both Bob and Roger occasionally brought up the prospect of my going into foundation work. They thought I had the right background and personality for it. I didn't know much about foundations, but found the idea intriguing. This became a recurrent theme in our conversations and was occasionally linked with the tantalizing prospect of a foundation presidency. However, nothing came of it and after a while it began to sound like background music. I finally put the matter out of my mind just as I had done with prospects for higher academic office.

~

In the fall of 1986, Roger asked me to meet with him and William Hewlett to discuss a grant request from the American University of Beirut submitted to the Hewlett Foundation. I had seen Bill Hewlett a few times at Stanford functions but had not actually met him. I was intrigued by the prospect and accompanied Roger to Bill's office at the old Hewlett Packard headquarters on Page Mill Road in Palo Alto, on the edge of the Stanford campus, close to where we lived. Hewlett's secretary escorted us into his office and he welcomed us warmly. I was surprised by the rather modest furnishings in the room, a far cry from what I had imagined. There were no Mies van der Rohe leather chairs, no solid rosewood desk, no Tolomeo lamps or fine art on the walls. I was even more struck by Bill's unassuming and friendly manner despite his legendary status in Silicon Valley. We spent a pleasant half hour together. I told them about the American University of Beirut, its historical background, and how it had emerged from the Lebanese civil war more or less intact although much weakened. The request from AUB was for $250,000 for general support. My recommendation was to give the money instead for self-evaluation and planning. Its leadership needed to take stock and plan for its postwar future—to turn disadvantage to opportunity. It was high time for AUB to take a long-range view rather than keep on living hand to mouth. Bill and Roger seemed pleased with what I said and they thanked me for my help.

A few weeks later, Roger let me know that they had taken my advice and AUB would get a grant for the purposes I had suggested. Shortly thereafter, I was contacted by Dr. Frederick Herter, the president of AUB, from his New York office. He invited me to be part of a steering committee that would oversee the self-study supported by the Hewlett grant. Nothing was said about my role in their getting the grant, but I was sure that the suggestion to include me in the steering committee had come from Roger Heyns. These were the committee meetings I referred to earlier that necessitated my taking the red-eye to New York during my mother's terminal illness. I agreed to do it out of a sense of obligation to AUB and in view of my role in its getting the grant. The committee included distinguished and interesting individuals like Najeeb Halaby. He was of Lebanese origin, a former navy test pilot, the then CEO of Pan Am, and the father of Queen Noor of Jordan. Thinking back on my days as a lowly freshman, I was pleased and grateful for the opportunity to serve AUB in such august company. My engagement with this evaluation process eventually led to my joining the board of trustees of AUB.

In the spring of 1989, I received another call from Roger Heyns. In his characteristic manner, he wanted to ask me for a big favor: Would I be willing to join the board of directors of the Hewlett Foundation? He went on to elaborate how Bill Hewlett would appreciate it, how useful I would be to the board, and how interesting I might find the experience. I said of course I would be happy to do it. It was the start of an association with Bill Hewlett and his family that would be highly significant for my life and career.

~

William and Flora Hewlett established their foundation with their oldest son, Walter, in 1966. Bill was in his fifties and an iconic figure in Silicon Valley. I had heard about how Bill Hewlett and David Packard started HP in a Palo Alto garage and built it into one of the greatest corporations in the country. However, it was not until years later when reading their joint biography, that I became aware of the scale of their accomplishments and the central role that HP had played in establishing the electronics industry in the Silicon Valley.[1]

Hewlett and Packard became the largest benefactors of Stanford next to its founders. Flora Hewlett was a Berkeley graduate with enduring ties to the University of California. After Flora's death, Bill and Roger essentially ran the foundation together. The board of directors consisted of half a dozen of Bill's friends along with Walter and his older sister, Eleanor Gimon. I joined the board along with Bill's younger daughter, Mary Jaffe—the second woman and the youngest person on the board. I was the first director who was neither a member of Bill's family nor his personal friend, and the youngest person on the board who was not a family member. The endowment of the Hewlett Foundation reached its peak at over $9 billion in 2007, when it gave away about $500 million a year. At Walter's initiative the foundation committed $400 million to Stanford in 2001 (the single largest gift made to the university), and over $100 million to the University of California at Berkeley six years later. I had had some experience with large sums of money in connection with Stanford's budget as vice provost. However, giving away of hundreds of millions of dollars entailed a different level of engagement. I wished my father had been alive to see me now; it would have meant a lot to him. As I realized the enormous leverage that vast resources made possible, my father's idea of "hiring ten doctors" now made more sense and my involvement in the foundation provided me a window into the world beyond academia.

It was fascinating to learn about the many aspects of philanthropy that I was unaware of. I learned that the size of a grant did not necessarily determine how much good it did, and how difficult it was to give away money effectively. For example, a community renewal program that started with great promise could not be replicated in the next community that truly needed it. A grant of $100 million of Hewlett and Annenberg money with other funds to improve K–12 education failed to produce significant outcomes that could be readily discerned. Admittedly, bringing about social change was difficult and it was necessary to take chances to achieve it. The success of such projects required careful staff work, but when a program officer became an advocate for a particular project there was a loss of objectivity and perspective. While these examples stand out in my mind, there were many other instances of highly effective grants that the foundation made over the years in the areas of the environment, population, education, conflict resolution, and the performing arts.

~

I became a trusted advisor to Bill and then Walter Hewlett. I was also an integral member of the Hewlett Foundation board along with members of the Hewlett family as well as others, like Arjay Miller, Bob Erburu, Jim Gaither, and Richard Levin, even though I had not held positions of responsibility comparable to theirs. (Miller was the former president of Ford Motor Company and dean of the Stanford Business School; Erburu was the CEO of the Times Mirror Co., which published the *Los Angeles Times*; Gaither was a senior partner in a prestigious law firm and chaired the Stanford Board of Trustees; and Levin was president of Yale.) My sixteen years on the board were to be a fascinating and deeply rewarding experience.

This was particularly true for my personal interactions with Bill Hewlett. Some of these were quite sensitive and involved Walter. After Roger Heyns retired, the board approached David Gardner, the former president of the University of California, to succeed him. David was reluctant at first but finally accepted the offer. As these negotiations were going on, I was surprised to learn that Walter had not met the prospective new president. Walter was vice-chairman of the board and the heir apparent to succeed his father as chairman, which meant that he would eventually have to deal directly with the new president. I brought this to Bill's attention. He was taken aback and said he took personal responsibility for the oversight. Two days later Roger had arranged for Walter to meet with David Gardner. This and other observations convinced me that Bill, who was approaching eighty, was not facing the prospect of his retirement from the board and grooming Walter to succeed him in the chair. I made an appointment to see Bill in his same old HP office. He didn't know why I had come. I began by saying that more kingdoms were lost for want of an orderly succession than any other cause. If Walter was next in line to chair the board, was he being prepared for that eventual transition? Bill was silent. I asked him if I should leave. With an emphatic gesture, he indicated that I should stay. He then said regretfully that he had not yet talked to Walter about it. Perhaps he should absent himself more often from board meetings so Walter would have more opportunities to chair them. I didn't think that would be a good idea. Walter needed to chair the meetings with Bill in the room, otherwise his father's shadow would always loom large over him. Bill nodded his agreement and thanked me. Soon thereafter, Walter succeeded his father and developed his own distinctive and effective style of leadership of the board. Unfortunately, Bill didn't stay well long enough to see this happen as he soon became incapacitated by a stroke.

My personal relationship with Bill became as important for me as my association with the foundation. I became Bill's old-age friend. Unlike most of his other friends, I wasn't an engineer, I didn't work at Hewlett

Figure 11.1 Bill Hewlett.

Packard, I wasn't an entrepreneur or a leader of industry, I hadn't served as a Stanford trustee, and I certainly wasn't immensely wealthy. Professionally, I would have had more in common with Bill's father, who was a distinguished professor of medicine at Stanford and who died when Bill was twelve years old. When I asked Bill what he would do differently were he to start his career over again, he said without hesitation that he would become a physician. Given his fascination with the workings of machines (and the body is a sort of machine) and his compassion for people, he would have made a great physician. I was also surprised that after his spectacularly successful career as an engineer and industrialist, he would opt for a career in medicine.

I first realized that Bill thought of me as a personal friend when I visited him during one of his earlier hospitalizations. As I was bidding him goodbye, he looked up from his pillow and said, "You help me gain perspective." I wasn't sure what he meant by that, but was pleased to hear it. Bill then suffered a stroke, from which he recovered only partially. When he could no longer drive, I would take him out for rides to the ocean and to his beloved San Felipe ranch. This magnificent property, jointly owned with Dave Packard, was one of the few remaining old Spanish ranches in California. It spread for miles on a mountain range on the edge of the Santa Clara Valley. We would get into a Jeep and Bill would direct me over the dirt roads that crisscrossed the hills. Then we would stop for a picnic lunch (prepared by Lucy) on one of the hilltops with a magnificent view overlooking the valley. I would tell Bill that he should have a house built at that spot and he would reply, "But I already have a house down below." I objected that "down below" wasn't the same thing, but he merely smiled.

It is said that as he got older, the gruff and curt Bill Hewlett evolved into a warmer, gentler man. This warmer, gentler man was the only Bill Hewlett I knew. It was difficult to witness his rapid decline. I noted in my diary, "A three-hour trip to the ocean. Bill has gotten much weaker. He is walking with difficulty. I barely managed to get him in and out the car. Occasional memory lapses." These outings became important for Bill. It was gratifying to hear Bill's second wife, Rosie, say, "The twinkle returns to his eyes when he sees you."

Bill and I were different in many ways. He had an insatiable natural curiosity about how things worked. I did not. Even as a boy, he would take apart locks and clocks and then put them together again. I couldn't imagine doing anything of the sort. I wasn't interested in how things worked as long as they worked. I never fixed anything until I was forced to become a handyman of sorts on the island in Finland. On the other hand, I was fascinated by the workings of the human mind, especially my own. Bill wasn't given to much introspection ("I'm an engineer," he would say by way of explanation). Nonetheless, he was a shrewd judge of character and masterful in dealing with people. He took a personal interest in his employees and with Dave Packard made "management by walking around" the "HP way." No wonder he became a much beloved figure for the tens of thousands of people working for HP. I would occasionally question Bill about the roots of his philanthropy: Why was he giving his fortune away? My question puzzled him. He had no answer beyond the fact that it was the right thing to do. He owed it to society. One had to give back. (One of his celebrated comments was, "Never stifle a generous impulse.") However, despite his immense wealth, Bill and his family lived rather simply. When driving Bill in one of the two family cars, I asked him what make it was. He said, with evident pride, it was a Mercury-something-or-other. A *Mercury?* Why not a Bentley, a Maserati, or a Hispano Suiza? Bill would smile and say nothing, or he would look at me indulgently as if I were a boy longing for a Tonka truck.

Given the materialistic and ostentatious culture I came from, it was hard for me to understand Bill's parsimonious ways—his worn-out jackets and beat-up favorite hat with the loose leather band. One day, when riding in my car, we stopped at a gas station. So as not to interrupt our conversation, I let the attendant pump the gas. And as I was paying him, Bill gave a gravely disapproving look for not using the self-service pump. I had never understood my father's frugal ways, but this was even more incomprehensible. Perhaps Bill thought it was not the money but the "principle," as my father used to say, but I never understood what that principle was. My father thought that money would spoil me when I was growing up. Actually it was the *lack* of money that spoiled me.

Bill's physical disabilities got progressively worse, but he endured them stoically. It was comforting that his mental capacities remained more or less intact until the next stroke, which took away his ability to speak and seriously clouded his mind. That made it very hard. During our outings we used to talk about our mutual interests: Stanford, the Hewlett Foundation, world history, and the state of the world. We talked about our children (but never about our wives). Now all that became impossible. I continued to try to talk to him but was never sure how much he understood of what I said. Once in a while, he would come out with an almost lucid sentence and then we would lose the connection. The end, when it came, was merciful.

William Hewlett's memorial service was held at the Stanford Memorial Church on January 21, 2001. The church was packed with family, friends, and prominent figures, including four Stanford presidents, trustees, faculty members, past and present HP employees, and leaders of Silicon Valley. I was honored to give the lead eulogy and was followed by Maggie Schneider (a friend of Mrs. Hewlett), Arjay Miller, David Woodley Packard, and finally Walter Hewlett, who gave a touching tribute to his father. It was the most important public address I was ever to give for the most remarkable man I had ever known and a cherished friend I would never forget.[2]

~

After my mother's death, Lucy continued to live in the townhouse on Williams Street that they had shared. We rented out my mother's former bedroom, but that turned out to be problematic. The first tenant was a young Chinese American who used the kitchen to cook his dinners. The strange smells from his exotic dishes bothered Lucy, and he left. The next one worked out better. He was a Japanese graduate student called Zen, who was very polite and deferential to Lucy. She became quite fond of him and often asked him to join her for dinner. Zen also developed a taste for Turkish coffee. Before leaving in the morning for class, he would sit quietly in the living room waiting for his coffee. When his parents came to visit him, he introduced them to Lucy. There was much smiling and bowing to each other, but they could not communicate otherwise. Zen was succeeded by Jenny, another Stanford graduate student. Lucy liked her a lot, and since Jenny was a single woman, she felt responsible for her. This led to a crisis when Jenny's boyfriend spent the night in her room. Lucy was shocked. I told her that she wasn't Jenny's mother and that Jenny was old enough to take care of herself. Lucy kept saying, "But they're not married; how can she spend the night with him?" The fact that I was not troubled by it became a problem in itself.

As finding tenants and dealing with them became increasingly burdensome, Stina and I looked for a place where Lucy could live and be taken care of eventually. We were fortunate to have her admitted to Lytton Gardens, a fine old-age facility in the heart of Palo Alto. Lucy lived there for the next decade to the end of her life. At first, she had a studio apartment with a small kitchen and lived independently. She was entitled to have her dinners in the dining room, but since the food was very bland, Lucy would take it to her apartment and dress

Figure 11.2 Lucy in her Palo Alto apartment. (Photograph by Margo Davis.)

it up. She often cooked for herself and more often for us. Stina and I visited or phoned Lucy almost every day and brought her to our house frequently. Nina and Kai saw her often when they were home on holidays. Nora kept in touch and Helen and Nello Sartoris continued to be her closest friends and companions.

Once she found her bearings, Lucy walked everywhere in Palo Alto, wearing her sneakers and baseball cap, which Nina and Kai had gotten for her. When going to the Stanford Shopping Center, she would cross the six-lane El Camino Real at whatever point she chose, traffic be damned. Our admonishments not to do that had no effect, especially when coming from me. What did I know? So far as she was concerned, I was still just a boy in her care. We knew she couldn't keep up her active life forever. A series of health problems increasingly troubled Lucy. Her stomach continued to bother her. During a routine gastroscopy, the surgeon accidentally perforated her stomach and she had to be rushed into emergency surgery. (It didn't occur to me to sue the surgeon.) Lucy then developed a blockage in her bile duct that required another operation, and she ended up with a tube in her abdomen to drain the bile. The tube itself would get blocked and had to be changed every month. After a dozen of these procedures, the surgeon decided not to replace the tube, and nothing untoward happened.

We had applied for U.S. citizenship for Lucy but the process took time. Finally I heard from the immigration office. Lucy needed to come for an interview. We went through the initial questions and then the immigration officer asked her if she was married. No, she was not. Did she have any children. Lucy was incensed. How could she have children if she was not married? The next question was if she had ever been a prostitute. *What?* The officer was staring at me waiting for me to translate the question. I told Lucy the man wanted to know if she had ever been in jail. She glared at him. Lucy didn't speak on the way home. She had lost faith in America. But her faith got restored at the next meeting. A year or so later, when it was time for her to go for the swearing-in ceremony, I called the immigration office to say that it would be difficult for Lucy to travel to San Jose because of her age and state of health. Was she handicapped? Yes, she was handicapped. That triggered a whole new procedure. Officers would come to swear her in where she lived. Two friendly middle-aged women came to her room. Would Lucy mind answering a few final questions. Not at all. But weren't they going to have coffee first? I looked at them imploringly and they agreed. Now Lucy was ready. What is your name? "Lucy." Where do you live? "Here." Who is the president of the United States? "Bill." If called upon, would you defend the United States against its enemies? "I old woman" said Lucy, "what I do?" I was going to say, "Knit a sweater for Bill" but I didn't want to take away from the solemnity of the occasion. No more questions. Lucy raised her right hand for the swearing ceremony and that was it. As the women congratulated her, a big smile broke out on Lucy's face and she said triumphantly, "I now American girl!"

~

What eventually made life more difficult was Lucy's progressive dementia. It began insidiously and never got to the point where she couldn't recognize us. Nonetheless, she could no longer manage on her own and had to be moved from independent living to the assisted care unit. She didn't want to leave the room where she had been so happy and couldn't understand why she should. I finally had to tell her that she just had to do it, we had no choice. Stina and I spent most of a day packing and moving Lucy's belongings to her new room. When we went back to see her the next morning, the room looked as if it had been hit by a tornado. Everything was out on the floor as Lucy, looking flushed and frantic, was trying to repack things. What was she doing? She said she was getting ready to move to her new room. She wouldn't believe that she was already in her new room. We put everything back in place and told her in no uncertain terms not to take them out again. In a week or so, she settled down and became resigned to her new location. The hardest part was that there was no kitchen—the center of her life had been taken away from her.

The next crisis came in the form of an incensed phone call. Lucy insisted that the woman next door had stolen her sweater. I tried to reason with her, but she was adamant. The woman was *wearing* her sweater—did I think she couldn't recognize a sweater she had knitted *herself?* Stina went to see the social worker, who arranged for Lucy to meet with the woman, who kindly showed them a picture of herself taken some time ago wearing the sweater in question. Lucy accepted it, but was not entirely persuaded until she found the "stolen" sweater among her things. She was duly abashed and apologetic but also pointed out that the sweaters were the same color and therefore easily confused. I agreed. Stina and I continued to visit Lucy almost daily. Our frequent visits didn't go unnoticed by the other residents, who had few visitors. An elderly Russian couple stopped me in the corridor and in heavily accented English expressed their admiration for my being so attentive to Lucy. I thanked them and said, "But she's like my mother." "All of us here," replied the woman, "are mothers and fathers."

The final blow came when Lucy had to be moved to the Alzheimer floor. She still had enough self-awareness to realize that most of the other residents wandering aimlessly had lost their minds. Why was she there with all those "crazy people," she asked indignantly. I had no answer. I tried to distract her by taking her out in a wheelchair to visit the neighborhood. One of our stops was at a coffee shop owned by an Iranian. I asked him if he could make us Turkish coffee. He said he would but then used the espresso steamer rather than simmering the coffee in an *ibrik* on a slow fire. It tasted awful, and Lucy and I agreed that the man was an idiot. One afternoon, as I was taking her around in her wheelchair, she asked in exasperation when my mother was going to return from her trip. I told her my mother had been dead for ten years. "What?" she cried, "She is dead?" and started to wail in the middle of the street. I had never told a bold-faced lie in my life, but this was a good time to start. The next time she asked me the same question, I said, "Very soon," and she was satisfied.

We left Lucy behind reluctantly in the early summer of 2001, when we took off for Finland. Kai was going to be in town and he promised to be with her as much as he could. True to his word, he visited her virtually daily. He would pick her up like a child, take her down the elevator without bothering about the wheelchair, and drive her around in his car. The nurses were amused and touched. Some of them might have wished he had carried them in his arms.

Late in August, Stina and I went on a Stanford alumni cruise around the Black Sea. As we were returning to Istanbul, the captain summoned us for an emergency meeting in the lounge. It was September 11. He gave a brief account of what had happened in New York but could provide few details, and being too far at sea, we could only see grainy images on television of the burning Twin Towers. We were stunned, and being on a cruise made everything seem even more unreal. We finally docked in Istanbul to find that airline connections to the United States were disrupted. Many of the passengers were stranded, but Stina and I had no trouble going back to Finland.

On September 26, Stina left the island to return home ahead of me. I was to spend a week in cherished solitude. On my first morning, I sat down to read art historian Robert Hughes's *The Shock of the New* (given to me by Nina for Christmas). An hour later, Stina called. Lucy was dying. Stina could handle the practical details but if I wanted to see Lucy before she died, I had to come right away. I took the next available flight to San Francisco. By the time I reached home, Lucy was no more. Nina also arrived from New York too late.

Lucy's memorial gathering at our house was a replay of my mother's memorial dinner. As one after another of the guests stood up to pay their respects, Kai and I sobbed. I was too undone to say more than a few words. Nina spoke movingly about what Lucy had meant to her.[3] It was amazing to see the depth of affection that Lucy had earned not only among our few relatives and her fewer friends, but many of our own friends as well. Her entire life had been spent in serving others. She was the only person I knew who had done more for others than others had done for her, including me. The little Armenian orphan who had been snatched from the jaws of death by Bedouins in the deserts of Deir Zor had found her final rest in America. She was buried at the Alta Mesa cemetery in Palo Alto, next to my mother.

~

Late in the spring of 1998, I had lunch with Walter Hewlett at the Stanford Faculty Club. He wanted to get my reactions to the prospect of establishing a smaller family foundation that would complement the work of the Hewlett Foundation. Over the years, Bill Hewlett had supported a considerable number of organizations and projects from his private funds. Some of these initially came as grant proposals to the Hewlett Foundation, but they didn't fit with its program areas, so he would fund them on his own. After Bill became incapacitated, his trust continued to provide about $20 million per year for such support until the time of his death. Walter realized that after his father passed away, this would no longer be possible to sustain. Having a smaller foundation would allow the Hewlett family to continue carrying on Bill's charitable donations as well as their own philanthropic interests. Moreover, a family foundation would provide Bill's twelve grandchildren with opportunities to become informed philanthropists and prepare them to serve on the board of Hewlett Foundation in due time.

Walter wondered what I thought about this prospect and the possibility of my being involved in it. Since I knew the family and was on the Hewlett Foundation board, I would be in a good position to act as a bridge. I thought it was a great idea. I was especially intrigued by the possibility of getting the younger generation engaged in philanthropy. I believed that philanthropy was most rewarding when it was part of one's whole life rather than when it came as a postscript toward its end. Anything truly worth doing was worth doing all your life, and it would be rewarding in different ways at different times. Grantees may not care who was giving them the money or why, but the impact on donors could vary considerably, depending on their philanthropic motivations.

When I returned from Finland in the fall, I met with Walter again and he now had a two-page statement titled "Some Ideas and Concepts for the New Foundation." It essentially outlined what he had told me earlier, with two new items: the new foundation's endowment would be $100 million and would come from Bill's trust, and I would be the president. When I expressed my surprise, Walter said that it was his wife, Esther, who had told him that I could run the foundation for them. Walter made clear that this was just an idea and it was going to take the right person to develop it into a working model and I was that right person to do it. Otherwise, nothing would come out of it.

I thanked Walter for his confidence. I would certainly like to help, but since I was still teaching full time, where would I find the time to do this? Walter thought it shouldn't take too much of my time since he wanted it to be kept simple and low key. Stanford permitted faculty members to spend a day a week consulting and I might be able do it on a one-day-a-week basis, at least while we set up the foundation. Susan Briggs, the attorney who handled Bill's trust, would help with the legal aspects of establishing the foundation and serve on the board.

I couldn't turn Walter down lightly, and I liked the prospect of getting involved in an exciting new venture. I agreed to do it but with some hesitation. I was again going to take charge of a vaguely defined venture for which my background had not prepared me. I consulted a few past and present foundation presidents, but their experience was not particularly relevant to what I was setting out to do. The primary challenge was to engage the twenty-one members of the extended family, in particular William and Flora Hewlett's twelve grandchildren, in the work of the foundation. To that end, I devised several mechanisms, which proved quite effective. The members of the Hewlett families, along with Susan Briggs and me, would constitute the Family Council. The council would act as an advisory group, and out of its membership would be elected the board, which I would chair and which would have fiduciary responsibility for approving the grants. Every grant request would have to be sponsored by a member of the Family Council, who would submit a statement justifying the grant. Smaller grants could be made at presidential discretion and reported to the board. We would have no predetermined program areas, thereby ensuring that grants would be most responsive to the interests of individual family members. I told them they could go in any direction they wished within reason in their grant-making choices. Their initial footsteps would gradually turn into paths, the paths into roads, and one day the roads would be paved and become program areas. To avoid the risk of being swamped by grant requests from all quarters, our grants would be made by invitation only. Every member of the Family Council would have the same set of privileges and responsibilities, from the youngest member, twelve-year-old Mary, to the oldest, Eleanor's husband, Jean-Paul Gimon. Family Council members, selected by lot, would serve on the board, with everyone rotating through over time.

~

We named the new organization the Flora Family Foundation (FFF) after Flora Hewlett—the beloved mother, mother-in-law, and grandmother of the extended family. She had been an integral part of Bill's philanthropic work from the outset and had also pursued her own charitable interests. I hired Ruth Sloan, who had been my assistant dean for administration at Stanford, as my general assistant on a part-time basis. We started working out of Ruth's home and then rented a single room in Palo Alto as our office until we moved into the new Hewlett Foundation building in 2002.

Our first Family Council meeting took place at a dinner at our house. I gave a short presentation and we discussed the organization and general purposes of the new foundation. It was an exciting occasion with a great deal of warmth and family solidarity. My next step was to meet with each family member individually to explore their philanthropic interests and the best ways to implement them. Our subsequent Family Council meetings would take place at the San Felipe ranch. These became important occasions to present reports on the past year, plan for the next, and provide an opportunity for the extended Hewlett family to get together.

In the beginning, it took some doing to have all of the family members, especially the younger generation, to sponsor grants. They had been brought up without a clear awareness of their family's wealth and were unprepared for the prospect of giving away tens of thousands of dollars just like that. One of the younger members pointed out that the grants he was sponsoring involved more money than what the parents of some of his friends made in a year. However, they caught on soon enough, and by the end of the first year our $5 million yearly budget was no longer adequate. I wrote to Walter suggesting that he consider adding to the principal. A few weeks later, we had another $100 million. My engagement went up from one day a week to two days and then to half time. I scaled down my responsibilities at Stanford but continued to teach most of my courses.

As our annual budget expanded to $10 million, it became clear that we needed more staff to administer it. I brought in Steve Toben, an able and respected program officer at the Hewlett Foundation whose term was coming to an end. He had been responsible for the programs in environment and conflict resolution and was highly knowledgeable about the field of philanthropy. I appointed him as my vice president and he brought with him Patricia Gump, a long-time staff member at the Hewlett Foundation, to serve as program administrator, and Annette Rado, an accountant, both of whom would replace Ruth Sloan.

Particularly in the earlier period, when we had ample assets and the most latitude, I was able to respond to whatever interesting prospect came my way. When I read about a natural disaster in the morning paper, I could go to the office and send a presidential discretionary grant to an agency that could respond to the emergency, or if I came across a good nonprofit organization, I would contact family members to generate support for it. Had I been wealthy, that is how I would have spent my money; now I could get almost the same level of satisfaction by spending someone else's money. Watching the younger generation develop into informed and imaginative philanthropists was the most rewarding part of our work.

The informal way in which I ran the foundation during its early years became more formalized over time. We suffered a serious loss of assets in 2001, when there was a sharp drop in the value of the HP stock, which accounted for most of our principal. To compensate for it, family council members were now limited to yearly allocations for sponsoring grants. Despite these limitations, morale remained high and interest didn't flag. I had cordial relationships with all of the family members, but Walter and Esther became my closest supporters and we became good friends. Esther was responsible for some of the most imaginative ideas that helped FFF set its agenda—notably the Gap program, which focused on the disparity between rich and poor countries. Obviously, these were immense problems and we could hardly make a dent. However, Gap provided good opportunities for supporting organizations like Ashoka and Oxfam, which worked effectively in the developing world. To see firsthand the problems in some of the poorest countries, we organized trips for small groups of Family Council members, starting with Mali in Africa. It was amazing how much we could learn even in a few short days. Given the enthusiastic response, we allocated a fifth of our grant budget to the Gap program, and international grant making came to account for 40 percent of the FFF grant budget (as against 3 percent for American foundations as a whole).

Given our lean staff, we came to rely on intermediary organizations that did great things in small ways. This was exemplified by a project in El Salvador, where a one-man-directed project consisted of digging wells with pick and shovel to provide fresh water to a village. We first made a small grant and were so impressed by the results that we then gave a larger grant, which made possible the purchase of a large drill mounted on a truck that could dig deeper wells. The grantee could now generate income by also digging wells for commercial outfits to make the project self-sustaining. Such projects produced immediate and tangible results that made it easier, particularly for younger family members, to learn from. It was much harder to see how much good we could do in connection with global problems, such as overpopulation and environmental degradation, where small grants would be like drops in the bucket.

Figure 11.3 FFF trip to Mali. To my left are Esther Hewlett and Jean-Paul Gimon. His daughter Marianne is at the far right. Two of our hosts are in the center.

This style of grant making would not save the world, but it could make a difference. Many of the most successful social entrepreneurs were women in the third world working individually or in small groups. Some of our grants entailed considerable risk, and it wasn't always possible to ascertain what they accomplished. For instance, during our trip to Mali, we met with a woman who had an interesting idea: she would buy lambs, let them graze in her backyard, and sell them as sheep during Islamic feast days. However, trying to figure out how this idea would actually work out and to estimate how much money it would generate drove me and Jean-Paul Gimon (who was a banker) to distraction. Jean-Paul, who was French, was amused when the Malinese woman congratulated him on how well he spoke French.

Judged by any reasonable measure, FFF was a notable success, especially in stimulating and sustaining the engagement of the younger generation in philanthropy. It established a new model for family foundations. However, judged by more stringent criteria for how effective our grants were, the results were more mixed (which probably would be true for virtually any other foundation). Some grants were not demonstrably effective; others helped the grantee but weren't essential for its work. However, there were also grants that had a critical impact.

A few of them were directly responsible for the creation and sustenance of an organization. Grants that were most effective were also usually the most personally satisfying to the sponsor; however, effectiveness and satisfaction did not always go together.

The grants that were most satisfying to me personally included support for *Cabinet*, the arts and culture magazine established by Sina Najafi, Nina's companion and later on our son-in-law. Supported by grants sponsored by Natalie Gimon (now Farman-Farma), *Cabinet* came to occupy a special niche in New York cultural circles and beyond. Other grants that I personally sponsored included the conservation of medieval churches in Armenia and support for the Hover men's choir, which recorded a sublime rendition of the mass of the Armenian Apostolic Church. Sometimes even a small grant to a large organization could have a compelling effect. For instance, the San Francisco Museum of Modern Art was about to eliminate several positions during a budget crisis that would have severely curtailed the work of its unit for art education. I happened to hear about it from a former student who was working there. The $25,000 grant SFMOMA received from FFF within a few days gave the education unit a reprieve and allowed time to raise larger funds to sustain its operations.

My engagement in philanthropy was one of the most satisfying experiences of my career. Through my work with the Hewlett Foundation and then FFF, I gained a better understanding of the problems of the world and the best ways to deal with them through philanthropic efforts. I also came to realize the difficulties in trying to bring about social change on a large scale and the blunders that foundations make in supporting projects that don't work out well while failing to support others that would have. I especially regretted the relatively low level of support that the humanities and the arts received from American foundations (let alone the federal government). I wasn't convinced that "strategic" philanthropy was always the best way to make grants. It is true that in addressing root problems, more could be achieved, but there was also something to be said for charity, pure and simple. The fact that a poor person would be hungry again tomorrow was no excuse for not feeding that person today.

At the end of five years, when I turned seventy in 2003, I decided I had fulfilled the task of setting up FFF and it was time for me to retire, at the same time that I was retiring from Stanford. There were no compelling reasons for me to retire in either case, and I could have easily stayed on for at least several more years. The main incentive to retire was the prospect of spending more time on the island in Finland. And since either job would have tied me down, I had to give up both at the same time. However, my retirement hardly gained me more time for leisure as I embarked on writing my next book, on guilt, and then came this memoir, which would take another two years.

~

My last task at FFF was to carry out an evaluation of our first five years. I conducted a survey of our grantees and wrote a short book giving the history of the founding of FFF and summarizing the findings of the self-evaluation. I hoped it would serve as a historical record and baseline for subsequent generations. Walter Hewlett wrote the introduction and Steve Toben a preface, in which they described my contributions to FFF.[4] Steve Toben succeeded me as president when I had already relinquished chairing the board to Susan Briggs. At about the same time I also retired from the Hewlett Foundation board after having reached its mandatory retirement age of seventy-two. My efforts on behalf of FFF were duly recognized by Family Council members, who presented me with a handsome album of photographs with generous acknowledgments of my contributions from every single member. Walter wanted to ensure that I did not underestimate the crucial role I had played in the establishment of FFF: "If you had said no," he wrote, "we very well might not have gone ahead with FFF. The idea without the right person is not enough. Very few individuals could have pulled it off, and the success of the concept really depended completely on the person who would be carrying it out. You took the skeleton of an idea and built it out into something fabulous. Herant running it, and our doing it at all, really went hand in glove. Please do not overlook your essential role in the very existence of FFF."

Much as I appreciated these generous sentiments, I basically thought of myself as an agent of the Hewlett family. It was their money and earnest engagement that made it all possible. My assessment of my own work in philanthropy was once again tinged with some doubts and ambivalence—like the feelings I had had in relation to my previous roles as psychiatrist and administrator. By all indications, I had done a good job in setting up FFF and serving on the Hewlett Foundation board, even though nothing in my background had prepared me for it. As Esther Hewlett kindly put it, FFF had brought together my talents as a teacher and my instincts as a philanthropist, which she thought is what I was at heart. I admit that there was some truth to this. However, I thought my engagement in philanthropy had come too late and I had accomplished too little. If I had made a career change into foundation work ten years earlier, I may have become the head of a large and nationally known ("real") foundation. Instead, I had ended up with yet another nebulous identity and another ephemeral accomplishment. I knew that was not quite true and that is not how I should feel, but that is how I felt nonetheless.

My encounters with wealthy individuals like Bill Hewlett and a few others, who also became good friends, changed my earlier conception of what it meant to be rich. My great-grandfather had owned ninety donkeys to ferry the produce to his house, and my family lived in one of the largest houses in Iskenderun, but all of that would count as small change compared with what these individuals were worth. Yet most of them lived relatively simply in contrast to the luxurious lifestyle they could have afforded. There was something quintessentially American in their way of being rich that reflected an almost puritan quality of self-denial. I found that both baffling and endearing.

Beyond the Hewlett Foundation and FFF, I have been most impressed by the selfless and creative ways in which some individuals make a difference. Three women friends in particular have exemplified this. The first is Helen Bing, who has done things that no one else would think of doing, and done it with great flair, such as brightening the bleak walls of the Stanford hospital with fine posters, having wild flowers planted on campus, and engaging in numerous other philanthropic ventures.

The second person is Esther Hewlett, who, apart from her key role in FFF, started her own organization (New Global Citizens) to foster philanthropy and public service among young people. The third is Laurene Powell Jobs, whose organization (College Track) has performed wonders in enabling students from disadvantaged backgrounds to go to college. All three women could have led contented lives as the wives of highly accomplished and wealthy men. Instead, they gave their "time, talent, and treasure," as Esther would put it, making their corner of the world a better place. Philanthropic contributions aside, we often fail to realize the critical role that such women play in the lives and accomplishments of their eminent spouses.

Philanthropy through giving money is not, of course, the only form of public service. The selfless volunteerism of Americans is expressed in the countless ways people help others. I never lived up to my youthful dream of devoting my entire life to the service of others. Yet throughout my life there has always been an element of caring for others—starting as a youth leader when I was young myself and later becoming a psychiatrist, teacher, author, administrator, and foundation executive. Another area where I was able to make a contribution was by serving on the governing boards of academic institutions. The first was Haigazian University College in Beirut. Founded by the Armenian Missionary Association of America (AMAA), it is an institution that mostly serves Armenians but also many others in Lebanon and the region as a whole. One Sunday morning I received a call from an old friend who was hoping I would join the board of trustees of the college, based in Los Angeles. I already had too much on my plate and was going to decline. However, Stina got wind of the conversation and told me I should do it—I was doing very little for the Armenian community and this was right up my alley. I groaned and accepted, and I am glad I did. During the two decades I served on the board, I came to play a central role as a friend and advisor to the chair and the president and secured substantial funding from various sources.

It was particularly significant for me to join the board of trustees of the American University of Beirut in 2001. I was the first person of Armenian descent to be on the board. Even though it was my connection with foundations that was probably the reason for my being asked to serve, the Armenian connection still mattered to me. Armenians were overrepresented at AUB among the faculty, the students, and the staff during the years I spent there, yet no Armenian had ever served on its board. It meant a great deal to me that I should be the first to do so. In this connection, perhaps my single most important contribution was initiating a program not for Armenians but for the better understanding of Islam. In the aftermath of September 11, I helped to obtain a grant to AUB from the Hewlett Foundation, with strong support from the president Paul Brest, to recruit Islamic scholars from the Middle East for short-term teaching appointments at American universities. The program was a resounding success, but visa restrictions and other security obstacles ultimately killed it. The eight years I spent on the AUB board until I reached the mandatory retirement age of seventy-five were deeply satisfying. About a third of the members of the board were native-born Americans. They included two former ambassadors to Arab countries in the Middle East, prominent business executives, and a number of academics, many of them from the Columbia Presbyterian Medical Center in New York. A second group consisted of naturalized American citizens, like myself, from the Arab world. The third group came from Lebanon and other Arab countries. I came to know many accomplished individuals whom I would not have met otherwise. Through them, I gained a deeper understanding of the issues in the region that I had lost touch with. After I retired from the board, I was happy to be invited to stay on as an emeritus trustee and continue to keep in touch with AUB.

~

The one area of public life I did not get involved in was politics. This was not a matter of indifference—it was an active rejection. The roots of this attitude were partly cultural and partly personal and in sharp contrast with the community I was a part of. We live in an area where people are engaged in many issues at the local, state, or national levels. They are deeply conscious of living in a democracy where each vote counts. My wife grew up in a country where the level of social consciousness and engagement is even higher. (Eighty percent of Finns vote in national elections.) Consequently, my reluctance to be more engaged socially was exasperating to Stina. I would tell her, by way of explanation if not excuse, that I grew up in a country and at a time where elections were rigged. It hardly mattered whether you voted or not. I recall one particular election when Nora's husband, Noubar, was running for the Lebanese parliament on the Armenian ticket. I was a college student and happened to be at a reception. As I excused myself for having to leave in order to vote (simply because the candidate was my relative), the host said not to bother. He could tell me who was going to win, and it was not the person I was about to vote for. I voted anyway out of loyalty to Noubar even if the outcome of the election was predetermined. Moreover, the fact that I belonged to an ethnic minority with little political leverage at the time was an added hindrance to my getting politically involved. What would have been the point? There was also the fear that if Armenians took sides and things went wrong, then we would end up getting blamed for it. I realized that these considerations no longer applied to my situation in the United States, but these attitudes were too deeply ingrained to ignore. As with earlier references to Lebanon, my observations come from fifty years ago, and much may have changed since.

My sense of being an outsider also fed into this cultural context. It is difficult to engage in politics if you feel like a foreigner rather than a native. I have been an American citizen for almost fifty years and feel like an authentic and integral part of Stanford and its surrounding immediate communities. But beyond that I am not sure where I fit. To be engaged in politics, one needs to be committed to a political perspective or have a personal commitment to a political leader. One needs to have a stake in the political system. I could not to find a political ideology or candidate for office that I could seriously care about. When I became an American citizen, my conservative background tended to turn me more toward what I thought were Republican views. However, given Stina's far more liberal politics and the fact that most of our friends were Democrats, I voted for the Democratic candidate in every presidential election without joining the Democratic party—until Barack Obama ran for president. I finally could take a personal interest in a candidate (aided and abetted by my exasperation with George W. Bush).

At a more personal level, two considerations added to my lack of engagement. My pragmatic perspective led me to not bother with issues that I couldn't do anything about. Moreover, when every disaster anywhere in the world confronts one with each morning's paper, it becomes overwhelming to be constantly engaged in the world. This doesn't mean I didn't care what happens to people outside of my orbit or empathize with their misery. But there were limits to what I could do. My compulsivity would also get in the way. If I could not solve a problem completely, I did not want to solve it only partially. In retrospect, I regret that I never took part in any marches for civil rights or against the Vietnam War. (I did take part in one march against the invasion of Iraq.) I now realize that one has to do what one can and let it go at that. As the saying goes, the best can be the enemy of the good. My need to be impartial and look at an issue from all angles has also tended to lead me to an unwillingness to take sides. (I have been called a "fence sitter," with some justification.) However, these liabilities have also had a positive side in helping me not to become a dogmatic, unquestioning advocate, let alone a bigot or a politically correct do-gooder. As somebody put it, I don't want to be like people who help save whales because they want to be seen as the sort of people who help save whales.

Two decades ago I was asked by Roger Heyns and Arjay Miller to join a small group of men who met for lunch every month or two to talk about ongoing issues of political interest. The group included John Gardner (former secretary of HEW under Lyndon Johnson, and the founder of Common Cause), Mel Lane (of *Sunset Magazine*), Wally Haas (of Levi Strauss), and Bill Reilly (former commissioner of the EPA), in addition to Roger and Arjay. These were men with a deep interest and engagement in public and political affairs. I was quite a bit younger than them and I had little to add to the discussion (peppered with statements like "Lyndon told me...,"), except when it came to problems in the Middle East, where I was the "expert." I greatly admired the social commitment of these men and their engagement with the world. I gradually became one of the senior members of the group as its older members died and were replaced with Walter Hewlett, Paul Brest, Laurie Hoagland, Jim Gaither, David Lyon, and my Stanford friends and colleagues Bob Greg and Dave Abernethy. These meetings exposed me to political discussions that supplemented my reading the *New York Times* at the breakfast table and the *Economist* in bed before going to sleep.

An extension of these discussions took place when Stina and I visited Arjay and his wife, Francis, when Arjay regaled us with stories about people like Henry Ford, Robert McNamara, and Warren Buffett, all of whom he had known for decades. When Arjay turned ninety, Paul Brest organized a lunch for some of his friends. A surprise guest was Warren Buffett, who flew in for the occasion. It was interesting to stand next to the one of the richest men in the world, who had given most his immense fortune to the Gates Foundation—the most selfless gesture in the history of philanthropy. In that casual setting, Warren Buffett did not stand out as an extraordinary person (and I noticed that I was wearing a more expensive-looking suit than he was). But Arjay said Buffett was a true financial genius beyond being a great stock-picker.

~

I officially retired from Stanford on September 1, 1999, at the age of sixty-six. However, I arranged to be recalled to active duty on a part-time basis for three more years. So I didn't stop teaching until I was nearly seventy. The decade that crossed into my retirement was relatively uneventful. I passed the landmark years of sixty, then sixty-five, and finally seventy and seventy-five at a steady pace. Like an airplane cruising at 35,000 feet, it was largely free of turbulence. Confucius said that he did not learn how to behave until he was seventy, and that would also be true for me.

Selling the house on campus that we had built and lived in for three decades was a major change in our life. We moved into a condominium complex at the Peter Coutts Circle that was initially built for junior faculty at Stanford but now has many retirees as well. The move was hard, especially when it came to dealing with our large number of books and oriental rugs. The rugs were tied with my personal life, and the books with my professional life. Giving them up meant giving up part of my past life. I tried to find a general rule to help me determine which books to keep and which ones to give away. I first picked those that I thought were more important, until I realized that the really important books were already in the library; the books I wanted to keep were those that were more personally significant for me. There was no way to weed out the rugs and kilims painlessly. We managed to find room for many of them on our condominium floors and walls. Nina was able to take some of them (including a large rug I used to play on as a child) and we set aside some others for Kai and Anni. The rest were put in storage. I told Stina that she could do with them as she wished after I was gone. We managed to re-create in our new unit the same sort of attractive space we had lived in earlier, but on a smaller scale. Moreover, it was easier to live in and to maintain. We no

Figure 11.4 With Stina. ("Growing old on the same pillow," as they say at Armenian weddings.)

longer had a dining room, where we rarely ate, or a living room where we did not live. Instead we now used fully the space we lived in. The experience of giving up so many things made me wonder what was to happen to the things we still had, but I decided I could no longer care about things beyond the grave. Our children would have to deal with them as we had with our parents' possessions. Besides, do we ever really own anything? Or are we merely its custodian for a while?

As we got older, I was happy that we could stay in remarkably good shape. After futile attempts at working out in a gym, I finally found a form of exercise that I would do diligently: walking five times every evening around our condominium circle, covering three miles in fifty minutes. Stina and I continued to be actively involved in our writing projects. While I worked on my memoir, she wrote a book based on her parents' correspondence during World War II and her own childhood recollections of the time she spent in Lapland and Sweden as a refugee.[5] We read and discussed everything each of us wrote. We could now spend much more time together. Our children and their spouses were healthy and happy, and we had no financial worries. It was a blessed way of aging together in good health and good humor (without feeling "old"), and looking forward to the time we might become grandparents.

~

My retirement from Stanford, the Hewlett Foundation board, the Flora Family Foundation, and the AUB board followed each other in short order. It was like turning the lights off in one room after another. There were memorable celebrations on each of these occasions. The outpouring of praise, respect, and affection erased whatever residual doubts I may have had over the value of what I had done during the preceding forty years—doubts that I should not have entertained in the first place.

The Human Biology Program organized a large gathering, during which several colleagues and former students spoke. Grace Yu, one of my star students, by then in medical school, gave a touching tribute.[6] This event was followed by a memorable dinner at the home of Anne and Russell Fernald; Russ was the director of the Human Biology Program and Anne was a child psychologist with whom I had co-taught in the Human Biology Core Course. It was attended by three former Stanford presidents, distinguished guests, colleagues, and good friends. Most importantly, I was surrounded by the members of my family. There were generous tributes and I made a brief response. I pointed out that during my four decades at Stanford, I had worked with five presidents, three provosts, many deans and faculty colleagues, and a great many students. I concluded with, "Ultimately, my primary sense of professional identity has been that of a teacher. I owe my students a debt of gratitude for making me who I am. Stanford's confidence in me has been boundless, perhaps misguided at times, and its generosity to me has verged on the irresponsible."

I received a gracious note from a former director of the Human Biology Program which showed how needless my worries had been

Figure 11.5 Human Biology retirement dinner.

over the value of my contributions to the program.[7] I was touched to receive a note from David Hamburg, who regretted not being able to attend my retirement dinner. It was a summation of my Stanford career—a career that I owed to him more than anyone else.[8]

My retirement dinner from the Hewlett Foundation board was held at the home of Paul and Iris Brest. It was a joint celebration for me and Bob Erburu, who was also retiring from the board. I was presented with a framed picture of Bill and Flora Hewlett with a generous tribute.[9] My retirement dinner from the Flora Family Foundation took place at Bill (Walter's younger brother) and Sally Hewlett's home. Virtually all of the extended family members were there, along with Susan and Ron Briggs, Steve and Janice Toben, and Stina. Their appreciation warmed my heart and I felt like a member of the extended Hewlett family.

I hadn't given much thought to what my actual retirement was going to be like. I kept immersing myself in new tasks that would keep me as busy as usual. When teaching my seminar on guilt and shame, I had been unable to find a book that covered various aspects of the experience of guilt from different disciplinary perspectives. I decided to write that book myself. Having taught the seminar for ten years and listened to my colleagues lecture, I thought I had the material more or less ready to put down on paper. I was wrong. It is one thing to teach an undergraduate seminar and another to write a book. I knew I had to do a lot of work in the areas that my colleagues had covered in their lectures. But I didn't anticipate the difficulty I was going to have finding sources that specifically dealt with guilt and shame from various religious and philosophical perspectives. I had to extract this material from the voluminous literature and lengthy discussions with my colleagues. The first draft of the manuscript was over 700 pages. No one would read it let alone publish it. Moreover, if it was going to be a book for a general readership, I had to make it more accessible. I spent two years rewriting and scaling the book down. It was well worth the effort. Finally, *Guilt: The Bite of Conscience* was published in an attractive volume by the Stanford University Press at the instigation of Mike Keller, the head of the Stanford Libraries and the press.[10]

I learned more from writing the book on guilt than from anything else I had written before. I gained a deeper insight into the psychology of guilt and its place in my own life. My sketchy knowledge of Judaism, Christianity, and Islam coalesced into more coherent entities. Whole new vistas were opened into Hinduism, Buddhism, and Confucianism. The moral philosophies of Aristotle, Kant, John Stuart Mill, and Nietzsche became more intelligible. It was the first time I understood how the legal system operated in ascertaining the guilt of individuals. This was the eighth book I had written or coauthored. The human sexuality textbook had had the widest impact (having been read by several hundred thousand people) but the guilt book was in some ways the most important book for me. (At least one colleague said it was my best book.) The scope of the subject matter and the style of its presentation came closest to reflecting who I was. I wasn't clear on how it would be received, how widely it would be read, and what impact it might have. However, the satisfaction I had already obtained from writing the book was sufficient reward. It also brought home to me my distinctive way of making a contribution. The book drew from a dozen areas of knowledge, many of which were outside my areas professional competence. I'm sure that there is someone, somewhere, who knows more about one or another aspect of guilt than I do. However, I know of no one who has written a book pulling all this together as I have done.

~

I had always wanted to teach at one of Stanford's overseas studies programs, and now was my chance to do it. I proposed teaching two courses at the Paris campus. One was a seminar on erotic art and literature in eighteenth-century France. It covered material from my sexuality class and was easy to put together. The second course required more extensive preparation. It was based on the antiquities at the Louvre from ancient Mesopotamia, Egypt, Greece, and Rome. The class would meet in the morning for lectures and discussion and then I would take the students to the Louvre for a two-hour tour to look at the antiquities we had discussed. I greatly enjoyed teaching the course, and it was a resounding success. The second time I taught it, in 2007, virtually all of the students in the program took it, making it the most popular course in the history of the Paris program.

We were delighted to live in Paris on both occasions. It was one thing to visit the great city and another to actually live and work there. I knew that having ample time to visit museums and historical sites would be exciting, but I had not anticipated the simple pleasures of going to a bakery to buy a croissant in the morning, or sitting at a sidewalk café in the evening to have a Campari with nuts or a kir with cheese and fig jam. These became traditions that we maintained after returning home. I then parlayed that teaching experience into another opportunity to go to Paris by compressing my Louvre course into a week-long program for small groups. The first group consisted of three families with children ranging from ten to sixteen who surprised me with how quickly and enthusiastically they delved into the material. The second group consisted of three couples of varying background, and they also became deeply engaged. I had the satisfaction again of being able to adapt what I taught to different audiences and circumstances and make it work.

David Hamburg and I had stayed in touch, and I now had the opportunity to become more actively engaged in his work. After David retired from Carnegie, he took up the critically important problem of the prevention of genocide by building on his long-standing interest in preventing deadly conflict. It was deeply gratifying for me to be able to be of some help to him in preparing his book and a related project whereby he interviewed eminent political and academic figures and produced a documentary together with his son Eric. The interview with me came to occupy a central part in the series by combining elements from my own family's past with the Armenian genocide. These opportunities brought us closer than ever before. David had always been enormously supportive and appreciative of our collaborations, and it was the least I could do for him. [11]

I had met Paul Brest when he was dean of the Stanford Law School but I came to know him better after he became president of the Hewlett Foundation. When I gave up my office at FFF, Paul offered me another office at the foundation that I could use as a resident scholar. It was an act of unsolicited kindness that made a huge difference to my life after retirement. I not only had a superb place to work in but enjoyed the companionship and help of the staff whose offices were close by. Without ever working for the Hewlett Foundation, I developed a sense of being a part of it, which I came to cherish.

Finally (and I don't know how to say this without sounding corny) my greatest debt is ultimately to America. I say America, instead of the United States, because that is the language of immigrants. There is nowhere in the world—be it my birthplace, the place of my ethnic origins, the country where I grew up, or even Europe, which I love—

where I would have been given the same amazing opportunities that I had in America. Social injustice and prejudice may still be a problem in the United States, but I was never subjected to either. As for Stanford, there is no university on the face of the earth that would have allowed me to do what I was able to do. I will never forget these debts and I will always be deeply grateful.

12

A Healing Place

It is early May. I am alone in my study at our summer house on the island off the coast of Finland. The leaden gray sea lies heavily with its soothing calm reflected by the overcast sky. Faint streaks of pale blue lighten the horizon like distant hopes. A lingering coolness hangs in the air with the receding echoes of the winter that held the island in its icy grip during months of darkness. Gulls wheel effortlessly through the vast silence. The distant chirping of a bird evokes the cry of a lost child. Pine trees stand erect like sentinels holding their breath in the morning mist. Birches with fresh green leaves of mouse ears stand self-consciously as recent arrivals not yet at home in the unfolding spring.

Sitting in front of the fireplace in my study, I stare at the softly burning birch logs. Their warmth permeates the room holding me in its embrace. The flames lap at each other voluptuously, then fall back exhausted. The fire struggles to stay alive. It flares up only to die down and break up into embers that turn to ashes that lie lifelessly in a cold heap. Is that how my life will also end?

This is a time of solitude that I cherish. Could I ever be truly alone? Could I ever get away from the burden of being myself? Even if I cannot, there is much comfort in having only myself to deal with. To let the rhythm of my body and mind carry me along with no concern for all else, free from the intrusion of the world that gets into my eyes, my ears, and between my teeth like fine sand.

Solitude feels inseparable from silence. St. Benedict, the founder of Western monasticism in the sixth century, made silence a cornerstone of his Rule. Trappist monks speak only when truly necessary. Idle talk disturbs their peace and receptivity to God's voice. The tradition of seeking God in the solitude and silence of the desert goes back to St. Anthony and the hermits of fourth-century Egypt. Even though God is no less present in the midst of a teeming city than in the wilderness, he is more accessible when there are no distractions. The Great Silence speaks to those who listen to him in silent solitude.

Silence descends in layers. I bask in the blissful quiet for the first few days. Then I listen to sounds I did not hear before. With each successive day, I go down to deeper levels of self-awareness. The residues of intrusive and troublesome thoughts sink to the bottom of my mind as deeper thoughts and feelings rise to the surface. These voices speak in hushed tones; they whisper no matter how urgent their message. The cities we live in are noisy places. Perhaps it has always been so. The braying of donkeys is as loud as the honking of cars. Ancient Rome must have been just as noisy as modern Beirut. Noise imposed on us is harder to take. We defensively shut it out as we retreat into cocoons of music clamped to our ears.

Sound turns to noise when it becomes disruptive, but sound itself can be soothing. Music of the right kind may be a perfect companion to solitude. I used to spend the last day on the island in the fall by listening to Haydn's *The Creation* oratorio while walking at the shore. But there are times when even the most sublime music disrupts my inner peace. Yet the sounds of nature are never disruptive. The murmur of the waves, the whispering of the wind, and the song of birds is never noisy; even the howling wind and the screeching gulls do not trouble me. A silent nature would echo the silence of death.

Solitude and silence must be chosen freely. Enforced solitude and silence are instruments of torture. It is the same for loneliness. I have been lonely and I have experienced solitude, so I know the difference. Being alone need not mean being lonely; and one can be lonely in a crowd. Loneliness is being unneeded and unwanted, being trapped in an aching void with longings that cannot be fulfilled. Solitude is different. It allows an intensity of thought and feeling one cannot experience with others. Reading becomes an intensified encounter with the absent author. Writing flows freely. Music is filtered of noise. Food tastes good no matter how simple the fare, and wine is savored without pretense.

Figure 12.1 Solitude and silence.

Solitude holds a mirror for me to see myself as I am. It opens windows into the innermost recesses of my soul. It makes the examined life worth living. Solitude and silence provide a respite from our cares. The injured body cannot heal if it is not allowed to rest. The injured soul also needs to rest through the peace of solitude. Distance from others is essential if we are going to get closer to them. I need to balance my need for solitude with the necessity of engagement with those I love. My longing for being alone does not mean running away from my family or abdicating my responsibilities to others. It makes me a better person—a better son, father, husband, and friend.

That is why for most of us solitude cannot be permanent. It cannot sustain itself indefinitely. Even the Christian hermits in the Egyptian desert came together in coenobitic communities of kindred souls. We need to be with those we love and even those we do not love, in order to live fulfilled lives. Solitude must be embedded in a life of action if it is not to lead to stagnation and atrophy. Yet without solitude an active life becomes a race to nowhere—a frantic journey with its attendant woes but none of its joys.

I wondered if others too have written on solitude and silence. Years ago, I had read Thoreau's *Walden* as part of my forlorn quest to read great literature. But it went over my head and I couldn't understand what made it a great book. Now I discovered that there is a whole literature on solitude and silence that spans many centuries and deals with their every conceivable aspect. As I read some of these books, I was struck by the similarities between my experiences with solitude and silence and what others had described in more systematic detail.[1] I was struck by the variety of ways solitude and silence may be experienced and pursued: to get closer to God, to flee from the world, to deepen self-knowledge, to unleash creativity, and to retreat into nature for respite and restoration. I came to understand why our island in Finland had come to occupy such an important place in my life.

~

The island of Bodö, where we have our summer place, is part of an archipelago located thirty miles southeast of Helsinki. No one lives in the forests of Bodö's interior. The houses sprinkled around its shores are mostly summer residences; only two local families live on it year round. Bodö (*ö* means island in Swedish) is one of the islands clustered around the larger community of Pörtö (Pirttisaari in Finnish).[2] Some two dozen houses surround its bay, which serves as its magnificent harbor, and constitute its small village. People have lived in the area since the sixteenth century, but the hamlet of Pörtö was established in the nineteenth century by ship pilots who maintained the lighthouse of Söderskär, at the outer edge of the archipelago, to steer ships through treacherous shoals on their way to St. Petersburg. When the lighthouse became automated, the pilots and then their sons became fishermen and builders and caretakers of boats.

There are no paved roads or cars on these islands. Footpaths crisscross the forest, and a few of them have become narrow dirt roads. Transportation is primarily by boat and over the ice when the sea is frozen. The main connection to the mainland is at the pier at Kalkstrand, which sits in the shadow of a limestone quarry. Bus service connects it to Helsinki and other locations in the area. Those who have larger boats can get easily to the mainland. Our open boat, *Sally Walker*, can also make it to Kalkstrand on calm days. It is seaworthy enough to handle rougher seas, but it would be a jumpy and wet ride. Our family has had a long-standing romance with *Sally Walker*. Nina originally came up with the name to mean "helper" and used it as a name for Kai for doing small chores. In the family vocabulary, the word turned to a verb ("I need some sally-walkering in the kitchen"). Stina then gave the name to the boat.

Sally Walker is a sixteen-foot prototype of a fishing boat from western Finland. We found it through a friend at a small boatyard in Turku. Since it never went into production, we have never seen another one like it. It came with a sluggish seven horsepower diesel engine, which required constant attention like an ailing relative. We put in a more efficient engine, and now it goes at a stately, ladylike pace of five knots per hour. *Sally Walker* can also be rigged with a wine-colored square sail, though we rarely do that anymore. Over the years, good old *Sally* has suffered quite a bit of abuse at our hands, as testified by several propellers bent out of shape in encounters with underwater rocks that always come up the winner. Many admirers have lusted after *Sally*. Her elegant lines, green exterior, brown interior, and teak trim endow *Sally* with more character than the fancier boats and the pedestrian aluminum tubs that run about frantically. *Sally Walker* has become like a family member, and we would not dream of parting with her.

For many years there was a small family-run grocery store on Pörtö, but it closed down when old Einar and his wife retired. Their son Lasse went into salmon fishing and occasionally serves as a water-taxi to take people to Kalkstrand. During the summer season, there is now *Christina*, a shop-boat outfitted like a grocery store. It brings in most of the provisions we need three times a week to the Bodö jetty. At the end of the summer season, one orders food from a supermarket on the mainland that is delivered weekly to the same jetty along with the post. Going to the shop-boat is one of the highlights of the week. *Christina* is often late, but waiting for her gives time to socialize, like the gatherings after a church service. The parting reward is an ice cream cone, which we savor before lugging the food home through the forest or, if weather permits, with *Sally Walker*. Island happiness is largely made up of such small pleasures.

There are coffee get-togethers in the old schoolhouse in the village on Saturday mornings during the high season. Two bigger lunch occasions provide a spectacular sea food buffet and a grill lunch. In September there is the fall market, when people sell homemade jams, smoked salmon, and an exquisite dark bread. There are woolen hats with tassels knit in a distinctive style, attractive jewelry, various knickknacks, and used books. The special treat is the salmon soup drenched with butter and cream. These events are organized by the indefatigable Baba (most people have nicknames), an amazingly energetic and enterprising woman who defies the years. The last event is a sit-down dinner at the schoolhouse near the end of October, which comes too late for us to attend.

The biggest event of the season is the annual *Sommarfest*. A three-man band from the mainland plays polkas, waltzes, and an occasional tango. After the older folks have left, swing and rock take over. Children run about, young people hang out, and older adults stand around talking to one another. In earlier times, these dances provided the opportunity for young people to meet. This still happens occasionally (that is where Kai met Anni). An important cultural event that was initiated by Stina brings in writers and musicians for an afternoon of presentations to a packed audience in the old schoolhouse.

Further out at sea, there are a dozen or so uninhabited skerries scattered around the island of Söderskär, on which the lighthouse still stands. These barren islands serve as nature preserves for migratory birds and are out of bounds until the middle of August. Their smooth granite surfaces, ranging from elephant gray to reddish brown, impart them with individual characters. Kokkomaa is a massive mound of granite slabs smoothed by the Ice Age and mostly free of vegetation. (A magnificent photograph taken by Anni's father, Christian Westerback, hangs in our living room at Stanford.) Inner and Outer Kittelskär are treasure troves of smooth pebbles and driftwood of every description washed ashore from the open sea. There is Jussinkari, where I found the massive tree trunk that like a totem pole now guards our houses. Every outing and picnic to these islands is memorable. I've spent many hours roaming around looking for unusually shaped rocks that have become my version of Chinese philosopher's stones. I rummage for driftwood and other cast-off items, which I put together into simple *assemblages* that have become the main vehicles of my modest creative talents.

~

Islands have occupied a special place in people's imagination. As vacation spots for rest and relaxation they have a distinctive character quite different from lakeside resorts or mountain retreats.[3] The attraction to forested land on promontories overlooking water may even have evolutionary roots. All this was very new to me. There were no islands to speak of off Beirut. My interest in islands started with Stina talking about the island in Finland when we were in Beirut. After I first saw it, I felt out of place, like a palm tree in the arctic. Then I fell in love with it.

Life on the island was very different from summering in the mountain resorts of Lebanon and I had to get used to it. I had had a sauna before, but this ritual now became an important part of my summer life. I came to enjoy the *löyly*, when water thrown on the hot stones of the wood stove produced a hissing steam that tingles your skin and goes into the recesses of your lungs. Sitting naked in the sweltering heat with family and friends created a special bond. Dipping into the frigid sea took some getting used to. I also began to take morning dips, which were tougher. The exposure to extremes of heat and cold led to a sense of exhilaration followed by a profound sense of relaxation. I found the taste of raw fish strange at first, but over time I came to love *senapssill* (herring in mustard sauce) with small new potatoes, and *gravlax* on toast, especially when fortified with chilled vodka (Absolut Citron, which I liked most until I discovered the Russian Beluga). I never acquired the Finnish passion for crayfish (a late-season delicacy), which required too much work for too small morsels. On the other hand I thought Finnish bread and sausages to be the best in the world.

Over the years I became adapted enough to feel more or less like a native, at least as a "summer guest." The island was to become the one place where I would feel most at home, especially when alone, with no one to remind me of who I was. When people used to live and die in the same place, as many people still do in many parts of the world, they had a clearer sense of place, which anchored their existence, framed their experience, and defined their identity. My grandparents' generation must have had such sense of place, which became diluted in my parents' generation and dissipated in mine. This, of course, is not peculiar to Armenians. Increasing mobility and the accelerated pace of change have led many people in the Western world to lose their sense of place, be it by choice or by force of circumstance.[4]

The island provided me with a special, albeit idealized, sense of place. Even though my presence on the island was seasonal and disconnected, the absences in between shrank as successive summers became linked in a continuous chain. Time became telescoped as we arrived in the late spring and it soon felt as if we had never left the previous fall. The seasonal associations with spring and fall became part of the experience of arriving and departing: spring, full of new life and anticipation, fall, of closing down and nostalgia. The fact that nature around us was undergoing its own transformations made us feel part of the greater cycle of life unfolding around us. When we arrived in the spring, the birds would be singing in a frenzy. In the fall, they fell silent and headed south. The birches turned golden and the maples blushed in envy; the heather bloomed in subdued tones and the bright

Figure 12.2 The island house, with my study at the right end.

red berries of the rowan trees glowed in the autumn sun. Toward the end of the stay, a chilly bite in the air enhanced the sweet sorrow of parting.

The sense of closeness to the island has become so deep in me that I hope to die there, or at least have my ashes scattered around a pyramidal rock that juts out of the sea off our shore. That rock will be my tombstone. I sometimes stop by it when canoeing in a calm sea and rest my hand at its top. There is nothing morbid about it. The prospect of ending there fills me with a sense of profound peace and joy. My remains will find their eternal rest there with the fresh scent of the sea and the sound of the waves while my soul soars free of the burden of my body.

~

Stina's association with Pörtö goes back to her maternal grandfather. Frans Blomqvist was an engineer and a methodical man looking for a place where his family could spend the summers. He got himself a sea chart and set out one morning in a row boat from Helsinki through the archipelago. He found Pörtö, where he managed to rent a couple of rooms from a local family. Years later, Stina's parents continued to summer there at various locations. Finally, her father rented some land on the nearby island of Bodö and set up two wooden huts to live in during the summer. These were yurt-shaped octagonal structures made up of wooden panels bolted together and surmounted by a conical roof. Originally, a wood stove in the center kept it warm. Lale had bought them from the Finnish army where they were used to house soldiers in the field during the winter.

Eventually, we were able to buy the land. We first built a sauna with a guest room where we lived for one summer. Then we had the main house built—a large log cabin—consisting of two bedrooms, a large living room and kitchen, with an attached room that became my study. We substantially renovated the round huts as well as another structure that Lale had built to store his tools. For many years we had no electricity or running water. But the main house is now heated by a wooden stove, fireplaces, and electric heaters. A pump installed into a drilled well now distributes water to the various structures and to the hot water heaters installed in the sauna and the kitchen. Other improvements have made island life increasingly easier and more comfortable, but we drew the line at getting a dishwasher or a washer and dryer. The dishes and the laundry are still done by hand and the clothes are hung out on lines to dry. Over the years, our island houses became attractive spaces for us to live in, as much as our house at Stanford. This was immensely important for Stina and, as it turned out, for the rest of our family as well. The island came to occupy a central place in each of our worlds and a wonderful legacy to leave to our children.

The part of the island where we live essentially consists of a massive slab of red and gray granite sloping gently to the sea. Its surface has been smoothed by the retreating ice sheets at the end of the last Ice Age, ten thousand years ago, leaving massive boulders stranded in its wake. A dense evergreen forest of pine, spruce, and juniper, interspersed with alder, rowan, birch, and maple trees, covers the land. The soil between the rocks is filled with blueberry and lingonberry bushes. Patches of moss cling to the trees, and a carpet of lichen in exquisite colors turns rock surfaces into a painter's palette. Here and there, a small bedraggled pine rises heroically out of the sheer rock, with its gnarled branches twisted in defiance of the relentless wind.

When I described all this to my father, none of it made much sense to him. People in Lebanon built villas in the mountains with carefully tended gardens surrounded by orchards, vineyards, and olive groves. Some of our American friends also found it difficult to visualize the island, except for those who were familiar with Maine's coastline. I loved the pine-covered mountains of Lebanon, which cascaded down to the sea, as I came to love the rugged California coast. But both of these landscapes lacked the intimacy of the Baltic sea that embraces the island. The deep blue Mediterranean was enticing like a Siren's song, and the Pacific Ocean awesome in its majesty, but I could only love them from a distance. The sea surrounding the island has a more immediate connection to my life that makes it feel much more intimate. Its cold waters are bracing, their moods subtle and changeable. The island faces south, with nothing between us and the horizon, to which the sea stretches free and unbounded. There are no tides in the Baltic, but the sea may rise or fall by several feet depending on the currents flowing through the Danish straits. This makes the movements of the sea unpredictable and interesting. One of my father-in-law's contraptions was a wooden box designed to measure the level of water, but he had made the hole in the graduated column too large and it went up and down, buffeted by the sea like an occult instrument gone berserk.

Standing on the shore, I watch the waves and try to discern their sounds. On calm days they undulate seductively toward me with barely a whisper. Pushed by winds, they speak in conversational tones. When they angrily crash on the shore, I stand with my head hanging low, taking in their reprimand. The roar of the sea in the background sounds like an orchestra playing Wagner, with the pounding of the tympani. The sound of the wind coming through the pine forest has the same clear whistling quality I remember from the pine forests of Lebanon. But when the birches and leafy trees become agitated, the sound is more muffled and confused. During a particularly spectacular thunderstorm the sea becomes covered with a white lace of foaming waves. Churned by Neptune's trident into a rage, sweeping whitecaps charge in a fury onto the granite shore, which does not blink an eye. The howling wind shakes the tall trees like rag dolls. Their trunks sway on the verge of snapping like matchsticks while their branches shake their leaves and pine needles into a helpless frenzy.

When we bought *Sally Walker*, our shore had no natural harbor to shelter it from southerly storms. Our distant neighbor Nisse Junasson, who stores boats for the winter, sold us a barge that he loaded with stones and sank it to the bottom to act as a shore break. That, however, was not the end. Huge rocks had to be piled on top of the barge to secure it against high seas and the pounding of the ice. The man who did this was as a taciturn and soft-spoken fellow with an amazing mastery of the giant forklift with which he hoisted up huge boulders from the bed of the sea and deposited them on the barge like an antique dealer handling an *objet d'art*. When every last rock was finally in place, he told me solemnly, "Now you have all rocks you need." He had produced a work of art and a guardian for our harbor for the ages to come.

On late summer evenings the sea is lulled into a glassy calm. Its surface glistens like a huge vat of oil gleaming in myriad shades of purple in the dying rays of the sun. Sometimes, a brilliant rainbow rises majestically from the sea and curves across the sky in a perfect arc. In the western sky, the horizon lights up with majestic brilliance. The light of Provence is deservedly celebrated, but the slanting rays of the North have their own allure, especially in the fall. I never realized how many shades of gray there were and how lovely they could be, how many bands of blue could be reflected by the sea, and how exuberantly its surface could sparkle under the sun. The sky and the sea engage in a seductive dance in this magical light. They match blue against blue and gray against gray, and on misty days they merge and recede to the horizon in a distant embrace.

Clouds are an intimate part of this romance between sea and sky. The low-lying cumulus clouds billow out into massive white balls, or break up into a field of cloudlets like a flock of sheep walking across the sky. The stratus clouds take no part in this drama. They just sit there covering the sky like a dull gray woolen blanket. Higher up, the wispy cirrus trace their delicate fingers across the sky unconcerned with what passes on below. One can see the same clouds in California, but I never looked at them until I realized that becoming fixated on the island should not blind me to what was also beautiful elsewhere.[5]

Light and time form another intimate pair on the island. Days and nights get respectively shorter and longer with the seasons as they do elsewhere, but the rhythm of change is more dramatic in the North. Daylight lingers on endlessly in the early spring, with the white night never going entirely dark. In the fall the days get rapidly shorter and darkness rushes in to take its place. On a clear moonless night, the pitch-black sky is lit up with countless stars, with the Milky Way spurting out like milk from Hera's breast. The dazzling brilliance of the full moon pours out in an immense river of light that turns the sea into a field of shimmering diamonds.

Some of the ineffably lovely sights are the most evanescent. In the fall, mushrooms in exquisite grays, reds, and yellows appear overnight, only to decay and die in a few days. Butterflies flit over the rock gardens lodged between the crevices of the granite; the lemon-yellow *Rhumni gonepteri*, the many-toned blue *Polyommatus* and, most enchanting of all, the *Vanessa io*—the peacock butterfly, with its deep brown wings and elegant circles at its tips touched with royal purple. They too are soon gone. So much beauty in such achingly short lives.

I had always looked at birds and flowers from a distance, but now I lived in their midst. Native flowers like the wild pansy, the fireweed, the stonecrop, and the twinflower (*Linnaea borealis*—the favorite flower of Linnaeus) surrounded me. Over the years, Lale had planted scores of wildflowers from all over Scandinavia and tended them lovingly. After he died, we could no longer sustain them, but we were grateful for the gratuitous gift of one or another flower blooming spontaneously.

I never became a serious birder like Lale, but I came to watch the island birds with fascination—especially the gulls, terns, eiders, and mergansers that were lorded over by a pair of arrogant swans. One year a red-breasted goose made a sensational appearance with its striking chestnut throat, breast, and face patch. From its habitat on the shores of the Caspian Sea, it had strayed far from home—just like me. Chaffinches, wagtails, and great tits filled the trees with birdsong. With the approach of autumn they would flit in and out of trees frenetically getting ready for the departure to their winter grounds. The silence they left behind was a reminder that we too must not overstay our welcome.

For several years a lame gull we called Limpy (even when it no longer limped) had been regularly showing up at dinnertime hoping for handouts. It had a way of making itself look submissive and helpless without losing its dignity. Even if it was all contrived, it was hard not to get up from the table and throw it a few scraps. (Stina would occasionally offer it food from the fridge while speaking to it tenderly in Swedish.) A more aggressive younger gull (which Nina nicknamed Skimpy) used to take the food away from Limpy, but after we started chasing it away it figured out it was better to make peace and share the food with our friend. When on occasion we put out larger quantities of leftovers, a dozen gulls would swoop down within seconds. I couldn't understand where they came from so quickly. And why did the gull who got to the food first emit shrill calls that attracted the other gulls, only to fight them off? Why not keep quiet and eat everything in peace by oneself? I suppose one has to be a gull to understand it.

On our kayaking excursions, we would occasionally see a sea eagle soaring in the sky or a pair of osprey, their screeching cries unworthy of their majesty. The most glorious sight of all was that of cranes heading south in V-formations in the early autumn. Their distant calls would fill me with the same nostalgia that I felt in my youth when the autumn fog enveloped the pine forests in the Lebanese mountains as I walked through them, lost in my daydreams.

~

When we built our summer house, I could have hardly known the central role the island would occupy in our lives. It made little sense for us to own a summer place halfway across the globe. We could more easily have found a spot on the Pacific Ocean or in the Sierras that would be accessible year round. Yet, the island provided our family with a space and an experience far more precious than simply having a summer retreat. It made it possible for Stina to return every year to a place she loved, to spend time with her parents and her sister, and keep up with her friends—all of which proved important for her personally and professionally. The island proved equally important for me and our children, in our own ways.

My time on the island was mostly taken up by my work and family. I had relatively little opportunity left for leisure and solitude, unless

I managed to spend a week or two on my own. The burden of taking care of the family primarily fell on Stina, especially when the children were young. This was also true for attending to guests, who came from Helsinki and elsewhere in Europe or the United States. Every morsel of food had to be brought in from the outside and cooked at home; there were no restaurants or places for take-out food to make life easier. When guests came from long distances they stayed with us for some time. It was a good of way of spending time with them. We also enjoyed having our children's friends because that gave us a chance to get to know more of their world.

My contribution to household maintenance ranged from minor fixing jobs to tarring the roof. There were no handymen available except for more major projects. I helped by going to the shop-boat, grilling fish and meat, washing dishes, and doing the laundry. I started by hardly knowing how to hammer a nail but learned quite a bit over time from Lale, from a friend who is a cabinet-maker, and through trial and error (mostly error). One task that I was most happy to do was chopping wood. I particularly enjoyed doing it on brisk chilly days and savored the smell of the freshly cut wood. With two fireplaces, a wood stove, and the sauna, there was a constant need for firewood. I went through several motor saws and almost lost a finger to one of them. I nailed the leather glove with the mangled index finger onto the woodshed as a warning. I put another mishap to good use. When I was splitting wood without goggles, because I was using only the small ax, a fragment hit my left upper eyelid. I realized I could have lost an eye. From then on, whenever I found myself fretting over a trivial matter, I would close my left eye for a while to give myself a sense of what it would have been like to have lost an eye, and that put my worries in perspective.

Fishing started as fun but became tedious. Casting caught very few fish and laying a net was a hassle, especially when the net got clogged with seaweed. Fishing for me was never a sport, and the idea of catch-and-release never made sense. I would have been just as happy if someone gave me the fish or let me buy it. I never fully overcame my aversion to killing fish or cleaning them. On one occasion, when alone on the island, I caught eight perch in the net. I needed the food, so I pan-fried one and made soup with another. I kept the rest alive in the

Figure 12.3 Happiness.

fish box at the shore. The next evening, when having dinner, I began to think about the captive fish. Did I really need to keep all of them? No, I did not. I went to the shore and released half of them. Then I began to wonder how the remaining fish felt being imprisoned in their own element. Did they have parents and children that would miss them? The question weighed on my conscience, so I let go the rest of them also. It meant eating sandwiches for the next three days before I could go into town and replenish my stores, but my inner peace was no longer ruffled by guilt.

I particularly enjoyed the time I spent with our children. Little Nina delighted me with her imagination and creativity. She thought up names for special rocks. There was *Elimeni*, a striking block of red granite with yellow lichen. Another large smooth rock shaped like a platform presided over the *Lejonplats* (the "lion's place"). Nina drew wonderful pictures and created works of art with little objects one would have hardly noticed. She was fascinated by nature and learned much about it from Lale and Nunni. When she was no more than three or four, we came across some inedible red berries that looked to me exceptionally luscious. As I was about to pick them, Nina piped up, "Daddy, I don't think you can eat these."

I was fascinated by Kai's single-minded dedication to all things associated with the sea. Propellers were the rage for a while. He would churn the sand with his little fingers to imitate their action. He would spend hours "sailing" small wooden boats that he pulled around attached to a stick with a string. When younger, he wouldn't go to bed until I carried him out on the porch, where he could see the blinking of the distant lighthouse at Kalbådagrund. The ferries that came from Helsinki fascinated him; he was usually the first (and cared the most) to identify them in the distance: *Standard*, *Natalia*, *Terhi*, *Katarina*, *King*. He befriended the captain, who would take him into the cabin, where he could hold the steering wheel and move the controls. One day, when I was taking the ferry to go to Helsinki for a few days, I hugged Kai to say goodbye and he started to cry ("Daddy come back quickly"). To cheer him up, I said, "Kai, don't cry. You're a captain." He looked at me through his tears and said, "I'm not a captain; I'm a little boy."

At the end of one summer, when I was alone, I made for Kai a replica of *Standard*. Taking precious time away from my writing, I spent many hours on it. A few days before leaving the island to return home, I ran out of the special wood I needed to make the last bench. I spent a whole day going into town to get it. I left the finished *Standard* with a note at a conspicuous place where Kai could easily see it when he came back the next summer. I had a strange thought that I might die and would never know about Kai's reaction to finding my labor of love.

Nina and Kai would curl up in my lap to listen to stories I read from children's books that I had never seen when I was a child. Through their eyes, I tried to capture a childhood I never had. Most of these books were in Swedish, including the famous *Moomin* characters of Tove Jansson and the adventures of *Pippi Longstocking* by Astrid Lindgren. The only book in English that I recall was *Goodnight Moon* by Margaret Wise Brown given to us by visiting friends. The island was an idyllic setting for Nina and Kai to spend their summers. With few other playmates around, they created their own world with a collection of small plastic Playbigs and Playmobils. These little figures evolved into two families of *mänskors* ("people"). They were called the Båtsmans and the Sjöbloms, and their distinctive language was some variety of heavily accented English. Their rituals included a big wedding celebrated between the two families. Characters such as Matti and Cook became like members of our own family. The *mänskors* flew back and forth between Finland and California, where their activities were limited mainly to excursions to the ocean. On one outing, Matti and Timo in a small toy boat drifted out. Kai went after them into the surf until Nina frantically called and pulled him back. Matti and Timo "drowned." It was a sad day. Stina bought replacements for them, and Nina and Kai staged a ceremony where they were "recarcassed," which transferred their souls into their new bodies. It was an elaborate ritual with chanting and speeches.

~

These were wonderful times for me as well even though my reading and writing got in the way. My wife and children were used to my working hard, but being on the island on vacation made it harder. I was always working on one book or another. As the pressure of teaching and administration built up, I came to rely increasingly on my time on the island to catch up on my writing. Until I learned, late in life, how to type and got dragged into the world of word processors, I dictated or wrote everything by hand in pencil. I shipped ahead the books and materials I thought I would need, but I had to go to the library in Helsinki if I needed additional sources. It took the whole day and I rarely found what I needed.

This is his how the fourth and fifth editions of *Fundamentals*, the two volumes of the Cohort Study, and my book on guilt were written. The last required extensive reading in religion and philosophy, which I found both fascinating and overwhelming. These were topics that I had been yearning to learn more about since my brief taste in my sophomore-year course on Western civilization. For moral philosophy, I intended to read only about Aristotle, Kant, J. S. Mill, and Nietzsche. However, once I started reading Bertrand Russell's *History of Western Philosophy*, I couldn't put the book down until I went through its 778 pages (and the index). My exhilaration comes through in the following entry in my island diary: "Another glorious day. I feel transported to another realm of life. I finished Russell's section on Catholic philosophy. It feels like water on parched land. I have soaked it all up. Dates, events, people are falling into place. Astounded by my ignorance and at having reached where I have despite it. I grieve over the wasted years—throughout my entire education, during the year in Washington, and the three years of being a University Fellow. I have been trading in trivia all these years; albeit not without profit. I have never felt so enriched and so free. And it is not just the surroundings—it is me."

There was a parallel process of reading and study in connection with my lectures for the Stanford Travel Study trips. The bulk of that also took place on the island. The readings were especially enjoyable since they covered a part of the world where I grew up without knowing much about its history, art, and culture. This also provided the chance to read some of the great classics in their entirety: Homer and Virgil; books on Byzantium and Islam as well as books on Iran, Egypt, Tunisia, Morocco, Mongolia, and Uzbekistan. Had I been doing this reading at leisure, it would have been an unalloyed pleasure; piled on top of my reading for work, it felt like a Roman feast of overindulgence.

Almost all of my reading was nonfiction. I had had a taste of reading novels during my childhood illness, but both my character and my careerist aspirations had pushed me into a utilitarian frame of mind. Facts were what mattered; reality won over imagination. Stina urged me to read for pleasure, which I initially resisted. Reading about ancient Rome, I argued, was just as pleasurable. I would read fiction after finishing the book I was writing. And even if I were to read something for fun, there would still have to be a plan and purpose to it; otherwise it was a waste of time. I also lacked the sense of boredom that motivates some people to read fiction. I never felt bored with myself, and my rich fantasy life was a good substitute for the literary fantasies of novelists. When I finally relented and began to read more fiction, I became totally absorbed in it. However, my reading tended to be limited to the "great works" of fiction. Why should I read novels by Tom, Dick, and Harry before I had read all of Dostoevsky and Tolstoy? However, once hooked, I would read everything by an author like E. M. Forster or C. P. Snow, finally delving into Anthony Trollope and Henry James. Alas, I discovered poetry too late in life and never fully caught on to it. I realize this may reflect poorly on the sort of person I am.

Figure 12.4 The passage of time on the island. (Top photograph by Kaius Hedenström.)

The impact of the island on our children's lives went beyond their childhood. We got the first windsurfing board for Kai when he was eleven, and windsurfing became a lifelong passion that led to his becoming a professional windsurfer and eventually going into the windsurfing business. Even as a child, he would spend hours in the frigid waters in the flimsiest of wetsuits teaching himself how to water-start. For a while, he could sail downwind but couldn't tack against the wind, so we had to get him back with *Sally Walker* to have him start over again.

The island also provided a free range for Nina's creative talents. As she became an accomplished conceptual artist, many of her works were created on the island or derived from it. One of the central themes of her work was based on interventions in nature, such as "repairing" torn spider webs with red thread (which the spider would then come and throw out). She used an egg and tadpoles to create *Artificial Insemination*, a work of particular imaginativeness. It was also a private joke, a parody by the daughter of the sex professor on the reproductive process.[6] She had enough works on these themes to qualify for a special show in Finland.

The island also provided Nina and Kai with a special opportunity to get to know their maternal grandparents, as well as their aunt Maj and her children and then grandchildren. Stina had always spoken Swedish to our children, and the exposure to Finland greatly expanded their cultural horizons. The island, more than our home on the Stanford campus, was to provide a sense of continuity to our family life. After Nina and Kai had left home, we were to spend more time together on the island than anywhere else. The island became the closest thing to an ancestral home they would ever have.

We spent over twenty summers with Stina's parents. They were delighted to be with their grandchildren and developed a special relationship with them. I watched with fascination the different ways in which Stina's mother and my mother interacted with the children. Part of the difference was due to the settings in which they interacted—the island in one case and my mother's house in Palo Alto in the other. Stina's parents were more didactic; they taught the children a great deal about the island's flora, fauna, and lore. My mother and Lucine were more emotionally engaged and indulgent, doing just about anything the children wanted (including playing "football" with a balloon and my mother as quarterback). Both sets of grandparents loved their grandchildren dearly, but they expressed their love differently. When Nunni kissed the children, which she did only occasionally, it was a discrete peck on the cheek; when my mother kissed them, which she did at every opportunity, she was all over them.

Nunni and Lale had tamed the island in unobtrusive and subtle ways. I marveled at how they had managed to move the rocks that served as foundations for the cottages or constructed steps. I started my own rock-moving project by building a walkway across the shallow bay to the promontory jutting into the sea. Even though I managed to line up the rocks in place with the aid of a crowbar, the ice dislodged my bridge every winter. Finally I gave up and learned to live with way the ice arranged them in place and they no longer moved about.

Lale was a fine naturalist with a vast knowledge of the vegetation and wildlife of the island. He was obsessed with the weather and had his own little meteorological station that recorded the changes in atmospheric pressure and measured the wind and the rainfall. He listened to every weather broadcast, even though they mostly repeated themselves, and then he told us what sort of weather they "promised." Since I grew up in a highly predictable climate, I never worried about whether the sun would shine or rain would fall. I could see it when it happened. It took me some time to learn that you needed to know what was coming before setting out with a boat in the Baltic. I was particularly eager to learn from Lale's handyman skills. He was exceptionally meticulous, and some of his constructions had a Rube Goldberg quality about them. For instance, he had repaired an old plastic lid, 10 by 8 inches, by putting in twenty-nine tiny brass screws, transforming it into a marvel of resourcefulness, a work of art, and a monument to compulsivity. He had carefully stashed away nails and screws of every conceivable size, most of which he would never use. I inherited these and added my own contributions to them.

Like Americans who lived through the Depression, Nunni and Lale had been marked by the economic privations of World War II. To the astonishment and admiration of the world, Finland fought against the formidable Soviet Union as well as Nazi Germany and survived as an independent nation, albeit at an exorbitant cost. When Lale was at the front, Nunni and the children were part of the exodus of Finnish refugees to the north of Finland and then to Sweden. Stina has provided a compelling account of these times in a book based on her parents' correspondence as I referred to earlier. The war experience reinforced the parsimonious and conservative elements in Nunni's and Lale's characters. Even though we owned the property, we deferred to their wishes in managing it. Nonetheless, the contrast with my seemingly profligate ways caused some tension. Lale was loath to have any trees cut down because they provided a barrier against the wind. I wanted open vistas to see the horizon. In deference to him, we settled for looking at the sea through trees. After Lale and Nunni were no longer with us, we tentatively and guiltily cut down a few branches from the spruces that formed a dense green curtain in front of the cottage where they lived. That broke the dam, and we cut more and more trees to open up a fabulous view of the sea. Lale laboriously cut wood for their stove using a saw and a small ax; I went at it with a power saw and a massive ax. I saved his old tools and used power tools instead. I suppose our children will also do things differently when we are gone.

My sister-in-law, Maj, was another important figure in our Finland world. Six years older than Stina, she was a well-known fashion designer who specialized in work clothes with her partner, Pi Sarpaneva. They helped spark my interest in the exquisite world of Finnish design and added another strand to my adulation of Finnish culture. Maj's flaming fox-red hair matched the flamboyant and dramatic aspects of her character (which turned an ordinary downpour into a "terrible storm"). She was fluent in English, but when she couldn't think of a word she made up an imaginative substitute for it. Thus a shepherd became a "sheep's waiter." Caviar turned into "unborn Russian fish babies." After her divorce, we saw much more of Maj with her companion Kaius Hedenström, a well-known press photographer who became like a member of our extended family. A fun-loving and generous soul, Kaius left a rich legacy of photographs of our family, particularly of our children.

We have watched with satisfaction Maj's three sons—Jan ("Snippe"), Kaj ("Utti"), and Mats ("Matti")—grow up and become established in their careers and form their own families. Their wives—Tuula ("Tupu"), Ira, and Sonja—with their respective professions, have been wonderful additions to the family. With Maj's nine grandchildren, we now have another generation to further enrich our lives. The latest additions to the family circle in Finland are our daughter-in-law Anni, her parents Christian ("Nista") and Irmeli ("Immu") Westerback, and her sisters Frida and Emmy and her family. In addition to our oldest friends in Finland, the Björkenheims, we now also have Thomas ("Bebo") and Satu Huber as part of this close group. Many of Stina's cousins and friends also became my friends. Her cousin Teddy, who plays the clarinet, his son Emil, who is an accomplished pianist, and I play trios together. The late Poppy Berghem, Birgitta ("Bitte") Bought, and their children have become like family. Prolific authors and intellectuals like Bitte and Merete Mazzarella, and of course Stina, have helped me gain a deeper understanding of the cultural and literary life of Finland. I now have as many relatives and good friends in Finland as I have in the rest of the world.

The summer of 1976 began uneventfully, but Stina's parents had to go back into town because of her father's worsening heart condition. Nina and I had started a project to make a list of all the wildflowers on the island planted by her grandfather. The list was to be in English, Swedish, and Latin binomials. We got the list up to twenty names, but there was one rare flower that bloomed only on occasional summers and that we couldn't identify from books. We decided to wait until Lale would come back and tell us what it was. Lale was hospitalized for the last time, and on July 28 Stina received a call informing her that her father had died. He was seventy-eight. She had to break the news to the children. They had suspected from the tone of her voice on the phone that something was wrong. They stood silently until Nina said, "Who is now going to tell us the name of the flower?" It was their first confrontation with the death of someone they loved. Nunni lived for another eleven years and continued spending the summers with us on the island. Finally, on February 12, 1987, she also passed away, at the age of eighty-three, a few months after my own mother died.

~

I began this chapter with reflections on solitude and silence. Even though these periods of retreat were all too short, they provided a much needed respite from my increasingly hectic life. The diary entry from September 2, 1981, provides some sense of that experience: "What a blessed, blissful time these days have been. I am like a ship in dry-dock being overhauled inside and out. Stores are being replenished, damages repaired, weak points bolstered. Scrubbed and polished.... There cannot be many places better than this for blissful peace and quiet. I am in the study. The crackling of the fire and the muffled sound of the wind are the only sounds intruding into the silence."

Being away from my wife and children made me realize how much I loved them: "I feel so enormously full of love towards my family. I think of them as one big lovable cluster. I also feel a profound and boundless affection towards them individually; a depth of feeling I could not have imagined." How could I sustain this blissful state, or as I wrote, "How is all this lovely mood, the island spirit, going to hold up after I return home?" (Or as a friend put it, in another context, "How do you take the mountain into the city?")

I have tried to do that back home at Stanford by using my evening walks around our condominium I referred to earlier as a time for solitude and silence. To provide some structure to my reflections, I rely on a simpler version of St. Ignatius of Loyola's spiritual exercises (*Consciousness Examen*). If there are issues troubling me, I try to deal with them. If there are matters that I need to resolve, I try to sort them out. Mostly, I walk quietly, occasionally looking at the stars and listening to the silence of God. Rather than praying in the ordinary sense, I engage in a conversation with God. It is a monologue, but I'm not just talking to myself. [7]

The most important reward of the periods of island solitude has been in bridging the distance and alienation from God that I felt in the press of my daily life. The following entry in my diary conveys the sense of both joy and frustration in trying to achieve this: "Enormously grateful for all of the Lord's blessings yet deep down, a certain sense of futility and sadness.... I have tried to wipe the slate clean of past resentments, grudges, anger, disappointments and ill-feelings. Having forgiven, God has forgiven me to the same measure.... The Lord is often on my mind, yet the joyful closeness to Him is not always there. I feel more like an obedient servant doing my job than the child of a loving father. Nonetheless, I feel strong, sustained, invulnerable—

whatever happens to me my spirit will survive.... Strangely, my life over this week has been imbued with God in a subliminal way like the evanescent morning mist. No revelations. No dramatic sense of joy, no tears of remorse or purging forgiveness. Not even the sense of child-to-father intimacy. Rather, a calm awareness of God's presence around and in me. One evening at the shore, I felt so full of awe and gratitude that I stepped out of my sandals, so real was God's holy presence. (No, there was no burning bush and I didn't think I was Moses.) St. Thomas Aquinas wrote, 'Man's ultimate happiness does not consist in acts of moral virtue; it consists in the contemplation of God.' I now know what that means."

~

I wrote earlier about my faith in God in connection with my brush with death as a child and my psychological struggles during adolescence that became entangled with my religious beliefs. However, what I have revealed so far about my faith has been like a thin thread running through my account. Writing about it in greater depth is going to be difficult. However, since God has been the central concern of my life, I cannot fail to address it more fully. This may come as a surprise even to readers who know me well. My relationship with God has been a deeply private matter that I have shared with hardly anyone. When several good friends read a draft of this memoir, the two revelations that surprised them most were my adolescent struggles with obsessional problems and my belief in God. They seemed out of character, and I can readily understand why.

I have had to overcome my reticence to delve further into this aspect of my life even more than when writing about my psychological problems or my ex-wife. Such personal revelations are embarrassing because they entail public exposure of our vulnerable self, which is the psychological basis of shame. These concerns get deeper when it comes to God. Universities, where I have lived and worked all my adult life, are bastions of secular thought. In the eyes of many of my colleagues, belief in God would be antithetical to rationality. It puts believers in the same camp as those who thought that the world was flat or created fully formed in 4004 BCE, as claimed by Bishop Ussher in the seventeenth century. I have the further liability of having fellow psychiatrists ascribe my faith in God to unresolved psychological needs and conflicts that I should have understood and overcome long ago.

There is yet another consideration that goes beyond what people might think of me. To allow others into the inner recesses of my faith entails an intrusion into my exclusive relationship with God. The seventh-century Sufi mystic Rabi'a ended one of her prayers with "Kings have locked their doors, Lovers are with their beloved. And here I am alone with You."[8] It is this feeling of exclusive intimacy that is at stake. Equally compelling is Rabi'a's desire to reach God without intermediaries, no matter how revered they may be. Devout Muslim as she was, she could say, "O Prophet of God, Who is there that doesn't love you? But love of the Real has so pervaded me that there is no place in my heart for love or hatred of another."[9] If this is true for Islam, it would even be more so for Christianity, which has been burdened with more than its share of intermediaries.

However, I have also realized that one's relationship with God must be shared with his other children whether they acknowledge and love him or not. This obligation—at long last—to share my beliefs with others and to testify for my faith has been a key reason for my writing this memoir. I did not state that at the outset because it would have been out of context, or perhaps I delayed it until I could overcome my reticence. Since religious life should not be a solitary journey, it has been a source of abiding regret that as a Christian I never became an integral part of a church in my adult life. After my relationship with the Armenian Evangelical Church in Beirut soured during my divorce, I no longer went to church except on special occasions. After we came to Stanford, I attended Memorial Church on and off. Since it was close to my office, I would often stop by to contemplate and pray amid its Byzantine mosaics and the light streaming through its stained-glass windows. I became good friends with its successive deans, going back to Davie Napier and on to Robert Gregg, Ernle Young, and Scotty McLennan. However, given the nondenominational and by necessity diffuse ethos of Memorial Church, I never became actively engaged with it and I hardly ever saw anyone I could recognize as a fellow member of the faculty. I went to the Catholic service once, but when the officiating priest announced that Communion was only for

Catholics, I never went back. Despite my personal bond with God, I missed being part of a Christian community. "I feel like an unused resource," I wrote in my diary. "I give nothing and get nothing from the church."

~

My purpose here is to convey some sense of my private spiritual experience, not to explain or expound theological doctrines. I am neither a theologian nor a scholar of religious studies, and I have nothing to teach anyone. My very use of language may be problematic. I use terms like God, faith, and spiritual in more or less their conventional sense without trying to explain what *I* mean by them. A colleague and friend who read this chapter referred to my "experience of God" rather than God. That is exactly right. My concern here is with my personal experience and not with a more general or abstract notion of God. Moreover, these terms have many layers of meaning and heavy doctrinal baggage. Trying to define them gets complicated very fast, very soon and will go over my head. Theologians and philosophers may need to deal with these complexities, but there is no room for them in an account of a personal spiritual journey. I will also avoid phrases like "I am spiritual but not religious" (which sounds like alcohol-free wine). And even though there may be a good deal of mysticism in my beliefs, I don't think of myself as a mystic. Nor do I believe in do-it-yourself religious faith. I have actually gained a great deal from various sources in addition to my own reading of the Bible.[10]

My faith established in childhood was not the outcome of a rational process. My understanding and beliefs derived from the faith of my childhood changed and matured over the years, but the core convictions remained the same. My belief in God is an inner experience for which I feel solely responsible to myself. Unlike other thoughts and feelings, which I may need to explain or justify to others, my faith is what it means for me whatever else it might mean to others. Ultimately it is that faith that is at the core of my spiritual life and religious convictions. The account of my spiritual journey may be an example of what William James in his magisterial book calls the varieties of religious experience. But it is not just one more such experience; it is *my experience*, which must be understood on its own terms.

I don't know where my childhood faith came from. So far as I can tell, my parents didn't inculcate it in me; no one "put" it there. I have vague memories of saying my bedtime prayers, which somebody (probably Lucine) must have taught me, but as I wrote earlier, I believe it was God who reached out and touched me. When going through psychiatric training, I learned about the psychological, sociological, and anthropological explanations of why and how religious beliefs are culturally generated and adopted by people in their own lives for their own purposes. Freud claimed that God was a projection of our repressed childhood longings for a supreme and benevolent father who would care for and protect us. Fifty years earlier, the German philosopher Ludwig Feuerbach had already expounded the idea that God was the personification of core human values like justice, mercy, and love—essentially an idealized reflection of our own moral nature. Feuerbach's and Freud's ideas became the basis for a model of God as the projection of conscious and unconscious human needs—an idea that continues to have wide acceptance in secular thought.[11]

I had to consider seriously whether my faith too had originated in response to my own psychological needs and conflicts, of which I had plenty. My problematic relationship with my father provided a lot of grist for the mill, particularly in light of Freud's views on the matter. In one entry of my diary, where I express my disappointment in my expectations of God, I noted, "As a teenager, this is how I felt towards my father. The parallels are striking." I could easily see how my relationship to my father could have colored my thoughts and feelings about God. However, I couldn't convince myself that my more fundamental belief in God was itself an artifact of the same unconscious process. Perhaps this was a form of denial, but then perhaps it was not. It didn't ring true to me and I had to accept what I thought to be the truth in order to be true to myself. While I couldn't establish to everyone else's satisfaction that my faith was not my own creation, I satisfied myself that it was not. This is not to say that psychological problems didn't contaminate my beliefs; in that sense, we all create our own gods. I didn't think, however, that that was all I had done, and I was the person I had to live with.

One reason why I have not tried to inculcate my own beliefs in my children has been to avoid the prospect that being indoctrinated would rob their beliefs of personal authenticity. Moreover, since religious instruction had not been part of Stina's upbringing, I was unsure how she would feel about it. I fervently hoped that my children would find God on their own (or God would find them as he found me). I was troubled, nonetheless, that keeping my own core values out of my children's lives would leave them open to the values of others instead. I also missed seeing them kneel at their bed saying their evening prayers. I missed having my family worship together. Going to Memorial Church on Christmas Eve was the closest we ever came to it. I consoled myself that my children could perhaps learn from my example without my having to preach to them.

~

Perhaps the key issue is not only how my childhood faith originated, but why it endured for the rest of my life. The continuity between the faith of my childhood and adult years is captured in the following diary entry: "There is such satisfaction and gratitude in knowing that throughout all these years I have remained firmly attached to God's lifeline. There is no piety or religiosity in any of this. Just the secure tug of being connected to God. He is now more distant, awesome and mysterious than the God of my childhood, yet no less loving, caring and forgiving."

There are two visions of God in this statement. There is the distant, awesome, and mysterious vision of God that reflects his *transcendent* (L. "climbing over") nature, placing him beyond human experience, beyond the limitations of the natural world, beyond everyday reality, and which is ultimately unknowable. Then there is the *immanent* (L. "remaining within") God who is in the world, knowable and engaged with the world as creator, lord, and judge. If I try to delve more deeply into these distinctions they become more opaque instead of clearer. It is comforting that those far better versed in these matters are not entirely clear about them either. In any case, it is the loving, caring, and forgiving God that I primarily care about.[12]

As I have grown older, God has become more detached from particular religious conceptions and traditions that preoccupied me earlier. Although I am a Christian (one of many kinds), I will present my beliefs at a level of generality that can be more easily understood and shared by others, whatever their own religious persuasions may be. Moreover, Christianity has become stripped for me of many of its historical and doctrinal accretions that have encumbered it through the centuries. I have had to redefine and refine my own beliefs just as religious traditions have gone through revisions and redefinitions. Some ancient beliefs and traditions that may have made good sense in their own time may no longer do so. At this point, it would take a great deal of self-deception and intellectual dishonesty for me to hang on to every tittle of traditional doctrine.

Nonetheless, I remain conflicted over what to reject and what to continue to hold on to in my beliefs. I feel reluctant to relinquish outright core Christian beliefs and instead take refuge in my inability to understand them. I agree that there is a need to "demythologize" the Bible and that it is hard to accept the "whole package," as a friend and ordained minister put it. But I also fear that this may lead to a slippery slope that reduces fundamental beliefs to their skeletal shadows. If we strip scripture from all that defies a naturalistic explanation, will we end up with a mishmash of history and myth? Would that still guide and comfort us and offer hope for eternity? Or would God get thrown out with the bathwater?

In my halting attempts at revision and redefinition, the more mystical nature of the Eastern Orthodox Church provided me with more breathing room. The Bible in Armenian is called the *Breath* or *Spirit* of God (*Astvaztasounch*, as in L. *spiritus*) instead of the *Word* of God. That has helped me avoid the straitjacket of its literal interpretation. Understanding the historical contexts in which the scriptures developed over the centuries is also an antidote against doctrinal absolutism. By the same token, while I respect earnest agnosticism, I have no use for the arrogance of atheism.

There is a great deal in Christianity that I still don't understand and probably never will. And while writing my book on guilt, I learned to integrate into my beliefs certain aspects of other religious and philosophical traditions. From Judaism, I learned the importance of the Law and the Prophetic call for justice and mercy. Islam taught me submission to God's will and gratitude for his blessings. From Hinduism I learned the importance of intention and context in determining the morality of actions, as well as working with one's natural dispositions rather than fighting against them; from Buddhism, I learned the importance of compassion for all forms of life and detachment from the snares of the world; from Confucianism, the role of shame (which includes what we call guilt) in knowing how to be human and to alleviate suffering in the world. Equally important have been the secular views of moral philosophers, in particular, Aristotle's ethics of virtue (which I find especially appealing), Kant's ethics of duty, Mill's ethics of utilitarianism, and Nietzsche's critique of the excesses of Western morality. Instead of diluting the essence of my faith, these perspectives have purified, enriched, and strengthened it. These insights from other religion traditions strike me by their similarities more than by their differences, and in their commitment to some version of the Golden Rule.[13]

The critical boundary for me is between believers and non-believers—rather than between adherents of particular religions or sects. The fundamental choice is between a purely materialistic view of the world that can only be "seen" empirically, and a view that relies on faith as evidence for things "unseen," as St. Paul expressed it. Words like *faith* may be problematic if they imply a mindless leap into the unknown. At the same time, it is futile to try to get to a level of specificity or clarity in matters that do not lend themselves to such analyses. In one of my classes, students wanted to know what the Buddha *exactly* meant when he said such and such. The guest lecturer told them, "Don't try to fine-tune it."[14] I too have had to learn to live with a certain degree of ambiguity, as well as maintaining the difference between what we cannot understand because of sloppy thinking and failing to understand what cannot be understood by the human mind.

I have never gone through a crisis of faith with regard to the existence of God. My problem has been in knowing God's will and purpose for my life. My account does not trace a spiritual journey from paganism to Christianity as does St. Augustine's monumental *Confessions* (and is dwarfed by comparison). If I was ever assailed by doubts, they had to do with how God was treating me. There is an entry in my diary written during a particularly distressing period in my life—a *cri de cœur*—that ends with, "Though I am grateful for all I have, I also have a sense of being let down by God. I have been faithful to the Lord; where is He when I need Him most? I know that

I have received more than I deserve in life, or could have dreamed about having, yet the rewards still seem paltry." This is the tip of the iceberg of a deeply troubling mystery of why a loving, omniscient, and omnipotent God would allow the suffering of the innocent and the existence of evil in the world. Known as theodicy, it is a question that has preoccupied far greater minds without their reaching a definitive resolution of the issue.

~

The conflict between science and religious faith is an extension of the older conflict between Christianity and Greek philosophy, which some of the finest minds in Western culture tried to resolve from Late Antiquity through the Middle Ages. The contrast between the religious, theistic (God as creator and supreme ruler) world of faith and atheistic, secular, naturalistic, and humanistic views comes into sharpest conflict with respect to the nature of the universe and all that is in it. There have been innumerable attempts to reconcile these viewpoints, or for one side to undermine the credibility of the other. Consequently, the nature of the universe remains ambiguous. It can be interpreted naturalistically or theistically. The conflict cannot be resolved in favor of one or the other side because the evidence could be understood either way and cannot be given quantifiable values. This makes belief in God a rational choice even if that has no bearing on the empirical reality of the existence of God.[15]

I have been fully aware that I could not resolve these enormously complex issues for myself. Yet I struggled with them, especially given my need for clarity, consistency, and certainty. Should I see God as part of an all-encompassing nature rather than supernatural and existing in a realm outside of nature? If religion makes bad science and science bad religion, should I simply keep them apart? I realize that science does not require a supernatural creator to understand the laws that govern the physical universe. Nor do we need faith to explain what can be demonstrated empirically. Therefore I have a basically secular view of the workings of the physical world, and I relegate my faith to a separate realm of reality. I think of God as the last resort when nothing else makes sense. I invoke him only when I am at the end of my rope rather than as an easy way out of a problem. Yet a secular understanding of the workings of the universe empirically does not negate my belief in the existence of God or stop me from experiencing the awe and exultation that I feel at the sight of a towering mountain or the wing of a butterfly. It is deeply comforting to think of God as the creator and sustainer of the universe, and all that is in it, but I reject creationism as a backdoor way of rejecting the theory of evolution. Watching the myriads of stars against the pitch-black sky is like being in a cosmic cathedral. The specter of a cold and impersonal universe is chilling. And the idea of our world and the universe heading toward an eventual extinction reduces everything to irrelevance, but if that is the reality then I assume that too must be part of God's plan. Whatever may happen to the universe (or however many of them there may be), God is not going out of business.

When people first saw the image of the planet Earth taken from outer space, it radically changed their view of the world. The transformative image for me was a photograph of our galaxy—the Milky Way—with the planet Earth a pinprick of light looking lost in billions of stars. Our world seemed utterly insignificant in their midst, let alone in the inconceivably immense cosmos; and my own place in it infinitesimally smaller than that of a grain of sand in the vast desert of humanity. Why would God care for our insignificant planet in the boundless and timeless cosmic scheme of things? Yet I believe that God does care, that the world does matter, that I too matter. Trying to reconcile this majestic, incomprehensible, indescribable, and unutterable Being (which Hindus refer to as "Oh Thou, before whom all words recoil") with my belief in a benevolent, forgiving, and loving father has not been easy, but the problem has troubled me less as I have grown older. Yet, one mystery in particular continues to bother me: Where did all the stuff in the universe—all the original matter and energy—come from? Neither the scientific nor the religious accounts of the origin of the universe address this issue. They both deal with changes and reordering of what already existed; they do not deal with the fact of creation itself—*ex nihilo*—from nothing. In the biblical account, God creates order from chaos that was already present. In the Big Bang theory, matter and energy expand and contract; they do not vanish and reemerge. A physicist friend told me questions that cannot be answered scientifically are trivial. That would make the most fundamental existential questions that cannot be answered trivial. I am also baffled by the models of the universe currently envisaged by cosmologists. What does it mean that "the universe does not have a single existence or history, but rather every possible version of the universe exists simultaneously"?[16] This sounds no less mysterious than the Hindu quotation above.[17]

~

There are other dilemmas closer to home in understanding the doings of the immanent God in our midst. What does God expect from me and what can I expect from God? What is the contract, the covenant that specifies our respective rights and obligations? I have been greatly helped by Rabi'a's "non-utilitarian" stance toward God when she wanted to extinguish hell and burn paradise because they got in the way of truly loving God for his own sake—as an end rather than as a means to an end. Such singular and uncompromising devotion to God should eliminate much of the begging and haggling that we engage in with God. Why drag God into the tawdry details of our mundane lives? Why not just stop with "Let your kingdom come and your will be done" and forget the small change? On the other hand, where else am I to turn when I am up to my eyeballs in trouble? The way I try to deal with this dilemma is to avoid using God to achieve some ulterior purpose while gratefully accepting his loving help and gratuitous favors. I like to think that God and I have "joint custody" of my self. I take care of its everyday needs and God manages the rest. I deal with the short term and God deals with the long term issues.

~

Ultimately it all comes down to the love of God—the first and greatest Commandment. I feel that love deeply but I cannot adequately describe or explain it, let alone act on it. Our ideas of love are shaped by our human relationships, but none of the human ways of loving can be applied to God except as metaphors, such as loving God as a parent. I long to be in the presence of God and being in communion with him in a deeply personal sense (beyond the fact that God is present everywhere at all times). God is part of everything, but he is not everything. (When I drink a cup of coffee, I am not drinking God.)

I feel utterly dependent on God like an infant at the mother's breast and I am filled with profound gratitude and indebtedness for every breath I take. I take immense satisfaction in being close to the source of all beauty, wisdom, and all that is worthwhile in the world. And an absolute certainty that God will never abandon me. I will in turn place God ahead of all else in my life and make him its ultimate concern—as far as I can. I want to live my life and act according to his will—as far as I can understand it and am capable of it. And when I fail, which I surely will repeatedly, God's compassion and forgiveness will sustain me. How do I reconcile this lofty conception of my faith in God with the realities of my life, which has had its share of pain and suffering (albeit less than the common lot of people generally), or in causing pain and suffering to others (hopefully less than others do more generally)? How has my faith in God helped me in being a better person? How do I close the gap between faith and life—believing and living? If I'm not as good a person as I should be, I am the one at fault, not my faith. Then what is the point of having a faith that I cannot live up to fully? Maybe the point is that without my faith, I would be a worse person than I am.

I am acutely aware that the beliefs, thoughts, and feelings about God expressed in these pages are full of uncertainties, ambiguities, and contradictions. The certainty of my faith is awash in a sea of uncertainty. As Thomas Merton says it more eloquently, "My Lord. I have no idea where I am going. I do not see the road ahead of me. I cannot know for certain where it will end. Nor do I really know myself, and the fact that I think I am following your will does not mean that I am actually doing so."[18]

My parlaying the nascent faith of a bedridden boy into a lifelong spiritual journey may be farfetched and naïve. In that case, I hope it is naïve in a childlike and not in a childish way. In either case, I am certain that the only way I will ever make it into the Kingdom of God would be as a child.[19]

~

Having been blessed with robust health in my advancing age, I stopped counting the years while waiting for the shoe to drop. However, I have had to face the eventual prospect of mortality, and

Figure 12.5 Our expanding family. To my left are Anni and Kai; to Stina's right, Nina and Sina.

that too has been bound with my belief in God and life beyond death. I have no idea about what that life would consist of. The notion of the resurrection of the body as we know it makes me shudder. Since so many people now die in old age, as I will, is Heaven going to be like an old-age home for eternity? If I become reunited with my children in the afterlife, at which ages will we be? Will people exist at all ages of their lives simultaneously (as in a parallel universe)? That makes no sense. If we would be given new bodies, it would be most interesting to see what those would be like. One may believe that God can suspend physical laws (part the Red Sea, allow Jesus to walk on water, take Muhammad on a night journey to heaven). However, there are some things God will not do because they are mutually exclusive and hence contradictory to his own rational nature. He will not have two and two add up to five.

Having come to terms with the prospect of death during my childhood illness, death held no terror for me and does not now. Several years ago, I could have died of a severe postsurgical hemorrhage. After the crisis was over and I had realized how close I had been to death's door, my feeling was not one of anxiety and fear but concern that I may not have left my affairs in good order for my wife and children to sort out.

As I bring this memoir to a close, it is tempting to end it with a summation of my life—an overall assessment of its achievements and failures, its joys and sorrows. We can all rest assured that I am not about to do that. For one thing, I know that such a review would rapidly escalate into a *tour d'horizon* that would take us far afield. Moreover, if I have been unable to convey how my life turned out in this account so far, it is too late to do it now. I will especially avoid the temptation to rate my life like giving a grade for a course. Given that my life is not over yet, I have not yet taken the final exam. As the ancient Greeks would say, we should call no one happy until we have learned the manner of that person's death.[20] Until then, much may happen in my life, and the life of my family and of others I care for, that would change the grade. Besides, success and happiness are relative. One is more or less successful or happy compared to others, not with respect to some abstract ideal. Perhaps the ultimate test is that when I compare myself to others, there is no one I personally know with whom I would have traded places. This is not because I have lived an exemplary life. There are certainly those who have done better and others worse than I have. When it comes to counting peaks, I'm reminded of what John Gardner (a truly wise man who had climbed many peaks) said to me: "There are no peaks. It's all a journey while it lasts." The peaks for me now are our expanding family.

The basic task of old age, concluded Erik Erikson, is the achievement of a sense of integrity as against a sense of failure. Integrity depends on accepting that one's life has been "good enough," not perfect; and mine has certainly been good enough. I am ready to face anything that comes my way during the sunset of my life, except for the suffering or death of my children. That would be my undoing.

The Iliad refers to miserable old age. So far, my older years have been far from miserable and some of the happiest times of my life. Does this mean that I don't miss the strength and vigor of my youth? Of course I do. However, what I want at this point to hold on to are health and fitness within the constraints of my age, a sharp and active mind, financial security, the love of my family, a clear conscience, and an intact bond with God. And the greatest prospect for new and boundless joy would be having grandchildren.

I know that sooner or later I am going to feel and act old, in addition to being old. What concerns me most is the prospect of disability and decrepitude. I would not want to go on living after the game is no longer worth the candle. I already perceive a certain detachment from the world that seems to be passing me by. That doesn't bother me. As for the last phase of my life, I take comfort in the words of our friend Sir Geoffrey Vickers, who died at the ripe old age of ninety-three. The program of his memorial service included the following short poem that he had written when he moved into a home for the elderly:

> I did not guess
> age had such bliss;
> simply to rest,
> be, still, desist.
> Let striving cease;
> body, mind, heart
> content to part
> in peace.

~

As I was finishing this book, we learned that Anni and Kai were going to have a child—our first grandson. Incredulity gave way to wonderment and then to jubilation and gratitude that we were going to live to see the day that we had longed for. My grandson Kian was born in Helsinki on December 28, 2012, and I got to hold him in my arms. It would take another book to adequately express my thoughts, feelings, hopes, and longings for him. I had dedicated some of my earlier books to Stina, Nina, and Kai. I wanted to dedicate this book to my parents and Lucy. They richly deserved it. And now I could also dedicate it to our first grandchild, and others that may follow, thereby linking the past and future generations anchoring my life.

Notes

CHAPTER I

1. Anatolia, also called Asia Minor, forms a bridge between Asia and Europe. The eastern third of Anatolia was part of ancient Armenia, but it has been part of Turkey since the Middle Ages. Anatolia represented only part of the vast Ottoman Empire, a large part of which was in Europe at its peak. Anatolia now constitutes most of what is Turkey, with only a small segment of the country left in Europe.
2. The custom goes back to classical times, when amulets were fashioned as outlandish phallic symbols of the Roman god Fascinus, from whom we derive the word "fascinate." The amulet "fascinated" away the evil eye, thus protecting the individual wearing it. Currently, more commonly used forms come in the form of blue beads, often in the shape of an eye—a good eye acting as a deterrent against the evil eye. Children are particularly vulnerable to the evil eye motivated by envy. One tries to avert this by not commenting on how attractive a child is.
3. The change from *Hrant* to *Herant* and my surname from *Khatchadourian* to *Katchadourian* took place when I entered the American University of Beirut to conform to its system of transliterating names into English. When I came to the United States, I realized that *Herant* and *Katchadourian* were easier to pronounce than the original versions, so I kept the new spellings. I will continue to use the original spellings when referring to my parents and the changed version when referring to myself and to my own family. I will do the same when I refer to my mother after she immigrated to the United States.
4. All foreign words are in Armenian unless specified otherwise.
5. A new generation of Turks is now discovering that their grandmothers were Armenians who survived the genocide as children and were raised by Turkish families as Turks. This is an issue that was not spoken of until

recently, and it may lead to a new Turkish perspective on the Armenian genocide and its more open discussion.

6. The terms Near, Middle, and Far East have been in use at least since the nineteenth century to designate various geographic regions of Asia. The point of reference is Europe, as in the term "Levant," referring to where the sun rises in the east. Near East is the oldest term, and it encompassed Turkey, Palestine, Syria, and Lebanon as parts of the Ottoman Empire rather than discrete political entities. The Far East referred to China, India, and Southeast Asia. The term Middle East originated in the British Foreign Office at the turn of the twentieth century and referred to the area stretching from Iraq to India (filling up the gap between the Near and the Far East). Its focus then shifted to the region around the Persian Gulf during World War II. Subsequently, the term "Middle East" became more inclusive and is now commonly applied to the entire region between Egypt in the west to Iran in the east, with Turkey in the north and the Arabian peninsula in the south.
7. I was greatly helped in understanding the convoluted history of the rise of Islamic fundamentalism by Lawrence Wright, *The Looming Tower: Al-Qaeda and the Road to 9/11* (New York: Vintage Books, 2006).
8. The use of suction cups (or "fire cupping") is recorded by the ancient Egyptians as early as 1559 BCE in the Ebers Papyrus. The practice became part of Islamic medicine and spread to other cultures. The idea behind it is that the cups create a vacuum that draws out the noxious elements causing the illness. Cupping also makes bloodletting more efficient when applied over cuts.
9. Philip S. Khoury, *Syria and the French Mandate: The Politics of Arab Nationalism, 1920–1945* (Princeton, NJ: Princeton University Press, 1987).
10. Majid Khadduri, "The Iskenderun Dispute," *American Journal of International Law* 39, no. 3 (1945): 406–425.
11. The Austrian Jewish writer Franz Werfel published a fictionalized account of these events in 1933 called *The Forty Days of Musa Dagh* (New York: Carroll & Graf Publishers, 1983). He may have intended to use the Armenian example to call attention to the dangers European Jews faced with the Nazis' coming to power. His warning fell on deaf ears. When Hitler was planning the extermination of the Jews, his advisors warned him that international public opinion would not stand for it. Hitler reassured them, saying, "Who, after all, speaks today of the annihilation of the Armenians?" Kevork B. Bardakjian, *Hitler and the Armenian Genocide* (Cambridge, MA: Zoryan Institute, 1985).

CHAPTER 2

1. *Accent Elimination* has been exhibited many times domestically and abroad and was acquired in 2008 by the Hood Museum of Art at Dartmouth College for its permanent collection.
2. The diaspora ("dispersion") originally referred to Jews living outside their historic homeland. Since the borders of Armenia changed so much over the centuries, what constituted being away from one's homeland became harder to define. For my family, and for many of the Armenians who emigrated to the United States after World War I, their homeland was not the modern country of Armenia but various towns in the Ottoman Empire, mostly in eastern Anatolia, which historically had been part of a greater Armenia. Following the breakup of the Soviet Union, there was a large influx of Armenians from Armenia into the United States, many of whom settled in Southern California.
3. Genealogy compiled by Jirayr Khatchadourian (unpublished MS in Armenian).
4. My cousin Violet Arslanian provided these details about her grandmother.
5. Elie H. Nazarian, *A History of the Nazarian Family* (1475–1988) (Beirut: privately published, 1988) (in Armenian).
6. On another visit, this time to the town of Kharpert (now Elazig, Turkey), we encountered the same obliteration of historical memory. Stina was writing a book about an American missionary, Theresa Huntington, who lived in Kharpert (American missionaries called it Harpoot) for six years before World War I as part of a missionary community that included a four-year college. With the coming of WWI the missionary endeavor ended and subsequently the compound literally vanished. I asked an old man about it, and he said he had heard about some Americans living there but they had "gone home." Stina Katchadourian, *Great Need over the Water: The Letters of Theresa Huntington Ziegler, Missionary to Turkey, 1898–1905* (Princeton, NJ: Gomidas Institute, 1999).
7. In *Great Need over the Water*, Stina discusses the American missionary enterprise in Turkey and its relationship to Armenians. For an extensive history of Aintab, see Kevork A. Sarafian, *Armenian History of Aintab*, 2 vols. (in Armenian) (Los Angeles: Union of Armenians of Aintab in America, 1953). My mother gave me this work inscribed for my fifty-first birthday.
8. *New York Times*, January 24, 2012, A-4.
9. Henry Morgenthau, *Ambassador Morgenthau's Story* (New York: Doubleday, Page & Co, 1919).

10. For more extended discussions of the Armenian genocide, see Ronald Gregor Suny, *Toward Ararat: Armenia in Modern History* (Bloomington: Indiana University Press, 1993); Richard Hovanessian, *Remembrance and Denial: The Case of the Armenian Genocide* (Detroit: Wayne State University Press, 1999); and Taner Akçam, *A Shameful Act: The Armenian Genocide and the Question of Turkish Responsibility* (New York: Metropolitan Books, 2007). For parallels between the Armenian genocide and the Holocaust, see Robert Melson, *Revolution and Genocide: On the Origins of the Armenian Genocide and the Holocaust* (Chicago: University of Chicago Press, 1996).
11. The Armenian American author William Saroyan writes defiantly, "I should like to see any power destroy this race, this small tribe of unimportant people, whose wars have all been fought and lost, whose structures have crumbled, literature is unread, music is unheard, and prayers are not answered. Go ahead, destroy Armenia. See if you can do it. Send them into the desert without bread and water. Burn their homes and churches. Then see if they will not laugh, sing and pray again. For when two of them meet anywhere in the world, see if they will not create a new Armenia." Quoted in Vartan Gregorian, *The Road to Home* (New York: Simon and Schuster, 2003), 189–190.
12. Stina Katchadourian, *Efronia: An Armenian Love Story. Based on a Memoir by Efronia Katchadourian* (Princeton, NJ: Gomidas Press, 2001).
13. I tried but failed to find Ramzi's death certificate in the archives of St. Catherine's House in London. I had Ramzi's surname as Namatollah and the year of his death as 1917. I couldn't find anything remotely connected to such a person. Other attempts to track him down were equally futile. During our visit to Iskenderun, we met the grandson of one of the past Persian consuls, but there was no Ramzi in their family. His elderly aunt recognized the name of my mother's friend Nouriyeh Hanem but not Ramzi's mother, Farouz Hanem. Ramzi's school records had been sent to Damascus to the mission office that ran the school, but we had no way of tracking them down. And I could find no trace of my mother's letters from Ramzi that she had buried in our garden, along with a locket with their engagement photograph. Consequently, there is no corroborative evidence that Ramzi ever existed. Yet it is inconceivable that my mother could have made him up.

CHAPTER 3

1. Edward Said, *Out of Place: A Memoir* (New York: Knopf, 1999).
2. Named after the British Major General Sir Edward Louis Spears, who was noted for his role as a liaison officer between British and French forces in

the two world wars. This and other street names like Rue Weygand and George Picot were a legacy of the French Mandate, adding to Beirut's metropolitan flavor.

3. I wonder if Dr. Calmette was related to the French bacteriologist Albert Calmette, who with his colleague Camille Guérin developed the BCG vaccine for tuberculosis, which was first used in 1921.
4. During the summer of 2010, I reread *Monte Cristo.* I could still remember some of the episodes such as the escape of Edmond Dantès from prison, but I had forgotten most of the rest. I marvel nonetheless at how I had managed to go through its 1,082 pages as a sick nine-year-old boy.
5. The event was probably the Deir Yassin massacre in a Palestinian village by Zionist Irgun paramilitary fighters, which took place between April 9 and 11, 1948—an event that contributed to the panic that led to the mass exodus of Palestinians. Thomas L. Friedman, *From Beirut to Jerusalem* (New York: Farrar, Straus and Giroux, 1989).
6. The Greek Orthodox St. Thecla was the daughter of a Seleucid prince and became a young disciple of St. Paul. Her life story is told in the apocryphal *Acts of Paul and Thecla*.
7. The first Christian Endeavor Society was formed in 1881 in Portland, Maine, for evangelizing among youth and harnessing their energies into Christian service. The organization expanded rapidly and had grown to 67,000 nondenominational chapters with 4 million members by 1906 (http://en.wikipedia.org/wiki/Young_People's_Society_of_Christian_Endeavour).
8. I am grateful to Brent Sockness for this observation and for pointing me to the book by William James discussed shortly.
9. This young woman's account is the opening case history of chapter 8, on the pathology of guilt in my book: Herant Katchadourian, *Guilt: The Bite of Conscience* (Stanford, CA: Stanford University Press, 2010).
10. William James, *The Varieties of Religious Experience: A Study in Human Nature* (New York: Collier Books, 1961 [1902]). See, in particular, chapters 4–7.
11. A variety of explanations have been offered as causes for obsessive-compulsive disorder (OCD), ranging from the psychoanalytic and other psychological theories to biochemical causes implicating the neurotransmitter serotonin; hence the use of drugs called serotonin uptake inhibitors, which enhance the level of serotonin in the body. None of these explanations has proven to be definitive. I still find the psychoanalytic explanations most interesting. In this view, obsessive-compulsive symptoms are generated by ego defense mechanisms that deal

with repressed aggression such as intellectualization, reaction formation, and undoing.

There is also an intriguing linkage between streptococcal throat infections and the onset of OCD in childhood. Since I did suffer from such an infection that led to my developing rheumatic fever, it may have also led to the development of these symptoms a few years later. However, ascribing a complex psychological set of symptoms to the effects of a throat infection seems to trivialize it.

CHAPTER 4

1. Lutfi M. Sa'di, "Al Hakim: C. V. A. Van Dyke," *Isis* 73 (1937): 31–32; "Life and Works of G. E. Post," *Isis* 77 (1938): 392–393. In this historical account, I have relied extensively on Stephen B. L. Penrose, *That They May Have Life: The Story of the American University of Beirut, 1866–1941* (Beirut: American University of Beirut, 1970). A more recent important source is Betty S. Anderson, *The American University of Beirut: Arab Nationalism and Liberal Education* (Austin, TX: University of Texas Press, 2011).
2. "This College is for all conditions and classes of men without regard to color, nationality, race and religion. A man white, black, or yellow; Christian, Jew, Mohammedan or heathen, may enter and enjoy all the advantages of this institution for three, four or eight years; and go on believing in one God, in many gods, or in no God. But it will be impossible for anyone to continue with us long without knowing what we believe to be the truth and our reasons for that belief." Frederick Bliss, *The Reminiscences of Daniel Bliss* (New York: Revell, 1920), 105.
3. Quoted in Penrose, *That They May Have Life*, 49.
4. Among other major donors, the Rockefeller Foundation provided substantial help in the development of the medical school.
5. I came to know personally six AUB presidents. Calvin Plimpton was my professor of medicine before he became president. I knew Samuel Kirkwood when I was on the faculty. I met Frederick Herter first when I served on a committee that he chaired. I met David Dodge and John Waterbury after I joined the board, and Peter Dorman when he became president in 2009.
6. Malcolm Kerr's wife, Ann, and his daughter, Susan, have written compelling accounts about him. Ann Zwicker Kerr, *Come with Me from Lebanon: An American Family Odyssey* (Syracuse, NY: Syracuse University

Press, 1994); Susan Kerr Van de Ven, *One Family's Response to Terrorism: A Daughter's Memoir* (Syracuse, NY: Syracuse University Press, 2008).

7. Most American medical schools require four years of college followed by four years of medical school. However, the AUB medical school consisted of five years preceded by three years of college. Consequently, my senior year in college was actually my first year in medical school. And my classmates shown here were my anatomy lab partners, as I will relate in the next chapter.
8. These events are discussed at length in Anderson, *The American University of Beirut*.
9. I generally refer to most individuals by their real names. However, Juliette and the names of other girlfriends I refer to are pseudonyms.

CHAPTER 5

1. Syphilis is thought to have been brought to Europe by Columbus's sailors. There is also contrary evidence that the illness already existed in Europe earlier but in a less virulent form.
2. Stanley E. Kerr, *The Lions of Marash: Personal Experiences with American Near East Relief, 1919–1922* (Albany: State University of New York Press, 1973).
3. One of the heresies of the early church advanced the notion that greater the sin, the greater will forgiveness bring glory to God. Martin Luther used the phrase to claim that since the forgiveness of God is infinite and human beings by their nature are prone to sin, we should act boldly in order to rejoice in Christ more fully. Dietrich Bonhoeffer called the abuse of the idea cheap grace—forgiveness without repentance.
4. This legend may be the origin of the Christian tradition of painting eggs at Easter.
5. I am using Sylvia's real name, since I cannot conceal her identify as my former wife.
6. I lost track of the album of our wedding pictures and thought it had been left behind and lost in Beirut during the civil war. Years later, Lucine told me that my mother had burned the wedding album after my divorce was finalized. I was annoyed at first, since that album was part of my life. But I forgave my mother for this one rash act in view of all the grief she suffered during my ill-fated marriage.
7. Mohammed Shafi Agwani, *The Lebanese Crisis, 1958: A Documentary Study* (Bombay: Asia Publications House, 1965).

CHAPTER 6

1. Pan Am flew its first Boeing 707 from New York to London a few months later, on October 26, 1958.
2. Jules Cohen and Stephanie Brown Clark, *John Romano and George Engel: Their Lives and Works* (Rochester, NY: Meliora Press, 2010). Although I knew a fair amount about Romano's background, this dual biography of Romano and Engel provided me with a far more comprehensive understanding of their lives.
3. The American Psychoanalytic Association, "Oral History Workshop," No. 39, New York, December 16, 1993; John Romano, "Conference on the Integration of Onchiota Conference Center," February 10–12, 1961. I am grateful to Sanford Gifford for bringing this source to my attention.
4. For Romano's earlier life, also see John Romano, "My Milwaukee, Part I: Earliest Years," *Wisconsin Psychiatrist* 33, no. 4 (1993): 20–22.
5. Nonetheless, the list of Romano's publications and presentations runs to 280 items. Many of these, however, are short papers and the texts of talks. Much of his original research antedates his taking on administrative responsibilities. Cohen and Clark, *John Romano and George Engel*, 199–212.
6. John Racy, "John Romano—Remembrance," Alumni Weekend, University of Rochester, October 16, 2009.
7. Cohen and Clark, *John Romano and George Engel*, 169.
8. I am grateful to Christine Tour-Sarkissian for bringing this matter to my attention.
9. Feldman wrote a delightful book full of his astute observations and gentle humor. Sandor Feldman, *Mannerisms of Speech and Gestures in Everyday Life* (New York: International Universities Press, 1969).
10. The motto of the English chivalric Order of the Garter.

CHAPTER 7

1. The name of patients referred to in this and subsequent chapters have been changed and identifying details omitted.
2. "Dear Dr. Katchadourian: This is a letter of gratitude, long overdue, after all of these years. I was one of your patients at the county Mental Hygiene Clinic from September 1961 through June of 1962. I was a callow lad of nineteen, a drop-out in a deep depression and usually unemployed. I spent my time writing novels, painting pictures, trying to understand a world filled with fear, hypochondriasis, nuclear missiles, pain and weltschmerz.

I had a basic good humor and intelligence, unrefined. I couldn't get in touch with my assets or who I was at the time....

"I know that you have had a very successful career at Stanford as a professor and a much-cherished teacher, and it is easy for me to see you in those roles, recalling your ability to communicate and provide a role model for youth. I find myself chuckling at your wit and gift for understatement.

"After your help (and you gave much), I joined the intelligence community of the armed forces and worked in various administrative positions, while working on a degree on the side.... I eventually obtained my degree and went to work as a special agent in four law enforcement agencies. I was married and raised two boys as I worked my way into management. I retired in 1993 and have since been teaching long-term substitute assignments with special education and other groups at the middle and high school levels.

"My family is raised now and on their own and I spend much of my time painting.... In closing, I will say that my life has had its ups and downs, like everyone ... but there is always the bedrock of my epiphany with you. You opened a much wider world to me, a world of color and light. You helped me to learn to float, to swim, to dive and to ride great big waves to sandy shores and to plant my feet against the race of the tide. I thank you." (Author's name withheld and some biographical details altered. Reproduced with permission.)

3. The Minnesota Multiphasic Personality Inventory (MMPI) is one of the most frequently used tests for studying personality structure and psychopathology.

CHAPTER 8

1. Actually, there is a rare genetic condition, called fatal familial insomnia, where progressive insomnia leads to other severe symptoms and eventually death within a year or two.
2. The project was based on value orientation studies developed by the Harvard anthropologists Florence Kluckhohn and Fred Strodtbeck. I was unclear about how these studies were related to mental illness, but it was part of Lyman's agenda and I assumed he knew why we were doing it.
3. The hotel that the Caves du Roy was a part of was destroyed during the second Lebanese civil war. I understand that there is now a disco at St. Tropez called Les Caves du Roy at the renowned Hotel Byblos. It is said

that the women there are beautiful and the men (unlike in Los Angeles or New York) actually know how to dance. This description would perfectly fit what Les Caves du Roy was in its heyday in Beirut.

4. Shorter accounts of this research have appeared as chapters in two books: "Culture and Psychopathology," in *Psychological Dimensions of Near Eastern Studies*, edited by L. Carl Brown and Norman Itzkowitz (Princeton, NJ: Darwin Press, 1977), 103–125; "Culture and Personality: The Case of Anjar," in *The Armenian Communities of the Northeastern Mediterranean: Musa Dagh, Kessab, Dort-Yol*, edited by Vahram Shemmassian (Costa Mesa, CA: Mazda Press, in press).
5. Lebanon provided the cedar wood and the craftsmen for the construction of the first Jerusalem temple, built by Solomon. There is a record of the transaction between Solomon and Hiram the King of Tyre. Assyrian stone reliefs show Phoenician ships hauling cedar logs for the construction of their great palaces. The cedar forests have long disappeared. The Romans were already concerned about the deforestation of the Lebanese mountains.
6. The word "serendipity" was coined in the eighteenth century based on ancient fairy tale "The Three Princes of Serendip," in which astute individuals discovered things by accident.

CHAPTER 9

1. The one possible competitor would be William Dement's course on sleep and dreams. For more details on the history of the human sexuality course, see my chapter "Sex Education in College: The Stanford Experience," in *Sex Education in the Eighties*, ed. Lorna Brown (New York: Plenum Press, 1981), 173–190.
2. David A. Hamburg, *No More Killing Fields* (Lanham, MD: Rowman and Littlefield, 2002). Also see David A. Hamburg, *Preventing Genocide: Practical Steps Toward Early Detection and Effective Action* (Boulder, CO: Paradigm, 2010). I am grateful to David Hamburg for the prominent place he provides for the Armenian genocide in his book, and for including me in a series of interviews with distinguished individuals.
3. Student evaluations were peppered with terms like "best lecture," "great course," and "best professor I have ever had at Stanford." Some of the more exotic comments included "Cool beans dude," "Katchadourian rules," "He is down with it man," "He's neato," "He's rich. Rich in knowledge, rich in culture, rich in personality."
4. The origins of sexology, the scientific study of sexuality, go back to the turn of the twentieth century. In 1919, Magnus Hirschfeld founded

the Institute for Sexology in Berlin, which fell victim to the Nazis, who considered the subject decadent; the fact that Hirschfeld was also Jewish and a homosexual made things worse. They burned his books.

5. For the citation, see Appendix A.

CHAPTER 10

1. The events of this period are described in Richard W. Lyman, *Stanford in Turmoil: Campus Unrest, 1966–1972* (Stanford, CA: Stanford University Press, 2009).
2. Herant Katchadourian, "The Psychiatrist as University Ombudsman" *Psychiatry* 36 (1973): 446–457.
3. I have related some of my experiences during this period in Herant Katchadourian, "The Psychiatrist in the University at Large," *Journal of Nervous and Mental Diseases* 154, no. 3 (1972): 221–227.
4. Herant Katchadourian, "Medical Perspectives on Adulthood," *Daedalus* 105 (Spring 1976): 29–56. This publication led to my contributing several sections to one of the standard textbooks of internal medicine: Herant Katchadourian, "The Life Cycle Perspective in Medicine," "Development to Adulthood," and "Adulthood," in *Cecil Textbook of Medicine*, 15th ed., ed. P. B. Beeson, W. McDermott, and J. B. Wyngaarden; revised in 16th ed. and 17th ed. (Philadelphia: Saunders), 15–21.
5. Herant Katchadourian, *Biology of Adolescence* (San Francisco: W. H. Freeman, 1977).
6. Steering Committee, Study of Education at Stanford, *Report to the University: Government of the University*, vol. 10 (Stanford, CA: Study of Education at Stanford, 1969).
7. I am grateful to my friend and colleague David Abernethy, who was my associate dean for these programs, for pointing out their greater significance.
8. This was not a trivial sum at the time. However, the closest current counterpart of my old position, which is that of the vice provost for Undergraduate Education, is now funded by the revenue from a $1 billion endowment raised by President John Hennessy for the improvement of undergraduate education.
9. Herant A. Katchadourian and John Boli, *Careerism and Intellectualism among College Students* (San Francisco: Jossey-Bass, 1985).
10. Herant A. Katchadourian and John Boli, *Cream of the Crop: The Impact of Elite Education on the Decade after College* (New York: Basic Books, 1994).

11. For the citation of the award, see Appendix B.
12. I am grateful to David Abernethy for helping me place these matters in better perspective.
13. Edith Södergran, *Love and Solitude: Selected Poems, 1916–1923*, bilingual edition, translated by Stina Katchadourian (Seattle: Fjord Press, 1992); Märta Tikkanen, *Love Story of the Century*, translated by Stina Katchadourian (Santa Barbara, CA: Capra Press, 1986); Stina Katchadourian, *Great Need over the Water: The Letters of Theresa Huntington Ziegler, Missionary to Turkey, 1898–1905* (Ann Arbor, MI: Gomidas Institute, 1999).
14. Herant Katchadourian, ed., *Human Sexuality: A Comparative and Developmental Perspective* (Berkeley, CA: University of California Press, 1979). Translated into Spanish as *Sexualidad humana: un estudio comparativo de su evolución* (México, D.F.: Fondo de Cultura Económica, 1983).
15. David A. Hamburg, *Today's Children: Creating a Future for a Generation in Crisis* (New York: Times Books, 1994).
16. *Your Changing Body*; *Sexuality*; *Reproduction* (Chicago: Everyday Learning Corporation, 1991).
17. Herant Katchadourian, *Human Sexuality: Sense and Nonsense* (Stanford, CA; Stanford Alumni Association, 1972); Herant Katchadourian, *Fifty: Midlife in Perspective* (Stanford, CA: Stanford Alumni Association, 1987).
18. See Appendix C.
19. When Kai was a teenager, he gave me a small trophy on my birthday that said, "World's Greatest Dad." The fact that it must have been produced in the thousands didn't detract from its value. More recently, Nina posted the following message on Facebook on Father's Day: "I love my dad. Expressive and incredibly loving, unconditionally supportive, fair, honest, and deeply optimistic, and always able to be both teacher and student. I lucked out in the father lottery big time."
20. I am grateful to Reva Tooley, whose questions about the application of my psychiatric skills beyond the point of my being a working psychiatrist elicited my own reflections in this regard.

CHAPTER 11

1. Michael S. Malone, *Bill and Dave: How Hewlett and Packard Built the World's Greatest Company* (New York: Portfolio, 2007).
2. Appendix D. During the reception following the memorial service, I received many compliments for my talk. Dick Lyman, who must have heard many eulogies as president of Stanford and then of the Rockefeller Foundation, said it was the best memorial tribute he had ever heard.

3. Appendix E.
4. Herant Katchadourian, *The Flora Family Foundation: The Early Years, 1998–2003* (Menlo Park, CA: Flora Family Foundation, 2004). Steve Toben sent a copy of this book to Virginia Esposito, the president of the National Center for Family Philanthropy. She wrote back, "I use the word sparingly because it is so over- and misused, but this report is genuinely unique. I read hundreds of reports, but don't believe I have ever read one that attempted to include the spirit of the founding, the history, the personal reflections, grants analysis, grantee feedback, the themes and lessons to be gleaned from all that, and messages for the stewards of the future. Whew!"
5. Stina Katchadourian, *The Lapp King's Daughter: A Family's Journey through Finland's Wars* (McKinleyville, CA: Fithian Press, 2010).
6. Appendix F.
7. "I will always appreciate the personal advice and counsel you gave me during my years as HB director, not to mention *all* the wonderful things you did for Human Biology and for Stanford. I hope you won't think it corny to say that you were the 'mold' for me and for so many of us in the professoriate. Intellect, statesman, spokesman, and genuine and sincere person, all wrapped into one. Thanks for setting such a shining example."
8. "I am very glad that we had at a least a chance to talk … about your magnificent career and my privilege of being a close observer of its manifold contributions.… I take much pride in the unusual circumstances that led to the great opportunity to recruit you, not once but twice: first in the 1950s to the NIMH and then in the 1960s to Stanford. On a personal level, it has been an experience of utmost satisfaction to work with you in a variety of enterprises from psychiatry to human biology to university governance to philanthropy. In every sphere of your activity, you made major contributions: in scholarship, statesmanship, mentorship and much more. Surely, you are one of the finest teachers I have ever known, and I have been lucky to know the best.… Generations of Stanford students are in your debt.… I will always cherish your wisdom, courage, generosity of spirit and delightful humor. I express my most heartfelt congratulations, my personal gratitude and enduring affection, and every good wish for the years ahead. Dave."
9. Appendix G.
10. Herant Katchadourian, *Guilt: The Bite of Conscience* (Stanford, CA: Stanford University Press, 2009).
11. David A. Hamburg, *Preventing Genocide: Practical Steps Toward Early Detection and Effective Action* (London: Paradigm Publishers, 2010). The

volume that preceded it was David A. Hamburg, *No More Killing Fields* (Lanham, MD: Rowman and Littlefield, 2002).

CHAPTER 12

1. There is a good deal of overlap between the books that deal with solitude and those that deal with silence, and some of them deal with both. Among books on silence, I found particularly interesting Sara Maitland's *A Book on Silence* (London: Granta, 2008). Maitland is a British writer who spent several years living alone in the Scottish Highlands and other isolated locations, including a remote island. Her account is mainly descriptive and evocative of the feelings she experienced during these periods. Philip Koch's *Solitude: A Philosophical Encounter* (Chicago: Open Court, 1994) is a very different sort of book. Koch is a philosopher at the University of Prince Edward Island in Canada. He also draws from his personal experiences with solitude, but the book is an extensive review and analysis of the literature on solitude. I found personally most compelling Thomas Merton's writings on silence and solitude, including *The Silent Life* (New York: Farrar, Straus and Giroux, 1957) and *Thoughts in Solitude* (New York: Farrar, Straus and Giroux, 1958).
2. I generally use the Swedish version of names because this is a Swedish-speaking part of Finland and also the language my wife speaks with her family and friends, even though they are also fluent in Finnish. Moreover, some of the names of islands are the same in both languages.
3. John R. Gillis, *Islands of the Mind* (New York: Palgrave Macmillan, 2004).
4. Stina has an unpublished essay on a sense of place that helped my own thinking in this connection.
5. Another experience opened my eyes to see beyond the obvious. When visiting Venice, I came down with a severe chill that confined me to bed. My window looked at a wall painted in a fading pink. There was a chimney and a television antenna at its top. A more boring sight was hard to imagine. Yet after staring at the wall for a whole day, I began to find it as interesting as any painting I was to see at the Gallerie dell'Accademia.
6. Actually, Nina had created a work with historical resonance. When Anton van Leeuwenhoek invented the microscope in the seventeenth century, he found sperm in semen wiggling around like tadpoles and called them *spermatozoa*.
7. I discuss this in some detail in my book *Guilt: The Bite of Conscience*, referred to earlier, pp. 224–225.

8. Jane Hirshfeld, ed., *Women in Praise of the Sacred: 43 Centuries of Spiritual Poetry by Women* (New York: Harper Collins, 1994).
9. Quoted in "Rabi'a: Her Words and Life in 'Attar's *Memorial of the Friends of God*," in Michael Sells, tr. and ed., *Early Islamic Mysticism* (New York: Paulist Press, 1996), 163.
10. Several books have been particularly instructive in the exploration of my personal faith. One is the classic work by William James, *The Varieties of Religious Experience: A Study in Human Nature*, cited earlier. Originally published in 1902, it focuses on personal religious experience and remains relevant to this day. Another is a wide-ranging modern philosophical and theological analysis of religion, including the nature of personal belief, by John Hick, *An Interpretation of Religion: Human Responses to the Transcendent* (London: Macmillan, 1989). The sweeping and extensive account in Diarmaid MacCulloch, *A History of Christianity: The First Three Thousand Years* (New York: Penguin Books, 2009), has given me a better grasp of the historical context of my faith. On the other hand, I have gained very little from shallow attempts to repudiate religious belief in God, such as *The God Delusion* by Richard Dawkins (New York: Crown, 2008). His arguments against the experiential basis for belief in God, which are of particular interest to me, sound naïve. Even the title of the book is a misnomer. It should be *The God Illusion*. An illusion is a commonly held false belief. It is a cultural phenomenon, whereas a delusion is a personally held false belief that often represents a mental disorder. Believers in God may be collectively mistaken in their belief, but they are not crazy.
11. Actually, psychological theories have no bearing on the fundamental question of whether God exists or not. Only if God does *not* exist and people still believe in him do these theories become relevant. By the same token, if God does exist and people do not believe in him, then we must seek psychological explanations for why they deny reality. It is just as easy to invent God as a substitute for a parent that we love as it is to refuse to believe in God because we would rather not have the parents that we hate. Whether God exists or not, apart from why people may or not believe in him, is an epistemological, not a psychological, issue.
12. I am grateful to David Abernethy for helping me clarify my thoughts in this section.
13. These issues are discussed in chapters 9 and 10 of *Guilt: The Bite of Conscience*. For an account of a similar experience at a greater depth, see Huston Smith, with Jeffery Paine, *Tales of Wonder: Adventures Chasing the Divine* (New York: Harper One, 2009).
14. Mark Mancall used the phrase in my seminar on guilt.

15. This issue is argued in comprehensive detail in Hick, *An Interpretation of Religion.*
16. Quoted in the review of Stephen Hawking and Leonard Mlodinow, *The Grand Design*, in *The Economist*, September 11, 2010, p. 101.
17. Huston Smith, *The World's Religions* (New York: HarperCollins, 1991), 69.
18. Thomas Merton, *Thoughts in Solitude* (New York: Farrar, Straus and Cudahy, 1956), 83.
19. Mark 10:15.
20. Herodotus, *The Histories*, translated by Carolyn Dewald and Robin Waterfield (New York: Oxford University Press, 1988), 16.

Appendixes

Appendix A. Citation for the Richard W. Lyman Award

For inventive and uncommonly effective initiatives to involve alumni and the broader Stanford community; for articulate advocacy on behalf of Stanford's underrepresented; for separating sense from nonsense with timely injections of wisdom, encouragement, and drollery in councils and classrooms both campus and worldwide; and for his entire career as an educator—part Old Testament prophet and part New World statesman—which bespeaks the institution's highest goals and standards.

Presented by the Stanford Alumni Association, June 17, 1984
Dixon Arnett, '60
Association President

Appendix B. Citation for the Lloyd W. Dinkelspiel Award for Outstanding Service to Undergraduate Education

For the clear vision, well-constructed innovations, and warm humanity he has brought to all of his service at Stanford as University Ombudsman, Vice Provost, and Dean of Undergraduate Studies; For a seemingly inexhaustible commitment to teaching undergraduates in a style which challenges, stimulates, and which makes even the largest lecture hall seem small and personal; For tireless efforts to reach beyond traditional boundaries of the classroom to touch hundreds of students' lives with the essential gifts of sympathy, perception and encouragement; For research on undergraduate education which has allowed this institution to build on its strengths and identify its weaknesses in thoughtful and informed ways, and which has contributed significantly to the national reexamination of undergraduate life,

Herant Katchadourian, Professor in the Program in Human Biology, is hereby designated a 1993 recipient of The Lloyd W. Dinkelspiel Award for Outstanding Service to Undergraduate Education.

Gerhard Casper, President of the University, June 13, 1993

Appendix C. Class Day Address

June 17, 1989

Harun al-Rashid, the fifth Caliph of the Abbasid dynasty, who reigned in the eighth century, was a man of great sagacity and discernment. His subjects, from within the vast empire stretching from the turquoise waters of the Mediterranean to the snow-covered peaks of the Himalayas, came to his court in Baghdad to exhibit their talents and curry his favor.

One day, as the Caliph was holding court, a man was ushered in his presence. Tall and gaunt, he was worn out and weary from the long journey he had undertaken in order to display what was rumored to be an extraordinary skill. After the Caliph motioned to proceed, the man bowed deeply, and with his sinuous and bony fingers he took out a long needle from a leather bag hanging from his belt. He stood still for a moment and then with a precise and bold flick of his wrist, he fired the needle and lodged it firmly in the wooden floor. He then took out a second needle, aimed it at the first, and planted it squarely in its eye. A third needle followed suit, landing in the eye of the second needle. Needle followed needle until nine of them rose precariously from the floor.

The court was astounded; all eyes turned to the Caliph. After a long silence, the Caliph turned to the Grand Vizier and said to him, "Give this man 100 gold dinars and a 100 strokes of the lash." The old vizier fixed the Caliph with a steady gaze and spoke out with a voice deferential yet challenging. "My Lord," he said, "you are the successor of the Prophet of Allah, the Compassionate, the Merciful, and the Defender of the Faith. It is not for me, your servant, to question your judgment, but why are you rewarding and punishing the man at the same time?" "I am rewarding him," said the Caliph, "for his great skill. And I am punishing him for using his great skill for such a trivial purpose."

Your Stanford education has equipped you also with many splendid skills—worth at least several needles. So the question I want to pose to you this morning is whether you will use these skills to deserve the gold or the lash. I do not fear that you may go out into the world and place yourself in the service of evil. But I do fear that you may not do enough good with the many talents you have. There are many ways in which you may squander and misuse your talents. Given the limitations of time, and in deference to the festive nature of the occasion, I will only dwell on two.

First, you are likely to try to do too much: work too hard, earn too much money, accumulate too many possessions, have too many husbands and wives, and consume more that your share of the world's dwindling resources. Too much action and doing will overwhelm reflection and being. Your life will appear full, but it will lack meaning.

I had a patient once; I will call him Jim. Because he looked particularly burdened, I asked him what problem was bothering him so much. "The problem," said Jim, "is too much world." Too much world is what is waiting for you out there. It is not likely to twist your reason or crush your spirit as it did with Jim. But it will fray and frazzle you and leave you depleted.

If spreading yourself too thin is the first pitfall, overly concentrating your talents and energies in one area is the second. Relentless specialization in your work will sink you deeper and deeper into an ever-narrowing hole wherein you will lose perspective on what you do. Focusing your passions and convictions onto one or another solitary cause or purpose—no matter how worthwhile—will entrap you into single-issue politics and confine your vision with the blinders of bigotry. Like a bull in the ring you will react reflexively to red flags rather than respond with reason and compassion to the full spectrum of the hues and shades of issues confronting you, as you must do if you are going to deal effectively with the awesome problems of an increasingly interdependent world poised on the brink of bankruptcy.

It is always a great pleasure for me to speak on these happy occasions. The pleasure is particularly acute this year because today I stand here not only as an honored teacher but also as a proud and happy parent, for our daughter, Nina, is also graduating from college and is sitting over there smiling at me.

Having spoken against the prospect of too much of anything, I want to make one exception. There can never be too much of the joy and the pride we feel at the accomplishments of our daughters and sons. Hence, l rejoice with you, fellow parents, and I congratulate you graduates—who are our children and friends, as well—and wish you well.

Appendix D. William R. Hewlett Memorial Address

January 21, 2001

I joined the board of the Hewlett Foundation when Bill chaired it, and I became his old-age friend when illness made it necessary for him to retire from public life.

There is no escaping the fact that Bill Hewlett was a very wealthy man. It would be difficult at first to dissociate the person from his purse. But it took very little to realize how unaffected he was by his immense fortune. Bill was virtually immune to the corrosive effects of money. He judged neither himself nor others on the basis on their financial worth. He and his family lived simply and comfortably, far from the ostentation and self-indulgence of the conventionally affluent.

I used to take Bill on long rides, usually to his beloved ranch, and we sometimes stopped at some hole-in-the-wall for a bit to eat. When it came time to pay, I would say, "Please let me take care of it; I don't think you can afford this place." He usually let me get away with it, with his distinctive twinkle in the eye, but on one occasion he insisted that he was going to pay the bill himself. But it turned out that he had no money. I asked him, "What's going to happen to you without friends like me?" "I don't know," he said. "I guess I'll be homeless."

Bill Hewlett set out on his engineering career doing what he loved to do most. And he was, of course, very good at it. The financial rewards of his work were almost incidental for him. He would have been just as happy laboring on as an engineer for the rest of his life as he was running HP with David Packard with its thousands of employees. I asked him once what he would have done if he had taken a different path. He said he would have become a physician. This would have given him a chance to tinker with the human body while helping people. In a sense it would have brought him closer to his father, a distinguished physician who died when Bill was just twelve years old.

Despite his relaxed informality, Bill was an intensely private man who did not wear his feelings on his sleeve. He could seem aloof. He cared deeply about his family, his friends, and his colleagues. He did not pamper them any more than he pampered himself. He expected from them no more and no less that he expected from himself—which was a lot. He was a kind but also a demanding man, and he could be tough. He did not achieve all he did by just being nice.

I cherish the times Bill and I spent together, and I learned a lot from him—always by example, never by preaching (for he never preached). I will never forget his humility, his wide-ranging intellectual curiosity, his love of nature, and his respect for people as individuals.

Some years ago when the Hewlett Foundation Board met in Ashland, we attended a Shakespeare play. Bill was in a wheelchair, and I sat with him in the section for the handicapped. Behind us was an elderly woman, the kind of person many people barely notice. Bill turned to her, smiled, and greeted her. The woman had no idea who was talking to her. But Bill's kind gesture may have made more of an impression on her than the immortal words of William Shakespeare.

The world is fortunate to have its share of gifted engineers, enlightened captains of industry, trustees of great institutions of learning and culture and of public service, and philanthropists on a grand scale. William Reddington Hewlett was all of those. But he was also one of a kind. I am going to miss him.

Appendix E. Nina Katchadourian's Remarks at Lucine's Memorial Gathering

When Kai was about eight years old, he became obsessed with airplanes. Applying his newly acquired knowledge of aeronautical engineering principles, he theorized that Lucy probably moved so quickly because she had less wind resistance. She always reminded me of a hummingbird, as she was so tiny and precise in her movements. Visiting her meant stepping into a miniature world. Everything about her was tiny: her little chairs, her miniature sweaters, even her extra-small little Turkish coffee cups. One time when I came to visit, she had bought herself new shoes, a small pair of black flats. I asked her where she got them, and she told me she had gone to Rapp's Shoe Store and found this pair in the children's department. She was also quite pleased that she had bargained the guy down: "Mister, I am very small." Instant discount.

She had incredible physical stamina for her size, even late into her life. Years ago, I stopped by just after she had done a run to the farmer's market and (somehow) hauled home a sack of red peppers to make into red pepper paste. When I walked in the door she was sitting on the floor, with the limberness of a little girl, her legs in a perfect V shape and a mountain of peppers as tall as she was on the floor in front of her. When Kai and I were really little, she let us ride on her back like a horse—and she must have been in her sixties at the time. She inherited the tiny play chairs that we outgrew, and there were many times I would come to visit and find her sitting in a little chair, even though all the big ones were available.

Her will, on the other hand, was not small. As a child, having been orphaned and then taken in by Bedouins, her chin was tattooed with traditional tribal markings. After she had been adopted into my grandparents' home, she scoured out these marks using her fingernails and detergent. This display of stubbornness amazed me, and I used to stare at her chin to try to see the remnants of those marks, but there were none. Several times over the past eight years, Lucy landed in the hospital, and we braced ourselves for the end. But time and time again, Lucy pulled through with some incredible strength she summoned from somewhere inside her tiny body. Visiting her in the hospital after she had her gall bladder removed, she yanked up her hospital gown to show me her massive scar. "Look like chopped lamb," she said, with a dismissive wave of her hand. Then she looked at me seriously and said, "Again, I live."

Watching television with Lucy was hilarious. In the afternoon she would often have the TV tuned to the soap opera channels and complain bitterly about all the kissing, which she thought was indiscreet and bad for children to see. As she ranted and raved about it, she also watched it and made fun of it, sometimes pantomiming along and saying, "Ehhhh, habibi!" making faces, and then quivering with giggles. We were watching the news together once, and the lead story had something to do with the Clinton–Lewinsky scandal. I got a bit concerned; had she been following this, with all the salacious details? She turned to me. "President Bill," she said (they seemed to be on a first-name basis), "President Bill have girlfriend." Pause. "Eh. He is man." She had valuable pre-feminist advice for my friends and me as well. Jessica Evans and I were visiting her once and over coffee she offered us her famous theory: "If boy have nothing in pocket, he no good."

She had a kind of wild, unlimited faith in my abilities. If we were watching *Family Feud*, she would encourage me (or almost admonish me, for not having done it sooner) to go on the show and win the money. "Why you not go, you can also, eh, go, make!" She saved a clipping for me from the Armenian paper announcing that a Van Gogh painting had sold for $18.3 million. "Look," she said, "you can also make." Her idea was that I could easily paint another picture just like that one and earn a few million for myself.

Lucy was always telling me to learn Armenian, something I never succeeded much in doing. But in an effort to communicate with her more "in her language," my English gradually bent more toward hers (as you've already heard!), adopting many of the same intonations and accents, phrases, and grammatical quirks. She had some wonderful mispronunciations that I always enjoyed: "Tortinas" for "tortillas," for example. I also loved the fact that this Mexican bread staple had been co-opted as the base for her lahmajouns. Or, if I was helping her cook, she might say, "Please put in freezon," meaning the refrigerator, but sounding like some strange hybrid of prison and freezer where the food was jailed in the cold.

She talked a lot about heaven, too, but pronounced it "Heavy." At first this word would appear in guilt-inducing phrases like, "Quickly, make marriage. I will go Heavy." And more recently, as she began to understand that her time on this earth really was winding down, it took on a more serious tone: "One day, I will go Heavy. I am not afraid—Heavy also good."

I think we can safely assume that she's in Heavy now, having crocheted herself a pair of little wings. Actually, she was probably prepared enough that she did that before she left, the same way she has already made me a wedding dress and about 10 different baby outfits for every possible combination of genders or multiple-birth scenarios.

On one of my last visits to her, I took her outside in a wheelchair and we stopped next to a cluster of planted flowers on University Avenue. It was sunny, and she was happy to be outside. We squinted in the sunlight. I sat on a bench and held her hand, and she looked around. "Good you come," she said and squeezed my hand. "My heart open." I think that's exactly the effect she had on so many of us.

Appendix F. Tribute by Grace Yu at Retirement Reception

November 9, 2002

I first met Herant eight years ago at Stanford when he gave the introductory lecture for the Human Biology Core Course. I was a college sophomore at the time, and I remember being struck by what an amazing teacher he was and his great sense of humor.... It wasn't long after that I asked him to be my academic advisor. Doing so changed my life. Many people speak about that one person, a mentor, who has believed in them even when they didn't believe in themselves, supported them through thick and thin, encouraged them to dream what they didn't think was possible for themselves. That one person who changed their life for the better. I hope Herant knows this, but he has been that person for me.

He has been an unfailing source of encouragement and support, cheering me on and believing in me more than I even believed in myself. It profoundly affected me in ways that are hard to describe. He also opened many doors for me: for scholarships, jobs, and academic programs. He encouraged me to study in London for a master's degree, helped me get into Stanford medical school, and provided financial assistance through TA- and RA-ships, without which I could not have afforded to continue my Stanford education. With certainty I can say that I wouldn't be where I am now in life without him and his personal and intellectual generosity.

Stanford students know Herant as a brilliant teacher. He can bring clarity to the most difficult concepts, and has won numerous teaching awards. He is famous at Stanford, and respected, admired, and loved by students. A few years back, on his first day teaching one of his most popular classes, he didn't know how many students would show up. Ten minutes before class, the students started to stream in. Soon, all the seats in the large auditorium were full. But the students kept coming. They were streaming down the stairwells, and when there were no seats left, they starting sitting on the stairwells. When all the seats and the stairwells were full, there were still people packed outside the auditorium, trying to fight their way in. In the end, some students resorted to sitting on stage, literally at Herant's feet while looking up at him. Needless to say, the class had to move to a much larger auditorium.

Students also know what a wonderful sense of humor he has. He is the best storyteller. I love the way his eyes twinkle as he tells a particularly funny story. He can keep a lecture hall entertained and eager to learn more. And for those who are lucky enough to get to know him even better, they discover he is an amazing human being. He has such a genuinely good heart, he wouldn't hurt a soul. Deeply kind and compassionate, he has given his energies and leadership to philanthropic foundations that serve those in need throughout the world. He is the best people person that I have ever met. Whether it be getting into a wonderful discussion about art history with an esteemed colleague or befriending a shy little three-year-old girl by bending down to be at eye level with her, Herant is a wonderful with people....

We are here tonight because Herant is a professor who has devoted his life to intellectual development—his and of all of his students. He is truly amazing.... He has amassed an astounding breadth and depth of interests and expertise. From history to psychiatry to religion and art, he is amazing.... Compassionate, deeply thoughtful, genuinely good-hearted, brilliant, and funny. Herant has been a wonderful mentor, teacher, role model, and friend.

Appendix G. Citation from the William and Flora Hewlett Foundation Board

Presented to Herant Katchadourian, Hewlett Foundation Board Member from 1989 to 2005. With Deep Appreciation for Sixteen Years of Service to the William and Flora Hewlett Foundation.

October 16, 2005

The members of the Board of Directors of the William and Flora Foundation express our deepest gratitude for sixteen years of generous service to the Foundation and your commitment to the ongoing legacy of William and Flora Hewlett. You have served as a friend, colleague and mentor to three generations of Hewlett family members. Your extraordinary impact on the Hewlett Foundation and the field of philanthropy will be felt for many years to come.